An Introduction to their Prediction

THERMOPHYSICAL PROPERTIES OF FLUIDS

Series on Chemical Engineering and Chemical Technology

ISSN: 1793-1304

Seires Editor: Bill William Arnot Wakeham *(University of Southampton, UK)*

An Introduction to their Prediction

THERMOPHYSICAL PROPERTIES OF FLUIDS

Marc J. Assael
Department of Chemical Engineering
Aristotle University, 54006 Thessaloniki, Greece

J. P. Martin Trusler
Department of Chemical Engineering
Imperial College, London SW 7 2BY, UK

Thomas F. Tsolakis
Department of Chemical Engineering
Imperial College, London SW 7 2BY, UK

Imperial College Press

ICP

Published by

Imperial College Press
57 Shelton Street
Covent Garden
London WC2H 9HE

Distributed by

World Scientific Publishing Co. Pte. Ltd.
5 Toh Tuck Link, Singapore 596224
USA office: 27 Warren Street, Suite 401-402, Hackensack, NJ 07601
UK office: 57 Shelton Street, Covent Garden, London WC2H 9HE

British Library Cataloguing-in-Publication Data
A catalogue record for this book is available from the British Library.

Series on Chemical Engineering and Chemical Technology — Vol. 1
THERMOPHYSICAL PROPERTIES OF FLUIDS
An Introduction to Their Prediction

ISBN 978-1-86094-009-5
ISBN 978-1-86094-019-4 (pbk)

for

Theodora and Ioanni-Alexandro
Mariam and Daniel

Contents

Preface

This book is concerned with methods for the prediction of thermophysical properties of fluids. It takes the reader from first principles through to practical methods of predicting thermodynamic properties, phase behaviour and transport properties of complex fluid mixtures over wide ranges of temperature, pressure and composition. The book is written at an introductory level and, while some familiarity with basic thermodynamics is assumed, no prior knowledge of thermophysical-property models is required.

The book is aimed at both undergraduate students and at professional engineers and researchers who need to predict thermophysical properties of fluids as a part of their work. The book would form an ideal basis for a dedicated undergraduate course in the thermophysical properties of fluids. The organisation of the material, the provision of numerous worked examples and the inclusion of simple computer programs are all intended to aid teaching and learning. In our experience, students taking their chemical-engineering design projects have real difficulties with the modelling of thermophysical properties; they either spend enormous amounts of time tying to find 'exact' values or presume that anything will do. These difficulties have not been entirely alleviated by the development of integrated flowsheeting packages as these offer a choice from what, to the uninitiated, appears to be a bewildering array of thermophysical-property models. We certainly intend that this book will help. Parts of the text would also form an ideal companion to an elementary course on statistical mechanics for science or engineering students.

Some of the models that we describe are standard issue with modern flow-sheeting software and are routinely used in the simulation of chemical process systems. This book will be useful to those seeking a better understanding of the capabilities and limitations of the various thermophysical-property models on offer in such packages. Indeed, a proper understanding of these issues is essential if meaningful simulations of process systems are to be performed and safe, effective and efficient plant designs developed. It will also be useful to anyone wishing to develop their own routines for predicting fluid properties.

The first chapter sets out some fundamental concepts of statistical mechanics at an introductory level with the objective of showing how fluid properties might, in principle, be calculated essentially from first principles. The two excellent references give at the end of that chapter should be consulted for a fuller treatment of the theory. In chapter two, we apply the statistical-mechanical theory to the evaluation of the properties of the perfect gas. Together with several kinds of experimental result, such calculations form the basis of our knowledge of perfect-gas properties. For practical purposes this information is usually correlated in the form of an empirical formula for the heat capacity of the prefect gas and we adopt a particularly simple polynomial. In Appendix A, coefficients are listed for 102 compounds. At the end of Chapter 2, we give our first example routine: a subroutine for the evaluation of perfect-gas properties. Real fluids differ from perfect gases because of inter-actions between the molecules and, in Chapter 3, we outline the nature of the intermolecular potential energy function which describes these interactions. In particular, we mention several simple models that incorporate some of the features of real systems. Chapter 4, on the virial equation of state, illustrates both the extent and the limitations of a purely theoretical approach. The virial equation has a sound basis in theory and it is also of significant practical utility for real systems but it has limitations. To overcome these limitations, some measure of empiricism is required. Chapter 5, on the principle of corresponding states, also has a theoretical intro-duction but it concentrates mainly on practical implementation of this powerful and predictive principle. Examples are given of the Lee-Kesler-Plöcker corresponding-states scheme together with a modification that permits application to mixtures containing strongly polar compounds. The workhorses of engineering thermo-dynamics are empirical equations of state and we devote Chapter 6 to that subject. We describe both the theoretical basis of van der Waals equation and the empirical modifications that led to modern cubic equations such as the Redlich-Kwong-Soave and the Peng-Robinson models. We also describe a variant of the more complicated Benedict-Webb-Rubin equation of state. Detailed formulae and example routines are provided that permit thermodynamic properties to be calculated for a single fluid phase of specified composition at specified temperature and pressure. These properties include the fugacity of each component in the mixture, calculation of which is an essential prerequisite for the solution of phase equilibrium problems. We next turn in Chapter 7 to activity-coefficient models that can be used to estimate the fugacity of a component in a liquid mixture in cases where standard equations of state are unreliable. Mixtures containing strongly polar or, especially, associating molecules are the main examples of systems that need to be treated in that way. Again, detailed formulae and example routines are given. Having established the

main techniques by which the fugacity of a component in a mixture may be determined as a function of temperature, pressure and composition, we turn in Chapter 8
to the central application in engineering thermodynamics: the solution of phase
equilibrium problems. We begin Chapter 8 with a brief review of simple phase
behaviour. Next, we describe in detail the calculation of the dew or bubble point of
a fluid mixture of specified composition at either a specified temperature or at a
specified pressure. Example routines based on both equations of state and activity-
coefficient models are given. We then turn to the application of these methods in
vapour-liquid flash calculation, again with examples based on different thermodynamic models, and finally we consider liquid-liquid phase equilibrium. Having
thus completed our treatment of thermodynamic properties, we turn to the transport
properties. Chapter 9 sets out the basic kinetic theory in relation to the viscosity and
thermal conductivity of fluids. We include a simplified derivation of the mean free
path and the transport coefficients for a gas composed of hard spheres as well as the
results of the more complete kinetic theory for real gases at low density. We also
describe the available theory for the transport properties of mixtures and dense
fluids. The practical implementation of these methods is described in Chapter 10 and
a number of examples are given. There are two appendices. The first is a database of
property values for a selection of pure compounds; information from this appendix is
used in many of the examples. The second appendix gives precise definitions of and
relations between the different terms into which thermodynamic properties are often
decomposed. A few key thermodynamic formulae pertaining to the fugacity of a
component in a mixture are also given in Appendix B

Throughout the book, we provide simple computer programs, written in
FORTRAN, that implement the models that we describe.[1] All of these programs are
designed to be as simple as possible to understand, modify and adapt; they therefore
employ the most elementary algorithms possible and, while these are usually reliable,
they are not always efficient. It would therefore be unwise to use our routines in any
large calculation that involves thousands or millions of property evaluations. Nor
should they be used in any property calculation the result of which is of critical
importance without first checking carefully (in truth, the same should be said of

---

[1] All routines given in this book are also available for collection by anonymous FTP from two
INTERNET sites:

sagres.ce.ic.ac.uk
transp.eng.auth.gr

Login as ANONYMOUS with use your e-mail address as the password and consult the file
README for the location of the FORTRAN files.

commercial packages). Nevertheless, these programs comprise a powerful and comprehensive set of tools with which to predict the properties of fluids of known composition at specified temperatures and pressure. They should constitute a useful teaching resource and provide those without access to large commercial software packages a means of estimating single-phase properties and phase equilibrium conditions. We must of course add a disclaimer: these routines are provided primarily as examples and we accept no liability whatsoever for any consequence that may follow from their application. Don't design a billion-dollar plant on the basis of results from our programs!

In compiling Appendix A, we have made liberal use of data tabulated by other authors. Although the sources are acknowledged in the references, we would mention in particular *The Properties of Gases and Liquids* (4th Ed.) by R. Reid, J. Prausnitz and B. Poling (McGraw-Hill, New York, 1988) from which we extracted many of the physical constants of pure compounds. We also made good use of *The Vapour Pressures of Pure Compounds* (2nd Ed.) by T. Boublik, V. Fried and E. Hala (Elsevier, Amsterdam, 1984), and *Computer Calculations for Multicomponent Vapour-Liquid and Liquid-Liquid Equilibria* by J. Prausnitz, T. Anderson, E. Grens, C. Eckert, R. Hsieh and J. O'Connel (Prentice Hall, London, 1980).

We are indebted to Dr John Dymond who agreed to read the entire text and provide us with his comments. We found these most useful and we thank him for his dedicated and extremely efficient efforts; he even read the comment lines in our programs! Any remaining errors (or perhaps we should say 'the remaining errors') are of course entirely our fault. We are also pleased to acknowledge many useful discussions with Professor William Wakeham. Finally, we acknowledge the considerable help that we have extracted from the lecturer notes of Dr Mike Carabine.

Marc J. Assael
J. P. Martin Trusler
Thomas F. Tsolakis

London and Thessaloniki
February, 1996.

List of Main Symbols

A	Helmholtz free energy		P	pressure
b	van der Waals co-volume		q	molecular partition function
B	second virial coefficient		Q	canonical partition function
c	speed of light, molecular speed		r	intermolecular separation, bond length
C	third virial coefficient		$\mathbf{r}_i$	position vector of molecule i
C_P	isobaric heat capacity		R	universal gas constant
C_V	isochoric heat capacity		R_X	roughness factor for property X
D	diffusion coefficient, fourth virial coefficient		s	specific entropy
			S	entropy
E	total energy		t	time
E_i	energy for quantum state i		T	thermodynamic temperature
f	fugacity, energy scaling parameter		u	fluid speed
			U	internal energy
f_{ij}	Mayer function		v	dimensioless volume
F	intermolecular force		V	volume
g	radial-distribution function		V_0	characteristic molar volume
g_{ij}	NRTL binary parameter		x_i	mole fraction of i
G	Gibbs free energy		X	general property
h	specific enthalpy, Planck's constant, volume scaling parameter		y_i	mole fraction of i
			Z	compression factor
H	enthalpy			
H^{cm}	centre-of-mass Hamiltonian			
I	moment of inertia			**Script Symbols**
k_B	Boltzmann constant		$\mathscr{F}$	Poynting factor
k_{ij}	binary interaction parameter		$\mathscr{H}$	Henry's constant
K_i	equilibrium ratio of component i		$\mathscr{L}$	configuration integral
m	mass		$\mathscr{N}$	number of systems in ensemble
M	molar mass		$\mathscr{P}$	probability
n_1	number of molecules of species 1		$\mathscr{U}$	total potential energy
N	number of molecules			
N_A	Avogadro's constant			**Greek Symbols**
p	momentum		α	NRTL parameter, activity
$\mathbf{p}$	momentum vector			

γ	activity coefficient
ε	depth of intermolecular potential, energy parameter
ε_i	molecular energy for state i
η	viscosity
Θ_r	characteristic rotational temperature
Θ_v	characteristic vibrational temperature
λ	thermal conductivity
λ_{ij}	Wilson and T-K-Wilson parameter
Λ	mean free path
Λ_{ij}	Wilson and T-K-Wilson parameter
μ	chemical potential
v	frequency
ρ	mass density
ρ_n	amount-of-substance density
σ	rotational symmetry number, hard-sphere diameter
τ_{ij}	NRTL parameter
Φ	potential energy function
Φ_i	fugacity coefficient of component i
ω	wave number, acentric factor
ϖ	orientation vector
Ω	collision integral, degeneracy of an energy level

Superscripts

b	boiling
c	critical
cm	centre of mass
E	excess
int	internal
L	liquid
mix	mixture
mol	molecular
pg	perfect gas
V	vapour
res	residual
tr	translational
∞	infinite dilution
σ	saturation

Subscripts

corr	correction term
e	electronic
E	Enskog
int	internal
ir	internal rotation
m	molar
mix	mixture
MD	molecular dynamics
n	nuclear spin
o	dilute gas
r	reduced, rotational
R	reference
t	translational
v	vibrational
w	water

1

Fundamentals

A fundamental treatment of the thermophysical properties of bulk matter must begin with the properties of individual atoms and molecules and the manner in which they interact. Such a description includes the structure of individual molecules, the manner in which they store energy and the forces that act between them. This information comprises the microscopic description of the system and, in principle, it may be combined with the laws of physics to obtain a complete description of the whole system as it evolves in time. For example, in the case of a collection of structureless atoms, the microscopic description would specify the size and mass of each atom and the forces between them as a function of their arrangement in space. If we then specified the initial position and velocity of each atom and the boundary conditions, solution of the classical laws of motion would provide a complete description of the system. The difficulty with such an approach is that far too many variables are involved in the microscopic description of a realistic number of atoms or molecules and the equations of motion could never be solved.

Our interest throughout this book lies only in certain bulk or macroscopic properties of matter. For example, we might be interested in the pressure exerted by a gas. By this we mean the average force exerted per unit area on the walls of the enclosure by the gas; we are not interested in the fluctuations of this quantity arising from the impact of individual molecules with the wall. Thus the interesting macroscopic properties of the system are certain long-time averages of the microscopic variables and there are far fewer of the former than of the latter. The complete description referred to above would actually provide a great deal of redundant information. For example, a system of N atoms requires $6N$ microscopic variables to describe the position and velocity of each atom but only four macroscopic variables are required to specify completely the thermodynamic properties of the system. This suggests that another approach to the problem may be more profitable in which we seek to establish only certain average properties of the system. Statistical mechanics provides just such an approach.

The rôle of statistical mechanics is that of a bridge between the microscopic and macroscopic descriptions of the system. The statistical mechanics of systems at equilibrium, from which the thermodynamic properties may be obtained, is based upon two postulates which are made to connect the idea of an ensemble of systems to the macroscopic properties. In the following sections we shall develop these ideas and show that a single statistical quantity, the canonical ensemble partition function Q, may be used to obtain all the observable thermodynamic properties of a system.

1.1. Microscopic States

We begin by recognising that, fundamentally, a microscopic description of a collection of molecules should be one based on quantum mechanics according to which the system can reside only in certain stationary states known as quantum states. Each state of the system may be labelled by a symbolic index i and will have a fixed energy E_i. When the system absorbs or gives up energy, it does so only by transition between quantum states. In principle, these quantum states and their corresponding energies may be obtained by solution of Schrödinger's equation. For a system of many molecules there will be an extremely large number of states and of possible energies.

Although it is generally impossible to obtain exact solutions of the underlying quantum-mechanical problem, we shall see later that the energy of the system may be separated into a number of terms each of which may be treated to a good approximation as independent of the others. In particular, the energy may be separated into the sum of individual molecular energies and an intermolecular potential energy. The former may be treated rather accurately using quantum mechanical models and the latter accounted for using a combination of theoretical and empirical results. Thus we may proceed on the assumption that the microscopic states of a system can be quantified at least in terms of the allowed energies E_i.

1.2. The Canonical Ensemble

The concept of an ensemble of systems is central to the development of statistical mechanical results. An ensemble is simply a hypothetical collection of a very large number $\mathcal{N}$ of systems each of which is a replica on a macroscopic scale of the actual thermodynamic system of interest. This means that every system in the ensemble shares common values of *some* of the macroscopic variables. An ensemble is classified according to variables which are common to all members of the ensemble. In what follows we shall deal only with the canonical ensemble in which each system has the same values of volume V, temperature T and number of molecules N. This ensemble is thus representative of a closed isothermal system. Although each member of the ensemble has the same values of N, V and T they are not identical in a microscopic sense. For example, although the total energy of the entire ensemble is fixed (because the ensemble is isolated), the energy of each individual system may fluctuate in time. We can however define an ensemble average of such mechanical variables as the instantaneous average of the specified variable over all members of the ensemble.

The ensemble is a useful devise only in connection with the two postulates of statistical mechanics. The first of these states that:

As the number of systems $\mathcal{N}$ in the ensemble becomes very large, the ensemble average of a mechanical variable approaches the long-time average of that variable in the real system.

By this postulate the time average is replaced by an average at one instant over all members of the ensemble. For example, the ensemble average of the energy is defined by

$$\langle E \rangle = \sum_i E_i \mathcal{P}_i , \tag{1.1}$$

where $\mathcal{P}_i$ is the probability that a system chosen at random from the ensemble will be found in the quantum state i with energy E_i. According to the first postulate, this ensemble average will approach the thermodynamic internal energy U of the real system as $\mathcal{N} \to \infty$:

$$U = \lim_{\mathcal{N} \to \infty} \sum_i E_i \mathcal{P}_i . \tag{1.2}$$

The second postulate of statistical mechanics states that:

The only dynamic variable upon which the quantum states of the entire canonical ensemble depend is the total ensemble energy E_t.

From this postulate we deduce that all states of the ensemble having the same energy are equally probable.

1.3. The Canonical Partition Function

Using the second of the two postulates made above, it can be shown [1,2] that the probability $\mathcal{P}_i$ that a system selected at random from the ensemble will be found in quantum state i varies exponentially with the energy E_i of that state. That is

$$\mathcal{P}_i(E_i) \propto \exp(-\beta E_i) , \qquad (1.3)$$

where β is a positive quantity which turns out to be related to the thermodynamic temperature of the system by

$$\beta = 1/k_B T . \qquad (1.4)$$

Here k_B is Boltzmann's constant. Since, however, there is unit probability that the system resides in *some* state we have that $\Sigma_i \mathcal{P}_i(E_i) = 1$ and

$$\mathcal{P}_i(E_i) = \frac{\exp(-E_i/k_B T)}{Q} , \qquad (1.5)$$

where

$$Q(N,V,T) = \sum_i \exp(-E_i/k_B T) . \qquad (1.6)$$

The quantity Q, known as the *canonical partition function*, plays a central rôle in statistical thermodynamics. It does not have a well-defined physical meaning but it serves as a useful statistical devise in terms of which all of the thermodynamic properties may be expressed. In the following sections we shall develop the connection between Q and these properties.

1.4. Thermodynamic Properties

We now examine the relation between the thermodynamic properties and the canonical partition function for the most general case.

1.4.1. Internal Energy

The internal energy is just the ensemble average system energy. In Eq.(1.1), the ensemble average was expressed in terms of the probability function while, in Eq.(1.5), the probability function was, in turn, expressed in terms of the partition function. Combining these equations we obtain the expression

$$U = \frac{1}{Q} \sum_i E_i \exp(-E_i/k_B T) \qquad (1.7)$$

which, in view of the definition of Q, Eq.(1.6), may be written as:

$$U = k_B T^2 \left(\frac{\partial \ln Q}{\partial T} \right)_{N,V} . \qquad (1.8)$$

This equation provides a direct relation between the internal energy and the canonical partition function.

1.4.2. Entropy

In order to obtain an expression for the entropy in terms of the partition function, we compare the relation between internal energy and the probability function with the second law of thermodynamics. According to classical thermodynamics, the fundamental equation for a change in the state of a system of fixed composition is

$$dU = T dS - P dV . \qquad (1.9)$$

Now, according to our statistical-mechanical arguments, when N is constant a change in the internal energy of the system can occur only if either the probability function or the energy levels change. Thus, from Eq.(1.2),

$$dU = \sum_i E_i \, d\mathcal{P}_i + \sum_i \mathcal{P}_i \, dE_i . \qquad (1.10)$$

Let us start with the second term of this equation. With N constant, the energy levels may change only if the volume changes and hence $dE_i = (\partial E_i/\partial V)_N dV$. Thus, comparing Eqs.(1.9) and (1.10), we see that

$$P \, dV = -\sum_i \mathcal{P}_i \, dE_i \qquad (1.11)$$

and

$$T \, dS = \sum_i E_i \, d\mathcal{P}_i . \qquad (1.12)$$

In order to obtain the entropy, we eliminate the energy levels E_i from Eq.(1.12) in favour of the partition function Q. We do this by obtaining an expression for E_i from the logarithm of Eq. (1.5) with the result

$$T \, dS = -k_B T \left(\sum_i \ln \mathscr{P}_i \, d\mathscr{P}_i + \ln Q \sum_i d\mathscr{P}_i \right) = -k_B T \sum_i \ln \mathscr{P}_i \, d\mathscr{P}_i \, , \qquad (1.13)$$

where we have used the fact that $\sum_i \mathscr{P}_i = 1$ and hence $\sum_i d\mathscr{P}_i = 0$.

Equation (1.13) can be also written as

$$dS = -k_B \, d\left(\sum_i \mathscr{P}_i \ln \mathscr{P}_i \right) \qquad (1.14)$$

and, since dS is an exact differential, we see that the right hand side of this equation is the product of a constant and an exact differential. We may therefore integrate Eq.(1.14) directly with the result

$$S = -k_B \sum_i \mathscr{P}_i \ln \mathscr{P}_i \, . \qquad (1.15)$$

Finally, using Eqs.(1.5) and (1.6) to eliminate $\mathscr{P}_i$ in favour of Q, we obtain

$$S = k_B T \left(\frac{\partial \ln Q}{\partial T} \right)_{N,V} + k_B \ln Q \, , \qquad (1.16)$$

which is the desired relation between S and Q.

1.4.3. Helmholtz Free Energy

We now have expressions for both U and S in terms of Q from which the Helmholtz free energy, A, can readily be obtained through the relation

$$A = U - TS \, . \qquad (1.17)$$

Combining Eqs.(1.8), (1.16), and (1.17), we find that A is given by the simple relation

$$A = -k_B T \ln Q \, . \qquad (1.18)$$

1.4.4. Other Thermodynamic Properties and Mixtures

Since A is the characteristic state function for the choice of N, V and T as the independent variables, all of the other thermodynamic properties follow from this quantity through the fundamental equation:

$$dA = -S\,dT - P\,dV + \bar{\mu}\,dN \;. \qquad (1.19)$$

Here, $\bar{\mu}$ is the chemical potential per molecule. The partial derivatives of A with respect to T, V and N then give respectively $-S$, $-P$ and $\bar{\mu}$. All of the other thermodynamic properties of the system can be obtained by standard thermodynamic manipulations of these quantities. In Table 1.1, we give expressions for internal energy U, entropy S, pressure P, Gibbs free energy G, enthalpy H, heat capacity at constant volume C_V, and chemical potential per molecule $\bar{\mu}$, all in terms of Q.

The development above has been given for a pure substance only. However, there is no difficulty in extending the treatment to a mixture of r components. We suppose that the mixture contains N_1 molecules of type 1, N_2 molecules of type 2, etc. The system energy levels E_i available to each ensemble member now depend

Table 1.1. Thermodynamic Properties in Terms of the Partition Function Q.

Property	In terms of A	In terms of Q
A		$= -k_B T \ln Q$
U	$= A - T(\partial A/\partial T)_{N,V}$	$= k_B T^2 (\partial \ln Q/\partial T)_{N,V}$
S	$= -(\partial A/\partial T)_{N,V}$	$= k_B T(\partial \ln Q/\partial T)_{N,V} + k_B \ln Q$
P	$= -(\partial A/\partial V)_{N,T}$	$= k_B T(\partial \ln Q/\partial V)_{N,T}$
G	$= A + PV$	$= -k_B T \ln Q + k_B TV(\partial \ln Q/\partial V)_{N,T}$
H	$= U + PV$	$= k_B T^2(\partial \ln Q/\partial T)_{N,V} + k_B TV(\partial \ln Q/\partial V)_{N,T}$
C_V	$= -T(\partial^2 A/\partial T^2)_{N,V}$	$= 2k_B T(\partial \ln Q/\partial T)_{N,V} + k_B T^2(\partial^2 \ln Q/\partial T^2)_{N,V}$
$\bar{\mu}$		$= -k_B T(\partial \ln Q/\partial N)_{T,V}$

upon the numbers $N_1, N_2, \cdots, N_r$, as well as V, and the canonical partition function is a function of the independent variables $N_1, N_2, ..., N_r, V$, and T. The relation between the Helmholtz free energy and the partition function, which was derived under an assumption of constant composition, is not affected by these considerations. Since the fundamental equation corresponding to the choice of $(V, T, N_1, N_2, \cdots N_r)$ as the independent variables is now

$$dA = -S\,dT - P\,dV + \sum_{i=1}^{r} \bar{\mu}_i\,dN_i\,, \tag{1.20}$$

where $\bar{\mu}_i$ is the chemical potential per molecule of type i, the equations of Table 1.1 remain valid (with the differentials at constant N interpreted as being at constant N_1, $N_2, \cdots N_r$).

1.5. The Molecular Partition Function

So far, we have considered an ensemble representative of a system of N molecules and we have formulated the partition function in terms of the energies E_i of the quantum states to which the system has access. We expect that the total energy of the system is, in general, made up of contributions from both the individual molecules and the interactions between them. However, for the special case in which the molecules do not interact, each is an independent subsystem and the energy of the system is the sum of separate contributions from each molecule.

Let us consider first a system containing just one molecule. By analogy with Eq.(1.6) we can define a *molecular (one-molecule) partition function* as

$$q = \sum_j \exp(-\varepsilon_j/k_B T)\,, \tag{1.21}$$

where ε_j is the energy of state j and the summation extends over all quantum states of the molecule.

If we now consider a system composed of N identical but somehow distinguishable molecules that do not interact with each other, we have N identical but independent sets of energy levels available. The canonical partition function will therefore be the product of the molecular partition functions for this system:

$$Q = q^N = \left(\sum_j \exp(-\varepsilon_j/k_B T) \right)^N\,. \tag{1.22}$$

However, in a pure fluid, the molecules are actually *indistinguishable* and configurations of the system which differ only by the assignment of molecular labels are really the same. Thus Eq.(1.22) greatly overestimates the partition function. This is illustrated in Example 1.1 for a system of two indistinguishable molecule.

If the number of possible quantum states is very large and we ignore quantum restrictions on the multiple occupation of energy levels then the value of Q given by Eq.(1.22) is too great simply by a factor equal to the number of possible assignments of labels to the N molecules [2]. Since this number is $N!$, the correct value of the partition function for N non-interacting indistinguishable molecules is

$$Q = (q^N/N!) = \frac{1}{N!}\left(\sum_j \exp(-\varepsilon_j/k_B T)\right)^N. \qquad (1.23)$$

EXAMPLE 1.1   *Canonical and Molecular Partition Functions*

As an illustration, we consider a system of two identical molecules which we label a and b and which each have two possible quantum states. The product of the two molecular partition functions is

$$q_a \cdot q_b = \left(e^{-\varepsilon_{a1}/k_B T} + e^{-\varepsilon_{a2}/k_B T}\right)\cdot\left(e^{-\varepsilon_{b1}/k_B T} + e^{-\varepsilon_{b2}/k_B T}\right)$$

and there are four terms in this product,

$$q_a \cdot q_b = e^{-(\varepsilon_{a1}+\varepsilon_{b1})/k_B T} + e^{-(\varepsilon_{a1}+\varepsilon_{b2})/k_B T} + e^{-(\varepsilon_{a2}+\varepsilon_{b1})/k_B T} + e^{-(\varepsilon_{a2}+\varepsilon_{b2})/k_B T}$$

corresponding to four possible states of the system of two molecules. However, the two states with one molecule in energy level 1 and the other in energy level 2 are indistinguishable and should not be counted separately. The complete partition function for this system should therefore be written

$$Q = e^{-(\varepsilon_1+\varepsilon_1)/k_B T} + e^{-(\varepsilon_1+\varepsilon_2)/k_B T} + e^{-(\varepsilon_2+\varepsilon_2)/k_B T}$$

As the number of quantum states to which the system has access becomes large, one can show that the product q^N over estimates Q by a factor $N!$.

❑

1.6. Factorisation of the Partition Function

The canonical partition function defined by Eq.(1.6) is, as it stands, impossibly complicated to evaluate for any real system. This is because it requires a summation over all possible quantum states of a system of N molecules and therefore involves an impractical number of terms. In order to make progress, we first break down the problem into a number of separate terms which may be treated independently.

The key to factoring the partition function is to recognise that the energy of the system is made up of several essentially independent contributions. First, to an excellent approximation, the *internal* energy of the molecules is independent of the translational kinetic energy and the intermolecular potential energy. Therefore, Q factorises into the product of two terms:

$$Q = Q^{int}(N,T) \; Q^{cm}(N,V,T) . \tag{1.24}$$

Here, Q^{int} is the part of the partition function associated with the internal energy of the molecules and Q^{cm} is the part corresponding to the position and momentum of the centres of mass. Since, by this separation, the internal energy of each molecule is independent of the position of any molecule, we see that Q^{int} is independent of the volume of the system. We further presume that the internal states of the molecules form N totally independent subsets (or subsystems) so that Q^{int} is simply the product of N identical terms:

$$Q^{int} = q_{int}^{N} = \left(\sum_{j} \exp(-\varepsilon_{j}^{int}/k_{B}T) \right)^{N} . \tag{1.25}$$

Thus, Q^{cm} is the only part of the partition function which depends on the density of the system. This separation is highly accurate for monatomic systems and for systems composed of rigid (as oppose to flexible) molecules. For other molecules it may be less accurate.

We will see that the evaluation of q_{int} from quantum mechanics is quite possible for simple molecules. However, when dealing with the centre-of-mass term, it is usual and appropriate to adopt a classical approach. This is justified by the fact that the translational energy levels are so close together that we may treat them as a continuum. This approximation leads to negligible errors for real systems at temperatures above 1 K [1]. The overall approach to the problem is therefore *semi-classical*.

In the semi-classical treatment of the partition function, the summation over translational energy levels is replaced by an integral over the position and momentum of each molecule to give

$$Q = Q^{\text{int}} \frac{1}{N! \, h^{3N}} \int \cdots \int \exp(-H^{\text{cm}}/k_{\text{B}}T) \, d\mathbf{r}^N \, d\mathbf{p}^N . \qquad (1.26)$$

In this equation, $d\mathbf{r}^N$ denotes the product $d\mathbf{r}_1 d\mathbf{r}_2 \cdots d\mathbf{r}_N$, $\mathbf{r}_i$ is the position vector of molecule i and $d\mathbf{r}_i = dx_i \, dy_i \, dz_i$. Similarly, $d\mathbf{p}^N = d\mathbf{p}_1 d\mathbf{p}_2 \cdots d\mathbf{p}_N$, $\mathbf{p}_i$ is the momentum vector for molecule i, $d\mathbf{p}_i = dp_{x_i} dp_{y_i} dp_{z_i}$, and p is linear momentum. The integration is performed over all possible positions and momenta of each molecule in the system. In addition, the factor $1/N!$ accounts for the indistinguishability of the molecules, and the factor $1/h^{3N}$, where h is the Planck constant, is a normalisation constant chosen such that the classical and quantum expressions become identical in the limit of high temperatures [1]. Finally, H^{cm} is the classical centre-of-mass Hamiltonian equal to the total (kinetic + potential) energy of all molecular centres of mass in the system of N molecules. This is given by

$$H^{\text{cm}} = \sum_{i=1}^{N} \frac{p_i^2}{2m} + \mathcal{U}(\mathbf{r}_1, \mathbf{r}_2, \cdots, \mathbf{r}_N) , \qquad (1.27)$$

where $p_i^2 = \mathbf{p}_i \cdot \mathbf{p}_i$, m is the mass of one molecule, and $\mathcal{U}(\mathbf{r}_1,\mathbf{r}_2,\cdots,\mathbf{r}_N)$ is the total intermolecular potential energy of the system.

The integration over the momenta is readily carried out analytically, yielding the factor $(2\pi m k_{\text{B}}T)^{3N/2}$, and hence the complete semi-classical partition function is given by

$$Q = \frac{1}{N!} \left(2\pi m k_{\text{B}}T/h^2\right)^{3N/2} Q^{\text{int}} \int \cdots \int \exp(-\mathcal{U}/k_{\text{B}}T) d\mathbf{r}^N . \qquad (1.28)$$

It is convenient to recognise that the partition function contains two distinct parts. The first includes information about isolated molecules and therefore depends only on the molecular properties of the system. The second part contains information about the interactions between the molecules of the system. We therefore write Q as the product

$$Q = Q^{\text{mol}} \mathcal{L} , \qquad (1.29)$$

where
$$Q^{\text{mol}} = \frac{1}{N!} (2\pi m k_{\text{B}}T/h^2)^{3N/2} Q^{\text{int}} \qquad (1.30)$$

and
$$\mathcal{L} = \int \cdots \int \exp(-\mathcal{U}/k_{\text{B}}T) d\mathbf{r}^N . \qquad (1.31)$$

Q^{mol} depends only upon one-molecule properties such as mass, moment of inertia, etc. The second term, $\mathcal{L}$, is called the *configuration integral* and depends upon the interactions between the molecules. This is presumed to be the only part of Q which depends upon the density.

1.7. Perfect-gas and Residual Properties

As we shall see later, the calculation of the configuration integral is an extremely difficult task. However, in the special case that the total intermolecular potential energy $\mathcal{U}$ is zero we have a perfect gas and $\mathcal{L}$ is simply V^N. The complete partition function of the perfect gas, Q^{pg}, is therefore

$$Q^{pg} = Q^{mol} V^N = \frac{1}{N!} (2\pi m k_B T/h^2)^{3N/2} Q^{int} V^N . \tag{1.32}$$

Thermodynamic properties obtained from this partition function, by means of the relations in Table 1.1, are called *perfect-gas properties* and will be discussed in detail in the next chapter. For a real fluid, we may write

$$Q = Q^{pg} Q^{res} , \tag{1.33}$$

where Q^{res} is the residual part of the partition function and is given by

$$Q^{res} = \frac{1}{V^N} \mathcal{L} . \tag{1.34}$$

It follows from Eq. (1.33) that the value of any thermodynamic property X can be written as the sum of a perfect-gas term and a residual term:

$$X = X^{pg} + X^{res} . \tag{1.35}$$

This separation is useful in practice because different techniques are applicable to the calculation of each term. We shall see that the perfect-gas partition function is, like both Q^{int} and Q^{mol}, amenable to nearly exact evaluation from molecular theory. The residual partition function is, like the configuration integral $\mathcal{L}$, very difficult to evaluate in practice because of the need to solve the N-body problem posed by Eq.(1.31). Only for gases at low or moderate densities can a fully-theoretical approach to residual properties be followed. It is important to recognise that the quantity X^{pg} which appears in Eq.(1.35) is a property of the hypothetical perfect gas which has *either* the same temperature and density *or* the same temperature and pressure as the real fluid. In the case of a property that is independent of density (or pressure) in the perfect gas, the choice of independent variables has no effect on the values of X^{pg} and X^{res}. However, this is not the case when the perfect-gas property is density (or pressure) dependent although, of course, the combination ($X^{pg} + X^{res}$) is independent of the choice of independent variables. For a more detailed discussion of the definition of residual properties and their relation to the configurational properties, the reader is referred to Appendix B.

1.8. Summary

In this chapter, equations have been derived for the thermodynamic properties in terms of the canonical partition function Q; these are given in Table 1.1. Expressions which form the basis for evaluation of Q itself have also been obtained in the form of Eqs.(1.29)-(1.31) or Eqs. (1.32)-(1.34). These will be developed in the later chapters into forms suitable for practical application to real gases and liquids. Evaluation of the configuration integral $\mathcal{L}$ is a difficult problem in general, and several methods will be discussed later. These include a series expansion in powers of the density (which yields the virial equation of state) and the principle of corresponding states. We shall also have to resort to empirical methods for the evaluation of residual properties in many circumstances; example of these include generalised equations of state and activity coefficient models.

The use of Table 1.1 together with Eqs.(1.29)-(1.31) rests on three important approximations:

1. That the factor $1/N!$ in the canonical partition function correctly accounts for the number of distinguishable states (i.e. that Maxwell-Boltzmann statistics are satisfactory); this is valid except at very low temperatures.
2. That the internal and centre-of-mass energies are separable; this is usually valid for monatomic fluids in both gas and liquid states but may lead to errors for polyatomic fluids at high densities.
3. That the translational part of the partition function Q^{cm} may be treated classically; this too is valid except at very low temperatures.

References

1. T.M. Reed and K.E. Gubbins, *Applied Statistical Mechanics* (McGraw-Hill, Kogakusha, 1973).
2. T.L. Hill, *An Introduction to Statistical Thermodynamics* (Addison Wesley, Reading, Mass., 1960).

2

The Perfect Gas

In the previous chapter, the fundamental equations relating the canonical partition function with the thermodynamic properties were derived. The simplest system to which these concepts may be applied is the perfect gas. Such a gas is composed of non-interacting molecules and the total intermolecular potential energy $\mathcal{U}$ is then independent of the positions of the molecules and may be taken as zero. This chapter is devoted to the thermodynamic properties of the perfect gas.

2.1. Principles

The perfect gas is a hypothetical substance for which the intermolecular potential energy is zero; that is $\mathcal{U} = 0$. This definition of the perfect gas implies that P-V-T properties conform exactly to the equation

$$PV = N k_{\mathrm{B}} T \tag{2.1}$$

and also that the internal energy U is only a function of temperature (and not of pressure or volume).

Real gases are composed of molecules between which the interactions fall off rapidly with increasing separation. When such a gas is very "dilute" (i.e. the density is low), the average molecular separation becomes large and the condition $\mathcal{U} \approx 0$ is fulfilled if no external fields are present. Thus all real gases exhibit perfect-gas behaviour in the limit of zero density and may sometimes be treated as perfect with sufficient accuracy at finite densities. Also, the thermodynamic properties of any fluid may be expressed as the sum of perfect-gas and residual terms and this separation often provides a convenient route for calculation.

As already discussed in Section 1.7, the canonical partition function in the case of the perfect gas is given from Eq.(1.32), as

$$Q^{\text{pg}} = Q^{\text{mol}} V^N = \frac{1}{N!} (2\pi m k_{\text{B}} T/h^2)^{3N/2} Q^{\text{int}} V^N , \qquad (2.2)$$

where Q^{int} was defined in Eq.(1.25) as

$$Q^{\text{int}} = q_{\text{int}}^N = \left(\sum_j \exp(-\varepsilon_j^{\text{int}}/k_{\text{B}}T) \right)^N . \qquad (2.3)$$

We consider a gas composed of polyatomic molecules. The internal energy of each molecule, and consequently q_{int}, can be further separated into independent parts due to:

- free rotation of the molecules (r),
- internal vibrations (v),
- internal rotation of molecular groups (ir),
- electronic motions (e), and
- nuclear spin (n).

This approximation is usually a good one with the exception of the separability of the free rotation and internal vibrations between which there is some coupling; this may be accounted for in a formal sense by inclusion of a correction term (corr). In formulating the partition function, it is usual to measure the internal energies ε of the molecule relative to the lowest accessible energy level ε_0. ε_0 is known as the ground-state energy or the *zero-point energy* and is associated with a degeneracy Ω_0 (i.e. there are Ω_0 quantum states with the same energy ε_0). With this separation, the internal energy of the molecule is given as

$$\varepsilon_{\text{int}} = \varepsilon_{\text{r}} + \varepsilon_{\text{v}} + \varepsilon_{\text{ir}} + \varepsilon_{\text{e}} + \varepsilon_{\text{n}} + \varepsilon_{\text{corr}} + \varepsilon_0 , \qquad (2.4)$$

from which it follows that

$$\begin{aligned} q_{\text{int}} &= (q_{\text{r}} \, q_{\text{v}} \, q_{\text{ir}} \, q_{\text{e}} \, q_{\text{n}} \, q_{\text{corr}}) \, \Omega_0 \, \exp(-\varepsilon_0/k_{\text{B}}T) \\ &= \left(\prod_i q_{(i)} \right) \Omega_0 \, \exp(-\varepsilon_0/k_{\text{B}}T) . \end{aligned} \qquad (2.5)$$

In the following sections, the translational contribution and each of the internal contributions to the partition function will be examined and the corresponding contributions to the thermodynamic properties will be calculated. When these contributions are combined, we have the thermodynamic properties of the perfect gas.

2.2. Translational Contributions

A monatomic molecule such as argon has only translational and electronic states; it cannot rotate or vibrate. It is true that the electrons can acquire a net angular momentum about the nucleus but such a state is of high energy and is classed as electronically excited, not as a rotational state. This and other electronically excited states are energetically so far above the ground states that they are important only at temperatures far above 1000 K, and are therefore ignored here. Furthermore, the electronic ground state is non-degenerate, so that $\Omega_0=1$, and if we take the overall zero of energy as the energy of the atom at rest then $\varepsilon_0 = 0$.

We will first examine the partition function in the case where only the translational contribution exists. The centre-of-mass contribution, Q^{cm}, to the partition function is given by Eq.(2.2) with $Q^{int} = 1$, as

$$Q^{cm} = \left(\frac{V^N}{N!}\right)\left(\frac{2\pi m k_B T}{h^2}\right)^{\frac{3N}{2}}. \tag{2.6}$$

Using Stirling's approximation, $\ln N! \approx N \ln N - N$, we obtain

$$Q^{cm} = \left(\frac{eV}{N}\right)^N\left(\frac{2\pi m k_B T}{h^2}\right)^{\frac{3N}{2}}. \tag{2.7}$$

This equation gives the translational contribution to the partition function; for a perfect monatomic gas, this is the only contribution. The translational contribution to the thermodynamic properties is readily evaluated using the expressions given in Table 1.1 and the results are given in Table 2.1.

We see, first, that we have obtained correctly the equation of state of the perfect gas and, second, that the heat capacity at constant volume is $(3/2)Nk_B$. It follows that the heat capacity at constant pressure is $(5/2)Nk_B$ and that the molar heat capacities are, at constant volume, $3R/2$ and, at constant pressure, $5R/2$. These conclusions are amply confirmed by experimental data for the inert gases at low densities.

The expression obtained for the entropy, known as the Sackur-Tetrode equation, is also open to experimental verification. In Example 2.1, the entropy of argon at its normal boiling point is calculated and compared with the experimentally determined value; the agreement is within the experiment error. Such a comparison illustrates the value of statistical thermodynamics when, as here, we know sufficient about the molecular mechanics of our system. The equation for the entropy given in

Table 2.1   Translational Contributions to the Perfect-gas Thermodynamic Properties.

$$A_t = -N k_B T \ln\left[\left(\frac{2\pi m k_B T}{h^2}\right)^{3/2} \frac{V e}{N}\right]$$

$$U_t = \frac{3}{2} N k_B T$$

$$S_t = N k_B \ln\left[\left(\frac{2\pi m k_B T}{h^2}\right)^{3/2} \frac{V e^{5/2}}{N}\right]$$

$$P_t = \frac{N k_B T}{V}$$

$$G_t = -N k_B T \ln\left[\left(\frac{2\pi m k_B T}{h^2}\right)^{3/2} \frac{V}{N}\right]$$

$$H_t = \frac{5}{2} N k_B T$$

$$C_{Vt} = \frac{3}{2} N k_B$$

EXAMPLE 2.1   *Calculation of the Perfect-Gas Entropy of Argon at its Normal Boiling Point.*

Argon possesses only translational contributions to the partition function and, from Table 2.1 with $V/N = k_B T/P$, the entropy is given as

$$S^{pg} = S_t = N k_B \ln\left[\left(\frac{2\pi m k_B T}{h^2}\right)^{3/2} \frac{k_B T}{P} e^{5/2}\right]$$

Since $m = 6.6337 \times 10^{-26}$ kg, we obtain $S^{pg} = 129.20$ J·K^{-1}·mol^{-1} at $T = 87.29$ K and $P = 0.10133$ MPa. The value of $S_m^{pg}(87.29$ K$) - S(T \rightarrow 0)$ obtained experimentally is 129.0 J·K^{-1}·mol^{-1} and the agreement with the calculated value is well within the experimental uncertainty.

❑

Table 2.1 requires only a knowledge of the mass of the argon atom, while the calorimetric value requires painstaking measurements of heat capacities, enthalpies of melting and vapourisation, and the second virial coefficient (to correct to the standard state of the perfect gas). However, in most cases, as we shall see below, the balance of accuracy and convenience is not so heavily weighted on the side of the statistical calculation.

2.3. Rotational Contributions

Only the free rotations of an idealised rigid molecule are considered here; complications that can arise from bond vibrations and internal rotations are in this section neglected. Although the neglect of bond vibrations does not lead to serious errors at ordinary temperatures, corrections for such effects as well as for internal rotations will be discussed in later sections of this chapter.

2.3.1. Linear Molecules

Diatomic molecules and some polyatomic molecules such as CO_2, C_2H_2, etc. are linear molecules. In this case, as two angles suffice to specify the molecular orientation in space, two degrees of rotational freedom exist. We also note that a linear molecule has two equal moments of inertia, I, one about each axis through the centre of mass perpendicular to the interatomic bonds (see Figure 2.1). The solution of Schrödinger's equation for a linear rigid rotor in a zero-potential field (p.433 of [1]) yields the allowed rotational energy levels and their corresponding degeneracy from which the rotational contribution to the molecular partition function is found to be

$$q_r = \frac{1}{\sigma} \sum_{j=0}^{\infty} (2j+1) \exp\left[-j(1+j)\Theta_r/T\right] . \qquad (2.8)$$

In this equation, Θ_r is the *rotational characteristic temperature* and is given in terms of the moment of inertia and other constants as

$$\Theta_r = \frac{h^2}{8\pi^2 I k_B} = \frac{hcB_e}{k_B} . \qquad (2.9)$$

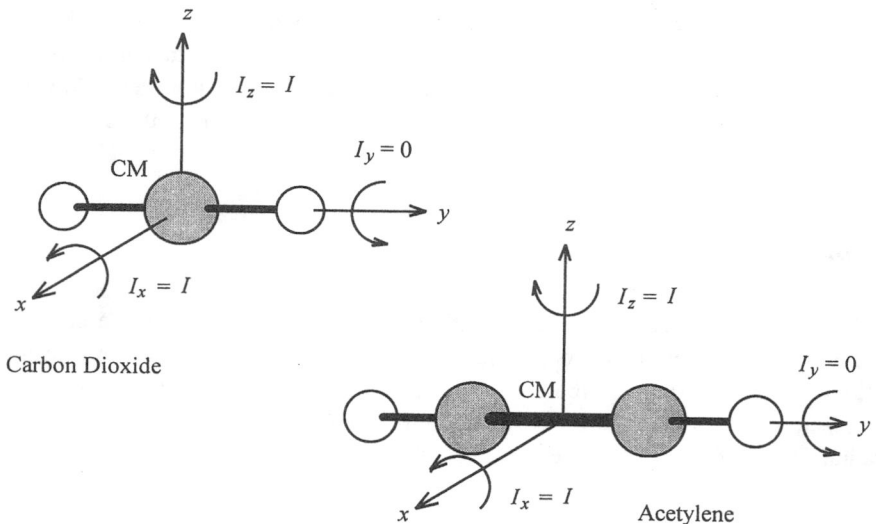

Figure 2.1 Rotation of rigid linear molecule, $I_x=I_z$, $I_y=0$ (CM = centre of mass).

The second form of Eq.(2.9) gives Θ_r in terms of the *rotational constant* B_e which may be determined spectroscopically. Values of Θ_r for some diatomic and poly-atomic gases are given in Table 2.2.

The symmetry number σ, which appears in the equation for q_r, takes the value 2 for a linear molecule with a centre of symmetry (e.g. all homonuclear diatomic molecules, acetylene, etc) and 1 for other molecules (e.g. CO, NO). This factor accounts for the fact that the number of distinguishable states for a molecule with a centre of symmetry is half that of a molecule which lacks this feature.

In Eq.(2.8), we can replace the summation by an integration when the actual temperature T is much larger than the characteristic rotational temperature Θ_r, as in such cases a large number of rotational levels are occupied. Hence

$$q_r = \frac{1}{\sigma}\int_0^\infty \exp(-x\,\Theta_r/T)\,dx ,\qquad (2.10)$$

where $x = j(1+j)$.

Table 2.2   Characteristic Rotational Temperatures

Diatomic [2]		Linear Triatomic [3]	
Molecule	Θ_r / K	Molecule	Θ_r / K
H_2	87.5	CO_2	0.561
N_2	2.888	N_2O	0.602
O_2	2.080	CS_2	0.157
Cl_2	0.351	COS	0.292
NO	2.460	HCN	2.127
HF	30.2		
HCl	15.22		
HBr	12.20		
HI	9.43		

Evaluating the integral, we obtain

$$q_r = \frac{8\pi^2 I\, k_B T}{\sigma\, h^2} = \frac{T}{\sigma\, \Theta_r} \tag{2.11}$$

and the corresponding rotational contributions to the thermodynamic properties are given in Table 2.3. For almost all substances, it is an excellent approximation to replace the summation over rotational energy levels by an integral. The only important exceptions are the isotopes of hydrogen and, possibly, some hydrides such as HF. These molecules have small moments of inertia and high rotational characteristic temperatures. For H_2, at 'ordinary' temperatures, only a few rotational levels

Table 2.3   Rotational Contributions to the Perfect-gas Thermodynamic Properties

	Linear molecules	Non-linear molecules
A_r	$-N k_B T \ln(T/\sigma\Theta_r)$	$-N k_B T \ln[(\pi^{1/2}/\sigma)(T^3/\Theta_A\Theta_B\Theta_C)^{1/2}]$
U_r	$N k_B T$	$(3/2) N k_B T$
S_r	$N k_B \ln(eT/\sigma\Theta_r)$	$N k_B \ln[(\pi^{1/2}/\sigma)e^{3/2}(T^3/\Theta_A\Theta_B\Theta_C)^{1/2}]$
P_r	0	0
G_r	$-N k_B T \ln(T/\sigma\Theta_r)$	$-N k_B T \ln[(\pi^{1/2}/\sigma)(T^3/\Theta_A\Theta_B\Theta_C)^{1/2}]$
H_r	$N k_B T$	$(3/2) N k_B T$
C_{Vr}	$N k_B$	$(3/2) N k_B$

are populated and it is necessary to evaluate q_r by direct summation. A further complication then arises for homonuclear molecules (e.g. H_2 and D_2 but not HD or HF) in the form quantum restrictions on the allowed rotational levels. This leads to two forms of molecular hydrogen, named ortho- and para-hydrogen, which have different heat capacities at low temperatures. A detailed treatment of this problem has been given elsewhere [2].

2.3.2. Non-linear Molecules

To specify the orientation in space of a non-linear molecule three angles are required. Hence, non-linear molecules have three principal moments of inertia, I_x, I_y, and I_z, about the body axes (see Figure 2.2). The origin of these axes is taken as the centre of mass, so that

$$\sum_i m_i x_i = \sum_i m_i y_i = \sum_i m_i z_i = 0 , \qquad (2.12)$$

where i refers to the various atoms of the molecule. The principal moments of inertia of a large selection of molecules have been tabulated elsewhere [3,4]. Values for a small selection of compounds are given in Table 2.4. For molecules not included in the tabulations the moments must be calculated from experimentally-determined bond distances and angles [5,6]. The calculation procedure is beyond the scope of this book but is described in detail in standard texts on classical mechanics [7].

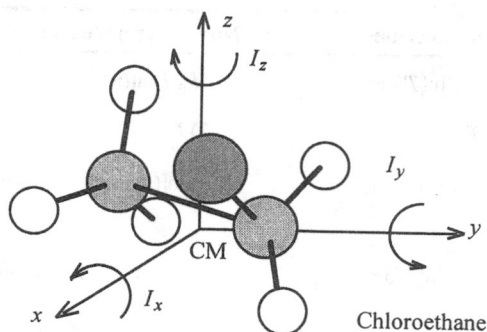

Figure 2.2 Rotations of a rigid polyatomic molecule with $I_x{\neq}I_y{\neq}I_z$ (CM = centre of mass).

Table 2.4 Principal Moments of Inertia of Some Non-linear Polyatomic Molecules.

| Spherical tops [5] | | Symmetrical tops [3] | | |
Molecule	$I_x / 10^{-47}$ kg·m^2	Molecule	$I_x / 10^{-47}$ kg·m^2	$I_z / 10^{-47}$ kg·m^2
CH$_4$	5.322	NH$_3$	2.816	4.43
CF$_4$	146.6	CH$_3$Cl	63.1	5.55
CCl$_4$	491	CH$_3$Br	87.5	5.40
SF$_6$	307			

| | Asymmetrical tops [3] | | |
Molecule	$I_x / 10^{-47}$ kg·m^2	$I_y / 10^{-47}$ kg·m^2	$I_z / 10^{-47}$ kg·m^2
H$_2$O	1.023	1.921	2.944
H$_2$S	2.694	3.097	5.927
SO$_2$	13.78	81.16	95.14
C$_2$H$_4$	33.84	28.09	5.75

Molecules for which the three moments of inertia are equal ($I_x = I_y = I_z$), for example CH$_4$ and SF$_6$, are known as *spherical tops*. Symmetric-top molecules have two moments of inertia equal, for example, CH$_3$Cl and C$_6$H$_6$; while non-linear molecules for which the three moments are all different are termed *asymmetric* tops (see, for example, Figure 2.2). This classification is followed in Table 2.4.

The procedure to obtain the rotational partition function is similar to that used for linear molecules but rather tedious; the result equivalent to Eq.(2.11) is

$$q_r = \frac{\pi^{1/2}}{\sigma} \left(\frac{T^3}{\Theta_A \, \Theta_B \, \Theta_C} \right)^{1/2} , \tag{2.13}$$

where the symmetry number σ is again the number of equivalent orientations in space (e.g. for CH$_4$, $\sigma=12$; rotating the molecule about each C-H bond produces three equivalent orientations, and so).

The rotational contributions to the perfect-gas thermodynamic properties of a non-linear molecule are exactly (3/2) times those given in Table 2.3 because we now have $q_r \propto T^{3/2}$ in place of $q_r \propto T$.

2.4. Vibrational Contributions

We consider a molecule composed of s atoms whose positions in space are described by $3s$ independent co-ordinates. It is convenient to divide these $3s$ co-ordinates as follows: three to describe the position of the centre of mass of the molecule; three to describe its orientation (or two if it is a linear molecule); and the remaining $3s$-6 (or $3s$-5) to describe the internal dispositions of the nuclei with respect to the centre of mass. Thus a diatomic molecule, which must be linear, has only one "internal co-ordinate", a linear triatomic molecule has four, and a non-linear triatomic molecule has three. One can show that the vibrations and internal rotations of a molecule may be separated into a number of independent modes equal in number to the number of internal co-ordinates. These independent modes, called the *normal modes*, have been described by, amongst other authors, Herzberg [2]. Here, we will consider only molecules without internal rotations and we will use the result that the vibrations of the molecule may then be analysed into $3s$-6 (or $3s$-5) normal modes. We also comment that the vibrational modes of the molecule give rise to characteristic absorption bands in the infrared and Raman spectra from which the molecular constants that we require in our analysis may be determined.

In the simplest treatment of molecular vibrations, the atoms are assumed to oscillate in a simple harmonic manner about their equilibrium positions (corresponding to the minimum potential energy of the molecule). The vibrational energy levels of the molecule are given in this approximation by

$$\varepsilon_{\rm v} = \sum_{i=1}^{3s-6}(n_i + \tfrac{1}{2})h\nu_i \ , \tag{2.14}$$

where n_i is the quantum number and ν_i the frequency of the ith normal mode. Allowed values of n_i are $n_i = 0, 1, 2, \cdots$ Since the vibrational energy is the sum of independent terms, it follows that the vibrational partition function is given by a product of terms, one for each of the normal modes:

$$q_{\rm v} = \prod_{i=1}^{3s-6} q_{{\rm v}i} \ . \tag{2.15}$$

The factor for each mode, excluding the zero-point energy, is given by the sum

$$q_{{\rm v}i} = \sum_{n=0}^{\infty}\exp(-n\Theta_{{\rm v}i}/T) \ , \tag{2.16}$$

Table 2.5 Characteristic Vibrational Temperatures

Diatomic [2]		Linear Triatomic [3]			
Molecule	Θ_{v} / K	Molecule	Θ_{v1} / K	Θ_{v2}^{*} / K	Θ_{v3} / K
H_2	6335	CO_2	1998	960	3380
O_2	2274	N_2O	1849	847	3199
N_2	3395	CS_2	945	571	2191
Cl_2	813	HCN	3006	1025	4766
CO	3120				
NO	2741				
HCl	4300				
HI	3322				

* The v2 mode of a linear triatomic molecule is a doubly-degenarate bending mode.

where the energy levels have been expressed in terms of the *characteristic vibrational temperature* Θ_{v} defined by

$$\Theta_{\mathrm{v}i} = \frac{h\nu_i}{k_{\mathrm{B}}} = \frac{hc\omega_{\mathrm{v}i}}{k_{\mathrm{B}}} . \tag{2.17}$$

The second form of Eq.(2.17) gives Θ_{v} in terms of the wavenumber $\omega_{\mathrm{v}i}$ of the corresponding infra-red or Raman absorption band. Some typical values of Θ_{v} are given in Table 2.5. Since the vibrational characteristic temperatures are generally comparable to, or larger than, the temperatures of interest, the summation in Eq.(2.16) cannot be replaced by an integral. However, being a geometric series, it can be evaluated exactly in closed form:

$$q_{\mathrm{v}i} = \sum_0 \exp(-n\Theta_{\mathrm{v}i}/T) = 1 + e^{-\Theta_{\mathrm{v}i}/T} + e^{-2\Theta_{\mathrm{v}i}/T} + \cdots$$

$$= \{1 - \exp(-\Theta_{\mathrm{v}i}/T)\}^{-1} . \tag{2.18}$$

The complete vibrational partition function is therefore given by

$$q_{\mathrm{v}} = \prod_{i=1}^{3s-6} \{1 - \exp(-\Theta_{\mathrm{v}i}/T)\}^{-1} . \tag{2.19}$$

In each of these equations, the upper limit for the number of terms should be replaced by (3s-5) for linear molecules. The vibrational contributions to the thermodynamic properties are the sum of contributions from each normal mode and the results are given in Table 2.6.

Since each vibrational mode is associated with a zero-point energy of $h\nu_i/2$, or $k_B\Theta_{vi}/2$, there is an additional constant contribution A_{v0} to the Helmholtz free energy which is not included in the equations of Table 2.6. This contribution is given simply by

$$A_{v0} \;=\; N \sum_{i=1}^{3s-6} (k_B\,\Theta_{vi}/2) \tag{2.20}$$

Since this is independent of T and V, but proportional to N, exactly that same term contributes also to U, H, and G, and the chemical potential is incremented by A_{v0}/N. Other properties are unaffected. In practice, these zero-point terms are of little consequence because only differences in A, U, H, G or μ are of physical significance.

Table 2.6 Vibrational Contributions to the Perfect-gas Thermodynamic Properties.

$$A_v \;=\; N k_B T \sum_{i=1}^{3s-6} \ln\{1 - \exp(-\Theta_{vi}/T)\}$$

$$U_v \;=\; N k_B \sum_{i=1}^{3s-6} \frac{\Theta_{vi}}{\{\exp(\Theta_{vi}/T) - 1\}}$$

$$S_v \;=\; N k_B \sum_{i=1}^{3s-6} \left[\frac{(\Theta_{vi}/T)}{\{\exp(\Theta_{vi}/T) - 1\}} - \ln\{1 - \exp(-\Theta_{vi}/T)\} \right]$$

$$P_v \;=\; 0$$

$$G_v \;=\; N k_B T \sum_{i=1}^{3s-6} \ln\{1 - \exp(-\Theta_{vi}/T)\}$$

$$H_v \;=\; N k_B \sum_{i=1}^{3s-6} \frac{\Theta_{vi}}{\{\exp(\Theta_{vi}/T) - 1\}}$$

$$C_{Vv} \;=\; N k_B \sum_{i=1}^{3s-6} \frac{(\Theta_{vi}/T)^2 \exp(\Theta_{vi}/T)}{\{\exp(\Theta_{vi}/T) - 1\}^2}$$

$$\mu_v \;=\; k_B T \sum_{i=1}^{3s-6} \ln\{1 - \exp(-\Theta_{vi}/T)\}$$

| **EXAMPLE 2.2** | *Calculation of Molar Entropy and Isobaric Molar Heat Capacity of CO_2 at $T=298.15$ K and $P=0.1$ MPa.* |

Molecular constants: CO_2 is a linear molecule with C=O bond length $r = 0.1159$ nm, $\sigma = 2$, and 4 vibrational modes with wave-numbers $\omega_i/\text{cm}^{-1} = 2349$, 1389, 667 and 667. Since the mass of an oxygen atom is $m_o = 2.656\times10^{-26}$ kg, the moment of inertia, $I = 2m_o r^2$, is 7.137×10^{-46} kg·m² and, from Eq.(2.9), $\Theta_r = (h^2/8\pi^2 I k_B) = 0.5643$ K. Since (hc/k_B) is 1.4388 cm·K, the characteristic vibrational temperatures $\Theta_{vi} = (hc/k_B)\omega_i$ of CO_2 are: $\Theta_{v1} = 3379.7$ K, $\Theta_{v2} = 1998.5$ K, $\Theta_{v3} = \Theta_{v4} = 959.7$ K (the reader may compare these values with those quoted in Table 2.5).

a)   Translational contributions. From Table 2.1:

$$C_{V,m,t} = \tfrac{3}{2} R \qquad\qquad = 12.472 \text{ J·mol}^{-1}\text{·K}^{-1}$$

$$S_{m,t} = R\ln\!\left[(2\pi\, m\, k_B T/h^2)^{3/2}(k_B T e^{5/2}/P)\right] \qquad = 155.943 \text{ J·mol}^{-1}\text{·K}^{-1}$$

where we have set $(V/N) = (k_B T/P)$ for the perfect gas and $N_A k_B = R$.

b)   Rotational contributions. From Table 2.3,

$$C_{V,m,r} = R \qquad\qquad = 8.314 \text{ J·mol}^{-1}\text{·K}^{-1}$$
$$S_{m,r} = R\ln(eT/\sigma\,\Theta) \qquad = 54.681 \text{ J·mol}^{-1}\text{·K}^{-1}$$

c)   Vibrational contributions. From Table 2.6,

$$C_{V,m,v} = R\sum_{i=1}^{3s-6}\left[(\Theta_{vi}/T)^2\, \exp(\Theta_{vi}/T)[\exp(\Theta_{vi}/T)-1]^{-2}\right]$$

$$= 0.013 + 0.460 + 2\times 3.742 \qquad\qquad = 7.957 \text{ J·mol}^{-1}\text{·K}^{-1}$$

$$S_{m,v} = R\sum_{i=1}^{3s-6}\left[\frac{(\Theta_{vi}/T)}{\exp(\Theta_{vi}/T)-1} - \ln[1-\exp(-\Theta_{vi}/T)]\right]$$

$$= 0.001 + 0.079 + 2\times 1.455 \qquad\qquad = 2.990 \text{ J·mol}^{-1}\text{·K}^{-1}.$$

The values of the required properties are therefore

$$C_{P,m}^{pg} = C_{V,m} + R = 37.06 \text{ J·mol}^{-1}\text{·K}^{-1} \quad\text{and}\quad S_m^{pg} = S_m = 213.61 \text{ J·mol}^{-1}\text{·K}^{-1}$$

The calorimetric values are $C_{P,m} = 37.14$ J·mol⁻¹·K⁻¹ and $S_m^{pg} = 213.9$ J·mol⁻¹·K⁻¹ [8]. The agreement is excellent and the calculated value of the entropy is probably more accurate than the calorimetric one.

❏

2.5. Internal Rotation Contributions

So for, we have treated the molecules as being essentially rigid except for well-defined molecular vibrations. This is indeed the case for many small molecules but, in more complex molecules, internal rotation of one part of the molecule relative to another often occurs. For example, in ethane the two CH_3 groups are able to rotate about the C-C bond. When such rotation is strongly hindered by a potential-energy barrier, the motion is essentially that of a torsional oscillator and it may be characterised best as a vibrational mode. However, the potential barrier to rotation is often quite small and almost free rotations occur. In such cases, one or more of the internal degrees of freedom is characterised as an internal rotation and there are correspondingly fewer truly-vibrational modes. The treatment of internal rotations will not be discussed here and the reader is referred to the literature on the subject (e.g. Janz [6]).

2.6. Nuclear Spin and Electronic Contributions

Due to the fact that the difference in energy between adjacent nuclear energy levels is extremely large ($\approx 10^{-5}$ J), almost all molecules are in the nuclear ground state except at extremely high temperatures ($T > 10^8$ K). The nuclear-spin partition function referred to the ground state at energy zero is

$$q_n = \sum_i \Omega_{ni} \exp(-\varepsilon_{ni}/k_B T) = \Omega_{n1} e^0 + \cdots \approx \Omega_{n1} \tag{2.21}$$

In the above equation, Ω_{ni} is the degeneracy of a given nuclear energy level. It can be seen that, since for practical purposes the nuclear-spin partition function q_n is a constant, it makes no contribution to U, P, H, or C_V and contributes a constant term to the A given by $A_{n0} = -Nk_B T \ln \Omega_{n1}$. The corresponding contributions to S, G and μ are A_{n0}/T, A_{n0} and A_{n0}/N respectively. Inclusion of these terms would therefore change the zero from which entropy and free energy are measured; but there would be no effect on differences in A, S, G or μ. Contributions from q_n may therefore be neglected.

Similarly, the electronic partition function is

$$q_e = \sum_i \Omega_{ei} \exp(-\varepsilon_{ei}/k_B T) = \Omega_{e1} + \Omega_{e2} e^{-\varepsilon_{e2}/k_B T} + \cdots$$

$$= \Omega_{e1} + \Omega_{e2} e^{-hc\omega_2/k_B T} + \cdots , \tag{2.22}$$

where ε_{ei} is the energy of the *i*th electronic state relative to ε_{e1}=0, *c* is the velocity of light and ω_2 is the wavenumber corresponding to the transition from the electronic ground state to the lowest excited state. Values of the wavenumber for electronic energy levels of atoms [10] and molecules [2,11] have been determined spectroscopically and tabulated together with the degeneracies of each level.

For the vast majority of molecules, the energy gap between the ground state and the first excited electronic state is large ($\approx 10^{-11}$ J), and almost all molecules are in the ground state unless the temperature is extremely high ($T>1000$ K). In such cases Eq.(2.22) reduces to the constant term $q_e = \Omega_{e1}$ and there is then no contribution to U, P, H, or C_V but A is incremented by $A_{e0}= -Nk_BT \ln\Omega_{e1}$, S by A_{e0}/T, G by A_{e0} and μ by A_{e0}/N. We shall generally neglect these terms but there are circumstances where electronic excitation is significant and at least the first two terms of Eq.(2.22) are required; Example 2.3 illustrates this.

EXAMPLE 2.3	*Calculation of the Partition Function of Chlorine Atoms at* $T = 1000$ K *and* $P = 1$ Pa.

Atomic constants: for the chlorine atom, $m = 5.887 \times 10^{-26}$ kg and the first two electronic levels are as follows: $\omega_{e1} = 0$, $\Omega_{e1} = 4$ and $\omega_{e2} = 881$ cm^{-1}, $\Omega_{e2} = 2$ [10]; higher excited states are essentially unoccupied at 1000 K.

For chlorine atoms, $q = q_t q_e$.

a)   Translational contribution. From Eq.(2.7),

$$q_t = \frac{k_B T}{P} \left(\frac{2\pi m k_B T}{h^2} \right)^{3/2} \qquad = 1.73 \times 10^{13},$$

where we have set $V = k_B T/P$ for a system of one atom.

b)   Electronic contribution. From Eq.(2.20),

$$q_e = \Omega_{e1} + \Omega_{e2}\, e^{-hc\omega_2/k_B T} + \cdots \qquad = 4.56$$

and therefore, $q = 1.03 \times 10^{32} \times 4.563$ $= 7.89 \times 10^{13}$.

□

2.7. Rotational-Vibrational Coupling Contributions

The assumption that molecular rotations and vibrations are independent is not strictly valid and may lead to errors at high temperatures. A more exact treatment [9] includes three corrections which arise from: (a) anharmonicity in the vibrations, (b) centrifugal stretching of the bonds due to rotations; and (c) periodical variation of the moments of inertia because of vibrations. However, for most molecules at normal temperatures the correction to the heat capacity and other properties is of the order of 0.1 per cent and is usually neglected. As the temperature rises, the correction may become more appreciable.

EXAMPLE 2.4	*Calculation of the Isobaric Molar Heat Capacity of CO in the Temperature Range 300 K to 1200 K.*

Assuming that (a) the rotation and vibration are independent contributions and (b) excited electronic states do not contribute at these temperatures, we have

$$C_{P,m} = R + C_{V,m} = R + (C_{V,m,t} + C_{V,m,r} + C_{V,m,v})$$

$$= R \left[\frac{5}{2} + \frac{(\Theta_v/T)^2 \exp(\Theta_v/T)}{\{\exp(\Theta_v/T)-1\}^2} \right] ,$$

where we have used the contributions to the molar heat capacity at constant volume from Tables 2.1, 2.3 and 2.6 and the fact that since CO is a linear diatomic molecule it has only 1 vibrational mode. The characteristic vibrational temperature, obtained from Table 2.5, is equal to 3120 K. We then obtain the results given in the following table.

T/K	$C_{V,m,t}/R$	$C_{V,m,r}/R$	$C_{V,m,v}/R$	$C_{P,m}$ / J·mol⁻¹·K⁻¹
300	1.5	1.0	0.003	29.128
600	1.5	1.0	0.151	30.355
900	1.5	1.0	0.414	32.539
1200	1.5	1.0	0.586	33.972

The correction to the heat capacity at constant pressure due to the rotational-vibrational coupling at 1200 K is only about 0.15 J·mol⁻¹·K⁻¹ [1], so that its neglect does not introduce serious error.

❏

2.8. Perfect-Gas Mixtures

We consider an r-component mixture in volume V at temperature T, having N_1, N_2, $\cdots$ N_r, molecules of components 1, 2, $\cdots$ and r respectively. We note that molecules of different species are distinguishable, while those of the same species are not, and furthermore that all molecules are independent. Each species will have its own set of energy levels which are functions of volume but unaffected by the presence of the other components.

Following a generalisation of the argument used to derive Eq.(1.23), it is readily shown that the canonical partition function of the system is simply the product of r terms, one for each component:

$$Q = \prod_{i=1}^{r} Q_i = \prod_{i=1}^{r} \frac{\left(q_i\right)^{N_i}}{N_i!} . \tag{2.23}$$

Here, $Q_i = (q_i)^{N_i}/N_i!$ is the canonical partition of component i alone in volume V at temperature T. This is given in terms of the one-molecule partition function q_i,

$$q_i(V,T) = \sum_j \exp(-\varepsilon_{ij}/k_B T) , \tag{2.24}$$

where ε_{ij} are the energy states for component i.

The Helmholtz free energy of the mixture is therefore given by

$$\begin{aligned} A &= -k_B T \ln Q \\ &= -N k_B T - \sum_{i=1}^{r} N_i k_B T \ln(q_i/N_i) \end{aligned} \tag{2.25}$$

where N is the total number of molecules of all types and Stirling's approximation has been applied to $\ln N_i!$. All of the other thermodynamic properties of the perfect-gas mixture may be obtained through appropriate manipulations of Eq.(2.25).

2.9. Summary

The expressions presented in this chapter for the various thermodynamic properties of the perfect gas are based on the assumption that the intramolecular energy is given by independent contributions from rotation, vibration, etc. according to Eq.(2.3). However, this approximation is relatively innocuous. In fact, the calculations are in excellent agreement with experiment for simple molecules and discrepancies, where they exist, usually arise from experimental error rather than unsound theory. The weakest point in the theory is in the treatment of internal rotations which often exhibit behaviour intermediate between free rotation and torsional oscillation.

The molecular constants required for the calculation of perfect-gas properties are obtained from spectroscopic measurements of rotational, vibrational and electronic energy levels. Such data are readily available for a wide variety of molecules [2,3,5,6,10,11] and, where they are not, bond-contribution methods exist for their estimation [12].

2.10. Computational Implementation

The perfect-gas properties are required as a separate contribution in the calculation of real-fluid properties and are therefore of more than academic interest. Although tables of perfect-gas properties, based on a combination of theoretical and experimental work, are available in the literature [13], it is convenient to have computer programs from which they may be evaluated routinely. Because of the difficulties with internal rotations and, to a lesser extent, vibration-rotation interaction, it is pragmatic to adopt empirical representations for some of the properties rather than to calculate everything directly from the partition function. That is not to say that the theory that we have derived is impractical, only that it is unsuitable for routine calculations in most cases because of the complicated treatment required for some terms. In the following, we give practical routines for the evaluation of perfect-gas properties.

2.10.1. Heat Capacities, Enthalpy and Entropy

Molar isobaric perfect-gas heat capacities obtained for a very wide range of compounds from theory and/or experiment [13], have been fitted [14,15] by the third-degree polynomial:

$$C_{P,m}^{pg} = C_{P,0} + C_{P,1}T + C_{P,2}T^2 + C_{P,3}T^3 \tag{2.26}$$

In Appendix A, the coefficients of these polynomials are given for a selection of compounds so that $C_{P,m}$ may be calculated. The molar perfect-gas enthalpy is then obtained from the relation

$$H_m^{pg}(T) = H_m(T_0) + \int_{T_0}^{T} C_{P,m}^{pg}\, dT \tag{2.27}$$

and the molar perfect-gas entropy from

$$S_m^{pg}(T,P) = S_m(T_0,P_0) + \int_{T_0}^{T}(C_{P,m}^{pg}/T)dT - R\ln(P/P_0). \tag{2.28}$$

In these equations, T_0 and P_0 are reference values of the temperature and pressure, $H_m(T_0)$ is a constant of integration for molar enthalpy and $S_m(T_0,P_0)$ is a constant of integration for molar entropy. Since only differences in enthalpy or entropy are of physical significance, one can chose the values of these quantities at will but one should be aware of the values used when comparing data from different sources. In this book we define the reference state as T_0 = 273.15 K and P_0= 0.10133 MPa; and we set $H_m(T_0) = 0$ and $S_m(T_0,P_0) = 0$. In other words, **enthalpy and entropy are set equal to zero at T = 273.15 K and P = 0.10133 MPa.**

PRFGAS (shown in Display 2.1) is a very simple subroutine that may be used to calculate the perfect-gas heat capacity, enthalpy and entropy, according to the aforementioned method. Although simple, it has wide application in property calculations.

```
C       PGCP=W1+W2*T+W3*(T**2)+W4*(T**3) in [J/mol/K]
C
C
C       - Enthalpy and Entropy are arbitrarily assumed to be equal to ZERO at 273.15 K
C         and 0.10133 MPa.
C
        T0=273.15
        P0=0.10133E+6
        R=8.31451
        PGCP=W1+W2*T+W3*T*T+W4*(T**3)
        PGCV=PGCP-R
        PGH=W1*(T-T0)+W2*((T**2)-(T0**2))/2.+W3*((T**3)-(T0**3))/3.+
       &W4*((T**4)-(T0**4))/4.
        PGS=W1*LOG(T/T0)+W2*(T-T0)+W3*((T**2)-(T0**2))/2.+W4*((T**3)-(T0**3))/3.-
       &R*LOG(P/P0))
        RETURN
        END
```

EXAMPLE 2.5   *Calculation of the Molar Isobaric Perfect-gas Heat Capacity of CO for the Temperature Range 300 K to 1200 K.*

Sample Input/Output is shown in Display 2.2. Values for the coefficients of Eq.(2.26) are taken from Appendix A. The calculated values agree, as expected, with values obtained from the partition function in Example 2.4 to within 1 per cent.

Display 2.2 PRFGAS Sample Input/Output

```
        PROGRAM EX0205
        DATA P,W1,W2,W3,W4/0.10133E+6,30.87,-1.285E-2,2.789E-5,-1.272E-8/
        WRITE (*,1000)
1000    FORMAT(2X,'TEMPERATURE   CP     CV     ENTHALPY   ENTROPY', +/,1X
       &'   [K]      [J/MOL/K]  [J/MOL]  [J/MOL/K]', +/,2X,'----------------------------
       &-------------------------------------------------------')
        DO 100 T=300,1200,300
        CALL PRFGAS(T,P,W1,W2,W3,W4,PGH,PGS,PGCV,PGCP)
100     WRITE (*,1020) T,PGCP,PGCV,PGH,PGS
1020    FORMAT(1X,5F10.2)
        END
```

TEMPERATURE [K]	CP [J/MOL/K]	CV [J/MOL/K]	ENTHALPY [J/MOL]	ENTROPY [J/MOL/K]
300.00	29.18	20.87	783.47	2.74
600.00	30.45	22.14	9680.42	23.24
900.00	32.62	24.31	19145.09	36.00
1200.00	33.63	25.32	29138.07	45.58

❑

EXAMPLE 2.6   *Calculation of the Molar Perfect-Gas Entropy of CO_2 in the Temperature Range 300 K to 1200 K at P=0.10133 MPa.*

The Sample Input shown in Display 2.2 is used together with values of the coefficients of Eq.(2.26) taken from Appendix A. As already stated, the values calculated by PRFGAS are relative to assumed zeros of enthalpy and entropy at T_0 = 273.15 K and P_0 = 0.10133 MPa. The values obtained are shown in the second column of the Table below. In the third column, the values of 'absolute' entropy obtained in Example 2.2. from the partition function are given. In order to permit comparison, we give in the last column the difference $S_m(T,P_0) - S_m(T_0,P_0)$ according to the partition-function method. The two sets of values agree to within 0.6 per cent.

T / K	$\dfrac{S_m(T,P_0) - S_m(T_0,P_0)}{J \cdot mol^{-1} \cdot K^{-1}}$	$\dfrac{S_m(T,P_0) - S_m(T \to 0,P_0)}{J \cdot mol^{-1} \cdot K^{-1}}$	$\dfrac{S_m(T,P_0) - S_m(T_0,P_0)}{J \cdot mol^{-1} \cdot K^{-1}}$
300	3.44	213.85	3.43
600	32.71	242.99	32.57
900	53.10	263.25	52.83
1200	68.89	278.91	68.49

❏

References

1. T.M. Reed and K.E. Gubbins, *Applied Statistical Mechanics* (McGraw-Hill Kogakusha, 1973).
2. G. Herzberg, *Molecular Spectra and Molecular Structure. Vol. 1. Spectra of Diatomic Molecules*, 2nd Ed. (Van Nostrand, Princeton, 1970).
3. Landolt-Bornstein, Band 1, *Atom-und Molekularphysik. Teil 2. Molekulen*, 1, p.328, (Springer-Verlag, Berlin, 1951).
4. *JANAF Thermochemical Tables*, 2nd ed. (Dow Chemical Co., Thermal Laboratory, Midland, Mich., 1965).
5. H.J.M. Bowen, *Tables of Interatomic Distances and Configuration in Molecules and Ions* (The Chemical Society, London, 1958; Suppl. 1965).
6. G.J. Janz, *Thermodynamic Properties of Organic Compounds*, rev. ed. (Academic, New York, 1967).
7. T.W.B. Kibble, *Classical Mechanics* (McGraw-Hill, London, 1966).
8. J.F. Masi and B. Petkof, *J. Res. Nat. Bur. Stand.* **48** (1952) 179.

9. K.S. Pitzer, *Quantum Chemistry* (Prentice-Hall, Englewood Cliffs, N.J., 1953).
10. C.E. Moore, *Atomic Energy States* (Nat. Bur. Stand. Circ. 467, vols.1-3, 1949-1958).
11. G. Herzberg, *Infrared and Raman Spectra of Polyatomic Molecules* (Van Nostrand, Princeton, N.J., 1945).
12. M.T. Howerton, *Engineering Thermodynamics* (Van Nostrand, Princeton, N.J., 1962).
13. *Selected Values of Properties of Hydrocarbons and Related Compounds* (Thermodynamic Research Center, Texas A&M University, 1977 and 1987).
14. T.P. Thinh, J.L. Duran, R.S. Ramalho and S. Kaliaguine, *Hydrocarbon Processing* **50** (1971) 98.
15. R.C. Reid, J.M. Prausnitz and B.E. Poling, *The Properties of Gases and Liquids*, 4th Ed. (McGraw Hill, 1988).

3

The Intermolecular Potential

In the first chapter, it was shown that all of the thermodynamic properties of a system may be obtained from the canonical partition function Q and that the effects of intermolecular forces enter the problem through the configuration integral $\mathcal{L}$. In the second chapter, the thermodynamic properties were calculated for the special case of the perfect gas; in that case, there are no intermolecular forces and $\mathcal{L}$ is simply V^N. We have also observed that the exact evaluation of the configuration integral for real systems with intermolecular forces is generally impossible and that some approximations must be made. Before examining possible approximations it is desirable to discuss briefly the nature of the forces between molecules which give rise to these difficulties. In this chapter, we therefore discuss intermolecular force, the associated intermolecular potential energy ϕ for a pair of molecules, and the total intermolecular potential energy $\mathcal{U}$ for a cluster of molecules. The most commonly encountered model pair potentials will also be presented.

3.1. Intermolecular Pair Potential Energy

We consider first the interaction of two neutral atoms a and b. The total energy $E_{tot}(r)$ of the pair of atoms at separation r is written as

$$E_{tot}(r) = E_a + E_b + \phi(r) . \tag{3.1}$$

Here, E_a and E_b are the energies of the isolated atoms, and $\phi(r)$ is the contribution to the total energy arising from interactions between them. We call $\phi(r)$ the *intermolecular pair potential energy function* and, in the present example, it depends only on the separation of the two atoms. Since this energy is equal to the work done in bringing the two atoms from infinite separation to the separation r, it is given in terms of the intermolecular force $F(r)$ by

$$\varphi(r) = \int_r^\infty F(r)\, dr \ . \tag{3.2}$$

By convention the force F is positive when repulsive and negative when attractive. For molecules, as distinct from atoms, the intermolecular potential energy depends, not only upon separation, but also upon the relative orientation.

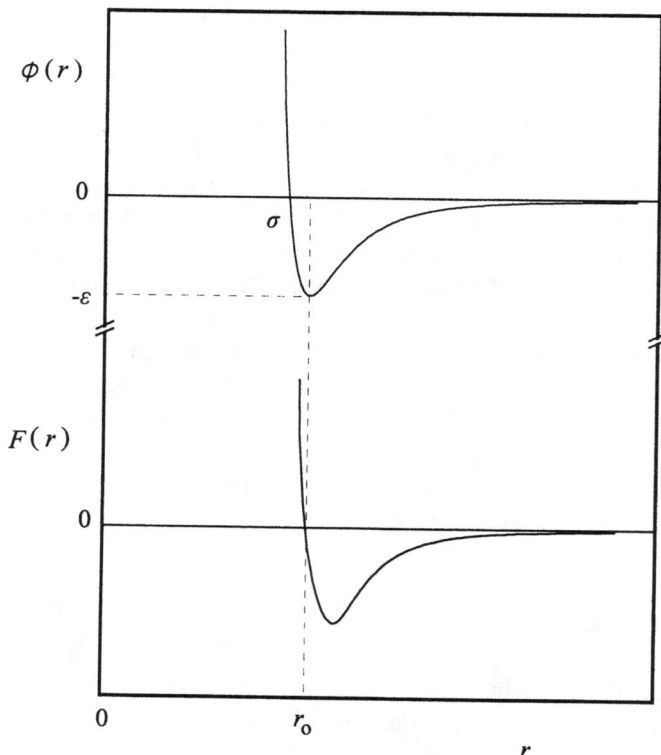

Figure 3.1 The intermolecular pair potential energy and force.

The general forms of $\varphi(r)$ and $F(r)$ are illustrated in Figure 3.1 [2]. We see that, at short range, a strong repulsion acts between the molecules while, at longer range, there is an attractive force which decays to zero as $r \to \infty$. Consequently, the potential energy $\varphi(r)$ is large and positive at small separations but negative at longer range. It is known that, for neutral atoms at least, there is only one minimum and no maximum in both $F(r)$ and $\varphi(r)$. The parameters σ, r_0 and ε usually employed to characterise the intermolecular pair potential energy are defined in Figure 3.1. σ is the separation at which the potential energy crosses zero, r_0 is the separation at which $\varphi(r)$ is minimum and $-\varepsilon$ is the minimum energy.

Intermolecular forces are known to have an electromagnetic origin [2] and the main contributions are well established. The strong repulsion which arises at small separations is associated with overlap of the electron clouds. When this happens, there is a reduction in the electron density in the overlap region leaving the positively charged nuclei incompletely shielded from each other. The resulting electrostatic repulsion is referred to as an *overlap force*. At greater separations, where attractive forces predominate, there is little overlap of electron clouds and the interaction arises in a different manner. Here, the attractive forces are associated with electrostatic interactions between the essentially-undistorted charge distributions which exist in the molecules.

There are in fact three distinct contributions to the attractive forces which will be discussed here only briefly; for a more detailed description the reader is referred to the specialised literature [2]. For polar molecules, such as HCl, the charge distribution in each molecule generates a permanent electric field and, when two such molecules are close, there is an *electrostatic force* which depends upon both separation and orientation. The force may be either positive or negative, depending upon the mutual orientation of the dipoles, but the net effect on the bulk properties of the fluid is that of an attractive force. Such interactions are not associated exclusively with dipole moments. Molecules, such as CO_2, which have a quadrupole moment also have electrostatic interactions of a similar nature. These interactions exist in general when both molecules have one or more non-zero multipole moments.

There is a second contribution to the attractive force which exists when at least one of the two molecules possess a permanent multipole moment. This is known as the *induction force* and it arises from the fact that the molecules are polarizable; that is, a dipole moment is induced in a molecule when it is placed in an electric field. Thus a permanent dipole moment in one molecule will induce a dipole moment in an adjacent molecule. The permanent and induced moments interact to give a force which is always attractive and, at long range, proportional to r^{-6}.

The third contribution to the attractive force, and the only one present when both molecules are non-polar, is known as the *dispersion force*. This arises from the fact that even non-polar molecules generate fluctuating electric fields associated with the motion of the electrons. These fluctuating fields around one molecule give rise to an induced dipole moment in a second nearby molecule and a corresponding energy of interaction. Like induction forces, dispersion forces are always attractive and, at long range, vary like r^{-6}.

3.2. Basic Approximations

As already discussed, the potential energy $\mathcal{U}$ of a cluster of molecules is a function of the intermolecular interactions, which in turn depend upon the type and number of molecules under consideration, the separation between each molecule and their mutual orientation. The term *configuration* is used to define the set of co-ordinates which describe the relative position and orientation of the molecules in a cluster.

When the distances between molecules are large, as is the case on average in a very dilute gas, the intermolecular interactions are sufficiently weak that the potential energy $\mathcal{U}$ of the system can be set to zero. However, when the system is compressed, the average intermolecular distances are reduced and the effects of attractive and repulsive forces between molecules give rise to a significant potential energy.

In order to estimate the potential energy of a configuration it is usual, and often necessary, to make some or all of the following simplifications:

1. The term *intermolecular pair-potential energy* is used to describe the potential energy involved in the interaction of an isolated pair of molecules. It is very convenient to express the total potential energy $\mathcal{U}$ of a cluster of molecules in terms of this pair potential φ. This leads to a very important assumption, the *pair additivity approximation*, by which the total potential energy of a system of molecules is set equal to the summation of all possible pair interaction energies. This implies that the interaction between a pair of molecules is unaffected by the proximity of other molecules.

2. The second important assumption is that the pair-potential energy depends only on the separation of the two molecules. This assumption is valid only for monatomic molecules where, due to the spherical symmetry, the centres of molecular interaction coincide with the centres of mass. For polyatomic molecules the absence of spherical symmetry causes *acentric interactions*.

3. Finally, since the intermolecular potential is known accurately for only a few simple systems, model functions need to be adopted in most cases. Typically, such models give φ as a function only of the separation between molecules but nevertheless the main qualitative features of molecular interactions are incorporated.

For a system of N spherical molecules, the general form of the potential energy $\mathcal{U}$ may be written as

$$\mathcal{U}(\mathbf{r}_1, \mathbf{r}_2, \ldots, \mathbf{r}_N) = \sum_{i=1}^{N-1} \sum_{j=i+1}^{N} \varphi_{ij} + \Delta \varphi_N \qquad (3.3)$$

where ϕ_{ij} is the potential energy of the isolated pair of molecules i and j, and $\Delta\phi_N$ is an increment to the potential energy, characteristic of the whole system, over and above the strictly pairwise interactions. According to the pair-additivity approximation, this reduces to

$$\mathcal{U}(\mathbf{r}_1,\mathbf{r}_2,...,\mathbf{r}_N) = \sum_{i=1}^{N-1}\sum_{j=i+1}^{N}\phi_{ij} = \sum_{i<j}\phi_{ij} \qquad (3.4)$$

This approximation implies that the N-body interactions (with $N > 2$) are negligible compared with the pairwise interactions. In fact, many-body forces are known to make a small but significant contribution to the potential energy when $N \geq 3$ and, for systems at higher density, the pair-additivity approximation can lead to significant errors. However, it is often possible to employ an effective pair potential which gives satisfactory results for the dense fluid whilst still providing a reasonable description of dilute-gas properties.

EXAMPLE 3.1   *Intermolecular Potential for Systems of three and four Molecules.*

For a set of three molecules there are three terms describing the pairwise interaction of the molecules and one term referring to the three-body interaction. Hence the potential energy according to Eq.(3.3) is written as

$$\mathcal{U}(\mathbf{r}_1,\mathbf{r}_2,\mathbf{r}_3) = \phi_{12} + \phi_{13} + \phi_{23} + \Delta\phi_3$$

Similarly for the four-molecules system

$$\mathcal{U}(\mathbf{r}_1,\mathbf{r}_2,\mathbf{r}_3,\mathbf{r}_4) = \phi_{12} + \phi_{13} + \phi_{14} + \phi_{23} + \phi_{24} + \phi_{34} + \Delta\phi_4$$

❑

3.3. Model Pair Potentials

The difficulties encountered in the theoretical evaluation of the intermolecular pair-potential energy led to the adoption of the following rather heuristic approach. The evaluation procedure starts with the assumption of an analytical form for the relationship between the potential energy φ and the distance r between molecules. Subsequently, macroscopic properties are calculated using the appropriate molecular theory. Comparisons between calculated and experimental values of these macroscopic properties provide a basis for the determination of the parameters in the assumed intermolecular potential-energy function. Finally, predictions may be made of thermodynamic properties of the fluid in regions where experimental information is unavailable.

In the following sections, we present some of the most widely used model potential-energy functions. For a more comprehensive discussion the reader is referred to specialised literature [2].

3.3.1. Hard-sphere Potential

In this model, the molecules are assumed to behave as smooth, elastic, hard spheres of diameter σ. It is apparent that the minimum possible distance between the molecules is then equal to σ and that the energy needed to bring the molecules closer together than $r = \sigma$ is infinite (see Figure 3.2). For separation $r > \sigma$, there is no interaction between the molecules. The mathematical form of the potential is given by the following discontinuous function

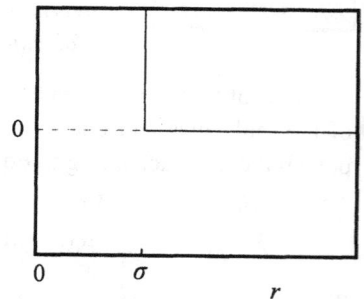

Figure 3.2 Hard-sphere potential.

$$\varphi(r) = \infty \qquad r < \sigma$$
$$\varphi(r) = 0 \qquad r \geq \sigma \tag{3.5}$$

Although this model is not very realistic, its does incorporate the basic idea that the molecules themselves occupy some of the system volume. The hard-sphere model is important in the theory of dense fluids.

3.3.2. Square-well Potential

This potential function is a more realistic one in the sense that it includes an attractive potential field, of depth ε and range $g\sigma$, surrounding the spherical hard core (see Figure 3.3). Commonly used values of g are between 1.5 and 2.0. The mathematical form of the model is:

$$\varphi(r) = \infty \qquad r < \sigma$$
$$\varphi(r) = -\varepsilon \qquad \sigma \leq r < g\sigma \qquad (3.6)$$
$$\varphi(r) = 0 \qquad r \geq g\sigma$$

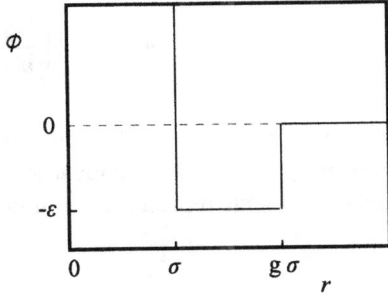

Figure 3.3 Square-well potential.

3.3.3. Lennard-Jones (12-6) Potential

The Lennard-Jones (12-6) potential illustrated in Figure 3.4, accounts for both attractive and repulsive energies and assumes that the interaction between the molecules occurs along the line joining their centres of mass. It is one of the most commonly used models due to its apparently-realistic form and its mathematical simplicity.

The functional form of the Lennard-Jones (12-6) pair-potential is given by

$$\varphi(r) = 4\varepsilon\left[\left(\frac{\sigma}{r}\right)^{12} - \left(\frac{\sigma}{r}\right)^{6}\right] \qquad (3.7)$$

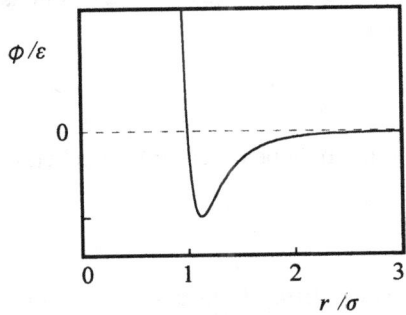

Figure 3.4 Lennard-Jones (12-6) potential.

Although this model looks realistic, it is not actually an accurate representation of any of the few intermolecular potentials that are known well. However, despite its approximate nature, the parameters can be chosen so as to give a useful representation of the bulk behaviour of many real systems.

A generalisation of the (12-6) model is sometimes used in which the repulsive exponent is permitted to take any value greater then 6. This function is given most conveniently in terms of the parameters ε and r_0:

$$\phi(r) = \varepsilon \left[\left(\frac{6}{n-6} \right) \left(\frac{r_0}{r} \right)^n - \left(\frac{n}{n-6} \right) \left(\frac{r_0}{r} \right)^6 \right], \tag{3.8}$$

where $r_0 = \sigma (n/6)^{1/(n-6)}$ is the separation at the minimum of the function. The extra flexibility afforded by the inclusion in $(n\text{-}6)$ model of the third parameter may permit a more accurate representation of the properties of real systems.

3.3.4. Interactions of Unlike Molecules

In a mixture there will be interactions between both like and unlike molecules and, often, the same model intermolecular potential-energy function is applied to all of them. In principle, the parameters for each of the potentials may be determined separately but there is often insufficient information available to do this for the unlike interactions and the question then arises of how best to predict these terms. The most common situation is that in which both like and unlike potentials are in the form $\phi_{ij} = \varepsilon_{ij} F(r_{ij}/\sigma_{ij})$ with a specified function F, such as the Lennard-Jones $(n\text{-}6)$ potential with n fixed. It is then a matter of predicting the values of the energy and length parameters, ε_{ij} and σ_{ij}, for the unlike interaction in terms of those for the pure components. Many mixing or combining rules have been proposed for this purpose but we mention just two. The first is the Lorentz rule for the length parameter which states

$$\sigma_{12} = (\sigma_{11} + \sigma_{22})/2 \; ; \tag{3.9}$$

this would be exact for hard spheres. The second is the Berthelot rule,

$$\varepsilon_{12} = (\varepsilon_{11} \varepsilon_{22})^{1/2} , \tag{3.10}$$

which arises from a simple treatment of the attractive dispersion energy. These combining rules form the basis of the treatment of unlike interaction terms in many corresponding-states and equation-of-state methods.

3.4. Non-spherical Systems

In the more general case of the interaction of polyatomic molecules, the angular dependence of the potential should be considered. It may be necessary to include up to five angular variables to describe the relative orientation of a pair of molecules explicitly. However, should we wish to do so, we can still think in terms of a one-dimensional function for any fixed orientation of the molecules. As an example, Figure 3.5 shows two sections through a model potential which has been proposed for the system $Ar\cdots CO_2$. In this case, the potential is quite strongly anisotropic and

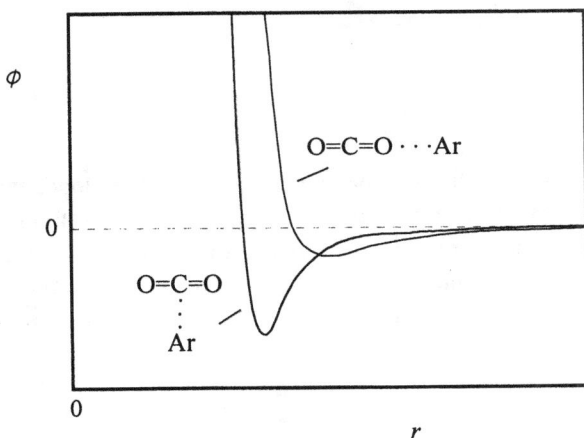

Figure 3.5 Sections through a pair potential for the system $Ar\cdots CO_2$ for a "T" shaped and a linear configuration.

the parameters σ and ε characterising the interaction along different paths of fixed orientation show marked variation. Clearly, the exact mathematical description of such potentials is very complicated. The key features of non-spherical molecules that give rise to anisotropic forces are:

1.   the non-spherical "core" geometry which dominates the anisotropy of the repulsive part of the potential; and
2.   the presence of electric multipoles, especially dipole or quadrupole moments, which give rise to anisotropic electrostatic forces that may be dominant at longer range.

This last point is of considerable importance and dipolar forces are often included in model intermolecular pair potentials where appropriate. The most common model which includes such forces is the *Stockmayer Potential* which consists of a central Lennard-Jones (12,6) potential plus the energy of interaction of two dipole moments:

$$\phi(r,\theta_1,\theta_2,\psi) = 4\varepsilon\left[\left(\frac{\sigma}{r}\right)^{12} - \left(\frac{\sigma}{r}\right)^6\right] - \frac{\mu^2}{4\pi\varepsilon_0 r^3}(2\cos\theta_1\cos\theta_2 - \sin\theta_1\sin\theta_2\cos\psi) \quad (3.11)$$

Here the angles θ_1, θ_2 and ψ define the mutual orientation of the dipole moments. θ_i is the angle made between the dipole moment on molecule i and the intermolecular axis, while ψ is the relative azimuthal angle between the two dipoles about the same axis.

3.5. Summary

Although the general form of the potential is well understood, as we move from a) monatomic interactions to b) rigid non-polar simple polyatomic molecules, c) polar simple polyatomic molecules and d) flexible molecules both polar and non-polar, its description becomes progressively more difficult. Approximations are therefore employed for all but the simplest systems and various models have been devise for this purpose.

References

1. J.D. van der Waals, *PhD thesis* (Leiden, 1873).
2. G.C. Maitland, M. Rigby, E.B. Smith and W.A. Wakeham, *Intermolecular Forces. Their Origin and Determination* (Clarendon Press, Oxford, 1981).

4

The Virial Equation

In Chapter 2, the thermodynamic properties of the perfect gas were discussed; we now turn our attention to real gases. Real systems are, of course, distinguished from ideal ones by the presence of intermolecular forces which enter the problem through the configuration integral $\mathcal{L}$. Unfortunately, even with the pair additivity approximation, evaluation of $\mathcal{L}$ is still an N-body problem which can be solved numerically for, at most, only a few thousand particles. This is clearly an impractical approach for real systems. In the present chapter, we examine an alternative treatment in which the configuration integral is expanded as a power series in the density about the zero-density limit. We show that the mth coefficient of this series is rigorously related to molecular interactions in clusters of m molecules. Hence, provided that the series converges satisfactorily, the intractable N-body problem is transformed into a soluble series of 1-body, 2-body, 3-body, $\cdots$ problems. Evaluation of the pressure from this series leads directly to the *virial equation of state*.

The coefficients of the virial series, known as *virial coefficients*, are functions of temperature and composition but not of density. Much of the importance of the virial equation of state lies in its rigorous theoretical foundation by which the virial coefficients appear not merely as empirical constants but with a precise relation to the intermolecular potential energy. Consequently, experimental values of the virial coefficients can be used to obtain information about intermolecular forces or, conversely, virial coefficients may be calculated from the intermolecular potential-energy function. Moreover, exact relations exist for the virial coefficients of a gaseous mixture in terms of like- and unlike-molecular interactions.

The virial series is valid (*i.e.* it converges) only for sufficiently low densities. The radius of convergence is not well established theoretically except for hard spheres for which it encompasses all fluid densities. In real systems, the empirical evidence suggests that the series converges up to approximately the critical density. It certainly does not converge either for the liquid phase or in the neighbourhood of the critical point. Furthermore, since not all of the coefficients of the virial series are known, the series is usually limited in practice to densities much below the critical.

Historically, the virial equation was originally proposed at 1885 on purely empirical grounds by Thiesen and later developed thoroughly by Onnes in 1901. Only in 1927, with the work started by Ursell, was it shown to evolve naturally from a statistical mechanical analysis of the forces between molecules.

4.1. The Virial Equation of State

It is known that the behaviour of any real gas approaches that of the perfect gas as the pressure, P, and the amount-of-substance density, ρ_n, approach zero. In this limit P/ρ_n is finite and equal to RT (see Table 2.1). Here, $R = N_A k_B$ is the universal gas constant, $\rho_n = N/(N_A V)$, and N_A is Avogadro's constant. The behaviour of a real gas at constant temperature can be expressed in a formal way by writing the compression factor, $P/\rho_n RT$, as an infinite power series in ρ_n with the requirement that $P/\rho_n RT = 1$ when $\rho_n = 0$. A Maclaurin expansion of $P/\rho_n RT$ about $\rho_n = 0$ serves this purpose and gives the well-known virial equation of state:

$$\frac{P}{\rho_n RT} = 1 + B\rho_n + C\rho_n^2 + D\rho_n^3 + \cdots . \tag{4.1}$$

It has already been mentioned that the virial coefficients B, C, $\cdots$ are functions of the temperature (and composition in a mixture) and that they may be related analytically to the potential energy of a cluster of molecules. Specifically, the second virial coefficient, B, arises from the interaction between a pair of molecules, the third virial coefficient, C, depends upon interactions in a cluster of three molecules, D involves a cluster of four molecules, and so on.

Expressions for the virial coefficients in terms of the intermolecular potential-energy function can be obtained in the following way. From Table 1.1 we have the pressure in terms of the canonical partition function, Q, as

$$P = k_B T \left(\frac{\partial \ln Q}{\partial V} \right)_{N,T} , \tag{4.2}$$

where Q is given by Eq.(1.28) as

$$Q = \frac{1}{N!} \left(2\pi m k_B T/h^2 \right)^{3N/2} Q^{\text{int}} \mathscr{L} , \tag{4.3}$$

and

$$\mathscr{L} = \int \ldots \int \exp(-\mathcal{U}/k_B T)\, d\mathbf{r}^N .$$ (4.4)

Examining Eq.(4.3), it is apparent that the only part which depends upon the volume (or the density) is the configuration integral $\mathscr{L}$; as explained in Section 1.6 Q^{int} is a function only of N and T. Thus Eq.(4.2) for the pressure may be written

$$P = k_B T \left(\frac{\partial \ln \mathscr{L}}{\partial V} \right)_{N,T} .$$ (4.5)

Comparison of Eqs.(4.1) and (4.5) will produce the required relations between the virial coefficients and the intermolecular potential energy. These expressions will be derived in the following sections.

4.1.1. The Mayer Function

It is convenient to develop the expressions for the virial coefficients in terms of the *Mayer function*, f_{ij}. This is a function of temperature and intermolecular distance defined by the relation

$$f_{ij} = \exp\left(-\phi_{ij}/k_B T\right) - 1 .$$ (4.6)

Here, ϕ_{ij} is the intermolecular potential-energy function so that, in the absence of intermolecular forces, f_{ij} is everywhere zero. For a realistic intermolecular potential, f_{ij} is positive in the attractive region of the intermolecular potential and $f_{ij} \to -1$ as $r \to 0$ and $\phi_{ij} \to \infty$. In Example 4.1 the Mayer function is calculated and plotted for the Lennard-Jones (12-6) potential at two temperatures.

According to the pair additivity approximation, Eq.(3.2), the potential energy of a set of N molecules may be written as the sum of all possible pair interaction energies:

$$\mathcal{U} = \sum_{i=1}^{N-1} \sum_{j=i+1}^{N} \phi_{ij} .$$ (4.7)

With this approximation, the Boltzmann factor $\exp(-\mathcal{U}/k_B T)$ which appears in the configuration integral may be written

$$\exp(-\mathcal{U}/k_BT) \;=\; \exp\left(-\sum_{i=1}^{N-1}\sum_{j=i+1}^{N}\Phi_{ij}/k_BT\right) \;=\; \prod_{i=1}^{N-1}\prod_{j=i+1}^{N}\exp(-\Phi_{ij}/k_BT) \quad (4.8)$$

so that, in terms of Mayer functions,

$$\exp(-\mathcal{U}/k_BT) \;=\; \prod_{i=1}^{N-1}\prod_{j=i+1}^{N}(1+f_{ij})\,. \quad (4.9)$$

EXAMPLE 4.1   *Calculate and Plot the Mayer Function for the case of a Lennard-Jones (12-6) Potential.*

The Lennard-Jones (12-6) potential is defined by Eq.(3.7) as a function of the intermolecular distance r as

$$\varphi(r) \;=\; 4\varepsilon\left[\left(\frac{\sigma}{r}\right)^{12} - \left(\frac{\sigma}{r}\right)^{6}\right],$$

where ε is the depth of the potential well and σ is the separation at which φ crosses zero. As an example, in Figure 4.1 we calculate and plot the Mayer function against (r/σ) for values of the reduced temperature k_BT/ε equal to 1 and 10.

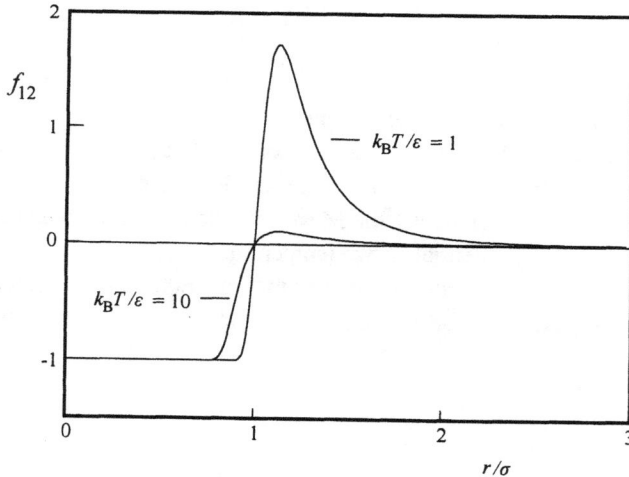

Figure 4.1   The Mayer function $f_{12} = \{\exp(-\Phi_{12}/k_BT)-1\}$ for a Lennard-Jones (12-6) potential.

4.1.2. *Expansion of the Configuration Integral*

We now express the configuration integral, Eq.(4.4), in terms of Mayer functions by means of Eq.(4.9) and multiply out the continued product to obtain

$$
\mathcal{L} = \int \cdots \int \left\{ \prod_{i=1}^{N-1} \prod_{j=i+1}^{N} (1 + f_{ij}) \right\} d\mathbf{r}^N
$$

$$
= \int \cdots \int \left\{ 1 + \sum_{i=1}^{N-1} \sum_{j=i+1}^{N} f_{ij} + \cdots \right\} d\mathbf{r}^N .
$$

(4.10)

One can show that the third term in the expansion involves summations over the product of two f_{ij}'s, the fourth term summations over the product of three f_{ij}'s, and so on. Since these higher terms involve the interaction of more than two molecules, they are not given exactly in the pair-additivity approximation.

We now integrate Eq. (4.10) term by term. The first term is readily evaluated as V^N. The second term involves interactions between all distinct pairs of molecules in the system and there are $N(N-1)/2$ such terms. However, since all the molecules interact with each other according to the same function ϕ, we can replace f_{ij} with, say, f_{12} and integrate over the co-ordinates $\mathbf{r}_3 \cdots \mathbf{r}_N$ one by one. Each such integration results in the factor V so that, approximating $N(N-1)/2$ by $N^2/2$,[1] we obtain

$$
\int \cdots \int \sum_{i=1}^{N-1} \sum_{j=i+1}^{N} f_{ij} \, d\mathbf{r}^N = \tfrac{1}{2} N^2 V^{N-2} \int\!\!\int f_{12} \, d\mathbf{r}_1 \, d\mathbf{r}_2 .
$$

(4.11)

Since f_{12} is actually a function only of the intermolecular separation $r_{12} = |\mathbf{r}_2 - \mathbf{r}_1|$, the integral can be further simplified by a change of variable to $\mathbf{r}_1$ and $\mathbf{r}_{12} = \mathbf{r}_2 - \mathbf{r}_1$. We can then integrate over the co-ordinates $\mathbf{r}_1$,

$$
\int\!\!\int f_{12} \, d\mathbf{r}_1 \, d\mathbf{r}_2 = V \int f_{12} \, d\mathbf{r}_{12}
$$

(4.12)

so that with $d\mathbf{r}_{12} = 4\pi r_{12}^2 \, dr_{12}$, Eq.(4.11) becomes

$$
\int \cdots \int \sum_{i=1}^{N-1} \sum_{j=i+1}^{N} f_{ij} \, d\mathbf{r}^N = 2\pi N^2 V^{N-1} \int_0^\infty f_{12} \, r_{12}^2 \, dr_{12} .
$$

(4.13)

---

[1] This approximation involves negligible error when N is large.

Finally, we obtain the configuration integral as

$$\mathcal{L} = V^N \left\{ 1 + 2\pi (N^2/V) \int_0^\infty f_{12}\, r_{12}^2\, dr_{12} + \cdots \right\}. \qquad (4.14)$$

The thermodynamic properties of the system all depend upon the logarithm of $\mathcal{L}$ and it is therefore useful to develop $\ln \mathcal{L}$ as a power series in $(1/V)$. This may be accomplished by noting that, at sufficiently low densities, the second and higher terms between brackets in Eq.(4.14) are small so that

$$\ln \mathcal{L} = N \ln(V) + 2\pi (N^2/V) \int_0^\infty f_{12}\, r_{12}^2\, dr_{12} + \cdots . \qquad (4.15)$$

4.2. The Virial Coefficients

Expressions for the virial coefficients can be obtained by inserting Eq.(4.15) in Eq.(4.5). Then, carrying out the differentiation with respect to volume, we obtain

$$P = \frac{Nk_B T}{V} \left\{ 1 - 2\pi (N/V) \int_0^\infty f_{12}\, r_{12}^2\, dr_{12} + \cdots \right\} \qquad (4.16)$$

and, comparing with Eq.(4.1), we see that the second virial coefficient is given by

$$B = 2\pi N_A \int_0^\infty \{ 1 - \exp(-\phi_{ij}/k_B T) \}\, r^2\, dr . \qquad (4.17)$$

Note that B has the dimensions of molar volume. One can show [1] that, in the pair additivity approximation, the third virial coefficient is given by

$$C = -\frac{8\pi^2}{3} N_A^2 \iiint f_{12} f_{13} f_{23}\, r_{12} r_{13} r_{23}\, dr_{12}\, dr_{13}\, dr_{23} \qquad (4.18)$$

while non-additivity corrections for the third virial coefficient (accounting for simultaneous interactions of more than two molecules) have been evaluated [1]. Expressions can also be obtained for the higher virial coefficients, although they rapidly become complicated by the increasing number of co-ordinates over which integrations must be performed. Coefficients up to the eighth have been evaluated for some potential energy functions. It should be remembered that the virial coefficients are function only of the temperature in a pure gas; they do not depend upon density.

In the special case of the hard-sphere potential, all of the virial coefficients are independent of temperature. The first eight virial coefficients have been evaluated [3] for this system and the results are given in Table 4.1.

Table 4.1 Virial Coefficients for the Hard-sphere Potential

$$B = 2\pi N_A \sigma^3/3 = b_0$$

$$C = (5/8)\, b_0^2$$

$$D = 0.28695\, b_0^3$$

$$E = 0.11025\, b_0^4$$

$$F = 0.03888\, b_0^5$$

$$G = 0.01307\, b_0^6$$

$$H = 0.00432\, b_0^7$$

| **EXAMPLE 4.2** | *Evaluation of B for the Square-well Potential* |

The square-well potential is defined by Eqs.(3.6) as,

$$\phi(r) = \infty \qquad r < \sigma$$
$$\phi(r) = -\varepsilon \qquad \sigma \leq r < g\sigma$$
$$\phi(r) = 0 \qquad r \geq g\sigma .$$

Therefore, substituting in Eq.(4.17) we obtain

$$B = 2\pi N_A \int_0^\sigma r^2\, dr + 2\pi N_A \int_\sigma^{g\sigma} \{1 - \exp(\varepsilon/k_B T)\}\, r^2\, dr$$

and hence

$$B = (2\pi N_A \sigma^3/3)\left[1 + (g^3 - 1)\{1 - \exp(\varepsilon/k_B T)\}\right].$$

The term $b_0 = (2\pi N_A \sigma^3/3)$ is the rigid sphere contribution, which, in the present case is modified by the presence of attractive forces through the term in square brackets.

❑

In Example 4.2, the second virial coefficient is evaluated for the square-well potential. The resulting expression differs from that of the hard-sphere model because of the presence of attractive forces. The difference is slight at high temperatures but, as the temperature tends to zero, B diverges towards $-\infty$ (in contrast to the hard-sphere virial coefficient b_0 which is temperature independent). With a suitable choice of parameters, the square-well model can represent the second virial coefficient of real fluids with remarkable accuracy. Its most obvious failure is that $B \rightarrow b_0$ in the high temperature limit whereas, for real molecules, the repulsion at short range is not infinitely step and B has a maximum at $k_B T/\varepsilon \approx 30$; above this temperature it declines. However, the maximum is only observable for helium and hydrogen (which have very shallow potential wells); for other molecules it occurs at inaccessibly high temperatures.

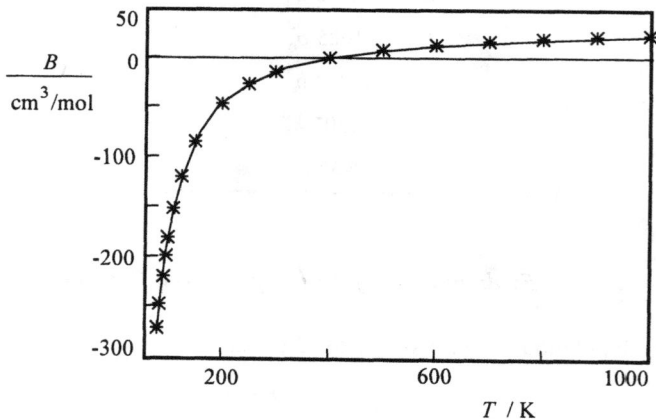

Figure 4.2 Second virial coefficient, B, as a function of temperature. Line from [2]; points calculated from Lennard-Jones (12-6) potential.

The success of the square-well model in correlating second virial coefficients is illustrated in Figure 4.2 where experimental values of B for argon [2] are compared with values calculated from a square-well potential with $\sigma = 0.309$ nm, $g = 1.689$ and $\varepsilon/k_B = 93.1$ K. Although this model is very different from the true potential for argon, it evidently offers an excellent representation of experimental second virial coefficients.

The general dependence of the third virial coefficient, C, on temperature is shown in Figure 4.3. Values for this coefficient are usually determined experimentally from gas-compressibility data by fitting the results at a given temperature

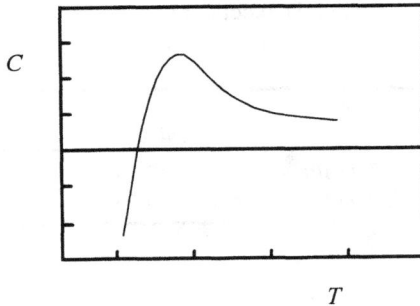

Figure 4.3 Typical temperature dependence of the third virial coefficient, *C*.

by a polynomial in density. The coefficients of this polynomial are then identified with the coefficients of the infinite virial series. Values for the virial coefficients obtained in this way usually depend upon both the degree of polynomial used and the density range of the compressibility data. The resulting uncertainties in the second virial coefficient may be small (usually less than 1 per cent) but are much larger for the third virial coefficient. An extensive collection of experimental data for the second and third virial coefficients of many pure gases and gaseous mixtures has been compiled by Dymond and Smith [2].

| **EXAMPLE 4.3** | *Calculation of the Density of CH_4 at a) $T = 293.15K$ and $P = 7.0$ MPa, and b) $T = 373.15$ K and $P = 9.0$ MPa.* |

To calculate the density of methane we use the virial equation of state truncated after the third virial coefficient:

$$P = \rho_n RT(1 + B\rho_n + C\rho_n^2).$$

Values of the second and third coefficients of methane have been given by Schamp *et al.* [4] and may be represented in the temperature range 270-470 K by the polynomials

$$B/(cm^3/mol) = -365.8277 + 2.0751\theta - 4.2936 \times 10^{-3}\theta^2 + 3.23556 \times 10^{-6}\theta^3$$

$$C(cm^6/mol^2) = 8741.32 - 40.439\theta + 8.6653 \times 10^{-2}\theta^2 - 7.1111 \times 10^{-5}\theta^3$$

where $\theta = T/K$. The results, expressed as the mass density, are shown in the Table. In the same Table, we give for comparison recent measurements of the density of methane [5] under these conditions, with an uncertainty of no more than 0.05 per cent. The agreement, as expected, is excellent with the maximum difference between calculated and experimental values well within the quoted uncertainty.

Conditions	Density / kg·m^{-3}	
	Calculated	Ref.[5]
a) 293.15 K, 7.0 MPa	52.29	52.31
b) 373.15 K, 9.0 MPa	48.73	48.73

□

4.2.1. Correlations of Virial Coefficients

Although experimental data on second virial coefficients are abundant, it is often necessary to estimate values of B for substances which have not been studied in sufficient detail. Several correlations have been developed for this purpose. One of the most common for non-polar gases is the extended corresponding-states method of Pitzer and Curl [6], given by Eq.(4.19) in Table 4.2. The parameter ω which appears in this correlation is the *acentric factor*. This will be discussed in detail in Section 5.2.1. but, for the present, we note that ω may be evaluated easily from vapour-pressure data through Eq.(5.24). Expressions for the dimensionless coefficients B_0 and B_1 were given by Pitzer and Curl [6] in their original formulation but we give, in Eqs.(4.20) and (4.21) of table 4.2, the more recent correlations in terms of the reduced temperature $T_r = T/T^c$ proposed by Tsonopoulos [7]. Third virial coefficients of non-polar gases have also been correlated using a similar model by Orbey and Vera [15] and the results are given in Table 4.2. Example 4.4 demonstrates their usefulness.

Second virial coefficients may be correlated rather well by one of several model intermolecular potentials such as the square-well or Lennard-Jones models. It is easy to show that, for any two-parameter potential, the reduced second and third virial coefficients, B/b_0 and C/b_0^2, are unique functions of the reduced temperature $k_B T/\varepsilon$ (here, $b_0 = 2\pi N_A \sigma^3 / 3$). Tables of reduced second and third virial coefficients have been compiled for several model intermolecular potentials [12] and values of the scaling parameters σ and ε in the Lennard-Jones (12,6) potential are available for a large number of systems [13]; selected values are given in Table A9 of Appendix A. Corrections to C for the effects of non-additivity of the intermolecular forces have also been tabulated [12].

Table 4.2   Correlations for the Second and Third Virial Coefficients [7,15]

$$B = \left(RT^c/P^c\right)\left(B_0 + \omega B_1\right) \tag{4.19}$$

$$B_0 = 0.1445 - 0.33/T_r - 0.1385/T_r^2 - 0.0121/T_r^3 - 0.000607/T_r^8 \tag{4.20}$$

$$B_1 = 0.0637 + 0.331/T_r^2 - 0.423/T_r^3 - 0.008/T_r^8 \tag{4.21}$$

$$C = \left(RT^c/P^c\right)^2 \left(C_0 + \omega C_1\right) \tag{4.22}$$

$$C_0 = 0.01407 + 0.02432/T_r^{2.8} - 0.00313/T_r^{10.5} \tag{4.23}$$

$$C_1 = -0.02676 + 0.0177/T_r^{2.8} + 0.040/T_r^3 - 0.003/T_r^6 - 0.00228/T_r^{10.5} \tag{4.24}$$

Intermolecular potential models may be applied to polar gases and correlations of B based on the Stockmayer Potential are available [13]. In addition to the parameters σ and ε, the dipole moment of the molecule is required for this model.

| EXAMPLE 4.4 |

Calculation of the Density of CH_4 at a) $T = 293.15K$ and $P = 7.0\ MPa$, and b) $T = 373.15\ K$ and $P = 9.0\ MPa$.

To demonstrate the use of the expressions shown in Table 4.2, critical constants can be obtained from Appendix A, while as it will be shown on Chapter 5, the acentric factor can easily be calculated from Eq.(5.36) from the normal boiling temperature and the critical temperature and pressure with the result that $\omega = 0.007$ for methane. Direct substitution in the equations of Table 4.2 gives the values for the virial coefficients for both temperatures and, solving the truncated virial equation, we obtain the results shown in the following Table. The mass densities are in almost as good agreement with the literature values [5], as are those from the previous example, where a correlation of virial coefficients for methane was used.

Conditions	Density / kg·m^{-3}		
	Calculated (individual B,C) *Example 4.3*	Calculated (reduced B,C) *Example 4.4*	Ref.[5]
a) 293.15 K, 7.0 MPa	52.29	52.36	52.31
b) 373.15 K, 9.0 MPa	48.73	48.67	48.73

❑

4.3. Gas Mixtures

The second virial coefficient of a multicomponent gas mixture is given exactly by a quadratic expression in the mole fractions:

$$B_{mix}(T) = \sum_{i=1}^{v} \sum_{j=1}^{v} x_i \, x_j \, B_{ij}(T) \ . \tag{4.25}$$

For a binary mixture of components 1 and 2, this reduces to

$$B_{mix}(T) = x_1^2 B_{11}(T) + 2x_1 x_2 B_{12}(T) + x_2^2 B_{22}(T) \ . \tag{4.26}$$

Here, x_i is the mole fractions of species i in the mixture of v components, B_{ii} is the second virial coefficients of the pure species i, and B_{ij} is called the interaction second virial coefficient. B_{ij} is defined as the second virial coefficient corresponding to the potential energy function $\varphi_{ij}(r)$ which describes the interaction of one molecule of species i with one of species j. B_{ij} is also referred to as the cross virial coefficient, the cross-term virial coefficient, or the mixed virial coefficient.

The third virial coefficient of a multicomponent mixture is cubic in the mole fractions and given by

$$C_{mix}(T) = \sum_{i=1}^{v} \sum_{j=1}^{v} \sum_{k=1}^{v} x_i \, x_j \, x_k \, C_{ijk}(T) \ ; \tag{4.27}$$

for a binary mixture of components 1 and 2 this reduces to

$$C_{mix}(T) = x_1^3 C_{111} + 3x_1^2 x_2 C_{112} + 3x_1 x_2^2 C_{122} + x_2^3 C_{222} \ . \tag{4.28}$$

Here, C_{iii} is the third virial coefficient of pure i, and C_{ijk} is the contribution that arises from the interaction of one molecule of species i, one molecule of species j and one molecule of species k.

Depending upon the availability of experimental P-V-T data, one of two general approaches may be adopted when dealing with multicomponent mixtures. For a specific set of components which have been studied in great detail, one can fit the experimental data to the virial equation truncated after, say, the third virial coefficient and obtain each of the possible pure-component and interaction virial coefficients. Strictly, this requires experimental data for each of the possible binary and ternary combinations of the components. An excellent example of this approach is offered by the GERG virial equation [9-11] for natural-gas type mixtures. For the 13 specified components, a total of 297 virial coefficients were required (B_{ii}, B_{ij}, C_{iii}, C_{iij}, C_{ijj} and C_{ijk}) and the resulting equation predicts the density of natural-gas mixtures with an uncertainty of approximately 0.1% (see also Section 4.6.2) at pressures up to 12 MPa and at temperatures between 265 K and 335 K.

If however, experimental measurements are not available then the following procedure can be adopted for a non-polar gas mixture.
1.   The acentric factor of the pure components can be obtained from Eq.(5.36) in terms of the critical constants and normal boiling temperature.
2.   The pure component virial coefficients can be calculated from the equations given in Table 4.2.
3.   In order to obtain the interaction second virial coefficients, mixing rules are applied from which an acentric factor ω_{ij} and a pseudo-critical temperature T_{ij}^{c} and a pseudo-critical pressure P_{ij}^{c} may be determined for each binary pair. B_{ij} for each binary pair may then be evaluated from the relations in Table 4.2.
4.   One of several empirical combining rules may be applied to obtain the interaction third virial coefficients from Table 4.2 in terms of the acentric factors and pseudo-critical constants of binary pairs.
The virial coefficients and the density may then be calculated.

Various mixing rules for the pseudo-critical constants have been proposed most of which are based on the Lorentz-Berthelot combining rules for the length and energy parameters in the intermolecular potential. Since, for a given functional form of the intermolecular potential, all configurational properties of a pair-wise additive system are functions of the reduced variables $k_B T/\varepsilon$ and V_m/σ^3, the Lorentz-Berthelot combining rules imply that T_i^c is the geometric mean of T_i^c and T_j^c, while $(V_{ij}^c)^{1/3}$ is the arithmetic mean of $(V_i^c)^{1/3}$ and $(V_j^c)^{1/3}$. The recommended mixing rules given in Table 4.3 are based on these combining rules together with the mixture acentric factor and the correlation of the critical compression factor proposed by Lee and Kesler [14]. These rules permit the evaluation of T_{ij}^c, P_{ij}^c and ω_{ij} for each binary pair in terms of the critical temperature and pressure and the acentric factor of each component. An optional binary interaction parameter k_{ij} also appears in the expression for T_{ij}^c. This can be adjusted to give agreement with experimental values of B_{ij} or, in the absence of sufficient data, set to zero. A correlation for k_{ij} in terms of critical molar volumes has been proposed for hydrocarbon mixtures [8].

Several methods have been proposed for the estimation of the interaction third virial coefficients C_{ijk}. Orbey and Vera [15] follow Chueh and Prausnitz [16] in proposing the relation

$$C_{ijk} = \left(C_{ij} C_{ik} C_{jk}\right)^{1/3} \tag{4.29}$$

in which C_{ij} is evaluated from the correlations in Table 4.2 with the acentric factor and pseudo-critical constants pertaining to the binary pair i and j. In a recent test against accurate experimental results for binary mixtures [17], this method was found to give satisfactory estimates of the interaction third virial coefficients. It was also concluded that the algorithm described here was sufficient to permit the prediction of gas density to within 0.1 per cent at densities to nearly two-thirds of the critical.

In Section 4.6.1, a computer subroutine will be described which implements these methods for a binary mixture.

Table 4.3 Lorentz-Berthelot Mixing Rules for Critical Properties

Mixture Acentric Factor

$$\omega_{ij} = (\omega_i + \omega_j)/2 \tag{4.30}$$

Mixture Critical Volume

$$Z_i^c = 0.2905 - 0.085\,\omega_i \tag{4.31}$$

$$V_i^c = Z_i^c\, R T_i^c / P_i^c \tag{4.32}$$

$$V_{ij}^c = \frac{1}{8}\left[\left(V_i^c\right)^{1/3} + \left(V_j^c\right)^{1/3}\right]^3 \tag{4.33}$$

Mixture Critical Temperature

$$T_{ij}^c = (1 - k_{ij})\sqrt{T_i^c T_j^c} \tag{4.34}$$

Mixture Critical Pressure

$$P_{ij}^c = Z_{ij}^c\, R T_{ij}^c \big/ V_{ij}^c \tag{4.35}$$

4.4. Virial Coefficients and Residual Properties

A residual thermodynamic property was defined in Chapter 1 as the difference between the actual value of the property and that in a hypothetical perfect gas at *either* the same temperature and density *or* at the same temperature and pressure. The two choices of independent variables do not in general lead to the same values for the residual properties because the perfect-gas property may depend upon density (see Appendix B). The usual choice in practice is to define the residual property as the difference between the actual value of the property and that in the perfect gas at the same temperature and pressure. Nevertheless, equations of state are usually written with temperature and density (or molar volume) as the independent variables and this leads to explicit expressions for the residual properties in terms of these variables. The thermodynamic relations required to derive expressions for the residual properties from an equation of state will be given in Chapter 6 (Table 6.1) and, in the case of the virial equation of state, one obtains the results given below in Table 4.4. We also give here expressions for the fugacity coefficient of a pure gas and for the partial fugacity coefficient of a component in a mixture.

Table 4.4 Residual Properties in Terms of Virial Coefficients

---

Residual Enthalpy, Entropy and Constant-Volume Heat Capacity

$$H_{\mathrm{m}}^{\mathrm{res}} = RT\left\{\left(B - T\frac{dB}{dT}\right)\rho_n + \left(C - \frac{1}{2}T\frac{dC}{dT}\right)\rho_n^2 + \left(D - \frac{1}{3}T\frac{dD}{dT}\right)\rho_n^3 + \cdots\right\} \qquad (4.36)$$

$$S_{\mathrm{m}}^{\mathrm{res}} = -R\left\{\left(B + T\frac{dB}{dT}\right)\rho_n + \frac{1}{2}\left(C + T\frac{dC}{dT}\right)\rho_n^2 + \frac{1}{3}\left(D + T\frac{dD}{dT}\right)\rho_n^3 + \cdots\right\} + R\ln Z \qquad (4.37)$$

$$C_{V,\mathrm{m}}^{\mathrm{res}} = -R\left\{\left(2T\frac{dB}{dT} + T^2\frac{d^2B}{dT^2}\right)\rho_n + \frac{1}{2}\left(2T\frac{dC}{dT} + T^2\frac{d^2C}{dT^2}\right)\rho_n^2 + \frac{1}{3}\left(2T\frac{dD}{dT} + T^2\frac{d^2D}{dT^2}\right)\rho_n^3 + \cdots\right\}$$

$$(4.38)$$

Fugacity Coefficient

$$\ln\phi = 2B\rho_n + \tfrac{3}{2}C\rho_n^2 + \tfrac{4}{3}D\rho_n^3 + \cdots - \ln Z \qquad (4.39)$$

Partial Fugacity Coefficient

$$\ln\phi_i = 2\rho_n\sum_k x_k B_{ki} + \tfrac{3}{2}\rho_n^2\sum_k\sum_l x_k x_l C_{kli} + \cdots - \ln Z \qquad (4.40)$$

| EXAMPLE 4.5 | *Calculate the Change of Entropy for* CH_4 *Gas at* $T=300K$ *between 1 MPa and 3 MPa* |

The entropy can be calculated from a perfect-gas and a residual contribution.
a) Perfect-Gas Entropy. Subroutine PRFGAS, described in Section 2.10.1, can be used. The subroutine calculates the perfect-gas entropy from the perfect-gas constant-pressure heat capacity coefficients given in Appendix A. It should be remembered that the reference state for the calculation is 273.15 K and 0.10133 MPa.
b) Residual Entropy. Equations for the second and third virial coefficients of methane are given in Example 4.3 and the residual molar entropy, S_m^{res}, can be calculated from Eq.(4.42). Values obtained in this way are given in the table.

P / MPa	Density / mol·m⁻³	S_m^{pg}	S_m^{res}	S_m / J·mol⁻¹·K⁻¹	ΔS_m	ΔS_m [LK]
1	407.8	-15.76	-0.38	-16.14	9.81	9.90
3	1265.5	-24.90	-1.15	-26.05		

In the table, the value calculated by the corresponding-states LEEKESL routine (described in Section 5.6.1) is also shown. The agreement is good.
❑

4.5. Summary

The principal reason for studying the virial equation of state is that it has a sound theoretical basis in statistical mechanics. The virial coefficients are experimental properties of real fluids at zero density, and they are related to intermolecular interactions by exact statistical-mechanical formulae.

Experimental values of the coefficients B and C are available for many systems [2]. They provide data for the quantitative study of various intermolecular pair and triplet potential-energy models and for the evaluation of the parameters which appear in them. Potential parameters thus estimated are useful in the statistical-mechanical calculations of other properties.

The virial equation is also of practical use. The density and all residual thermodynamic properties may be calculated, often with good accuracy, for both pure gases and gaseous mixtures. The calculations may be performed using virial coefficients obtained either directly from experiment, such as those given in the compilation of Dymond and Smith [2], or from generalised correlations and mixing rules, such as those given in Tables 4.2 and 4.3.

4.6. Computational Implementation

4.6.1. Density of Binary Hydrocarbon Mixtures

In Section 4.3, the estimation of the density of binary hydrocarbon mixtures was discussed with reference to the three-term truncated virial equation. The method involves: (a) calculation of the second and third virial coefficients of each of the pure components from the equations in Table 4.2; and (b) evaluation of the cross virial coefficients using the same relations with critical constants from the mixing rules of Table 4.3.

Subroutine PITZCURL calculates the density of a binary mixture of hydrocarbon gases according to the aforementioned method. The subroutine is listed in Display 4.1, while its use is demonstrated in Example 4.6. As already mentioned, the acentric factor required is obtained from Eq.(5.37).

```
                                        DISPLAY 4.1  PITZCURL List
      SUBROUTINE PITZCURL(T,P,WMOL,TB,TC,PC,D)
C
C     The Subroutine calculates the density of binary mixtures of hydrocarbon gases
C     using a three-term truncated virial equation. The coefficients are expressed
C     in the Pitzer-Curl form and are calculated from the critical properties,
C     and the acentric factor. OMEG  The acentric factor is calculated from the Lee-Kesler
C     equation in terms of the critical constants and the boiling temperature
C     - For the cross-coefficients the Lorentz-Berthelot mixing rules are employed
C     - Data required are: mole fraction of component 1 (WMOL), for each component
C       normal boiling temperature, TB(I) in [K], critical temperature, TC(I) in [K] and
C       critical pressure, PC(I) in [Pa]
C
      DIMENSION OMEG(3),B(3),C(3),TB(3),TC(3),PC(3)
      DATA R,DEN/8.3145,100./
C     Calculation of Acentric Factors
      DO 100 I=1,2
      TBR=TB(I)/TC(I)
      A1=LOG(0.10133E+6/PC(I))-5.92714+6.09648/TBR+1.28862*LOG(TBR)-
     &0.169347*TBR**6
      A2=15.2518-15.6875/TBR-13.4721*LOG(TBR)+0.43577*TBR**6
  100 OMEG(I)=A1/A2
C     Mixing Rules
      OMEG(3)=(OMEG(1)+OMEG(2))/2
      VC1=(0.291-0.08*OMEG(1))*R*TC(1)/PC(1)
      VC2=(0.291-0.08*OMEG(2))*R*TC(2)/PC(2)
      VC=(((VC1**(1./3.))+(VC2**(1./3.)))**3)/8
      WK=1.-((VC1*VC2)**.5)/VC
      TC(3)=(1.-WK)*((TC(1)*TC(2))**.5)
```

```
      PC(3)=(0.291-0.04*(OMEG(1)+OMEG(2)))*R*TC(3)/VC
C     .....Calculation of Virial Coefficients (BB and CC)
      DO 200 I=1,3
      TR=T/TC(I)
      B0=0.1445-(0.33/TR)-0.1385*(TR**(-2))-0.0121*(TR**(-3))-0.000607*(TR**(-8))
      B1=0.0637+0.331*(TR**(-2))-0.423*(TR**(-3))-0.008*(TR**(-8))
      B(I)=(R*TC(I)/PC(I))*(B0+OMEG(I)*B1)
      C0=0.01407+0.02432*(TR**(-2.8))-0.00313*(TR**(-10.5))
      C1=-0.02676+0.0177*(TR**(-2.8))+0.04*(TR**(-3))-0.003*(TR**(-6))-
     &0.00228*(TR**(-10.5))
 200  C(I)=((R*TC(I)/PC(I))**2)*(C0+OMEG(I)*C1)
      XMOL=1.-WMOL
      BB=(B(1)*WMOL**2)+(B(2)*XMOL**2)+(2.*WMOL*XMOL*B(3))
      CC=(C(1)*WMOL**3)+(C(2)*XMOL**3)+(3.*WMOL*XMOL*C(3))
C     .....Calculation of Density (D)
 300  D=P/(R*T*(1+BB*DEN+CC*DEN*DEN))
      CHK=ABS(D-DEN)*100./D
      IF (CHK.LT.0.01) GO TO 400
      DEN=D
      GO TO 300
 400  D=D/1000
      RETURN
      END
```

EXAMPLE 4.6   Calculation of the Density of $(0.85\ CH_4+0.15\ C_2H_6)$ at $T=288.15\ K$ and $P=6\ MPa$.

Normal boiling points, and critical constants for methane and ethane are obtained from Appendix A. The calling program is shown in Display 4.2

```
                                    DISPLAY 4.2  PITZCURL Sample Input/Output
      PROGRAM EX0406
      DIMENSION TB(3),TC(3),PC(3)
      DATA WMOL,T,P/0.85,288.15,6.E+6/
      DATA TB(1),TC(1),PC(1)/111.6,190.55,4.599E+6/
      DATA TB(2),TC(2),PC(2)/184.6,305.33,4.871E+6/
      CALL PITZCURL(T,P,WMOL,TB,TC,PC,D)
      WRITE(*,1000) D
1000  FORMAT(1X,'DENSITY (KMOL/M3) = ',F5.3)
      END

DENSITY (KMOL/M3) = 2.973
```

In the following table, we compare the results for methane at 288.15 K and 6 MPa obtained from the routine PITZCURL with those produced by subroutines

LEEKESL (a corresponding-states method that will be presented in Section 5.6.1) and VGERG (estimated accuracy 0.1%). The agreement is within 0.4 per cent.

Density / kmol·m^{-3} of mixture at $T=288.15$ K and $P=6$ MPa		
(PITZCURL)	(LEEKESL)	(VGERG)
2.973	2.967	2.978

□

4.6.2. *Density of a Multicomponent Mixture*

As discussed in Section 4.3, the GERG virial equation [9-11] is an excellent example of the use of a three-term truncated virial equation for a multicomponent mixture. The equation predicts the density of mixtures of up to 13 common components of natural gas with an estimated uncertainty of 0.1%.

In Display 4.3, subroutine VGERG, based on the GERG equation [9-11] is presented. The ranges of applicability are shown in the listing. Application of GERG subroutine is illustrated in Example 4.7.

```
                                           DISPLAY 4.3  VGERG List
        SUBROUTINE VGERG
        &(T,P,CF,DE,C1,CN2,CO2,C2,H2,C3,CO,C4,HE,C5,C6,C7,C8)
C
C       Calculation of Density of Natural Gas Mixture (up to 13 Components)
C       according to: Jaeschke et al. SPE Prod. Engng. Aug. 343 (1991).
C       - Limits of mole fractions [%]:
C               Methane                 (C1)    >=   50   %
C               Ethane                  (C2)    <=   20   %
C               Propane                 (C3)    <=   5    %
C               Butanes                 (C4)    <=   1.5  %
C               Pentanes + Benzene      (C5)    <=   0.5  %
C               Hexanes + Ethylbenzene  (C6)    <=   0.1  %
C               Heptanes + Toluene      (C7)    <=   0.1  %
C               Octanes + Higher        (C8)    <=   0.1  %
C               Nitrogen                (CN2)   <=   50   %
C               Carbon Dioxide + Ethene (CO2)   <=   30   %
C               Hydrogen                (H2)    <=   10   %
C               Carbon Monoxide         (CO)    <=   3    %
C               Helium                  (HE)    <=   0.5  %
C       - Temperature range [T]  265 - 335 K,   Pressure range [P] 0 - 120 BARS
C
C       Expected Accuracy 0.1%
C
```

```
      DIMENSION B(37),C(62)
      COMMON /COEFF/ X(111),Y(186)
      R=.0831451
      DO 100 I=1,37
100   B(I)=X(I)+X(37+I)*T+X(74+I)*T**2
      DO 120 J=1,62
120   C(J)=Y(J)+Y(62+J)*T+Y(124+J)*T**2
      BB=C1**2*B(1)+CN2**2*B(13)+CO2**2*B(20)+C2**2*B(26)+H2**2*B(31)+
     &C3**2*B(32)
      BB=BB+CO**2*B(34)+C4**2*B(35)+HE**2*B(36)+C5**2*B(37)
      BB=BB+2*C1*(CN2*B(2)+CO2*B(3)+C2*B(4)+H2*B(5)+C3*B(6)+CO*B(7)
     &+C4*B(8))
      BB=BB+2*C1*(C5*B(9)+C6*B(10)+C7*B(11)+C8*B(12))
      BB=BB+2*CN2*(CO2*B(14)+C2*B(15)+H2*B(16)+C3*B(17)
     &+CO*B(18)+C4*B(19))
      BB=BB+2*CO2*(C2*B(21)+H2*B(22)+C3*B(23)+C4*B(24)+C5*B(25))
      BB=BB+2*C2*(H2*B(27)+C3*B(28)+C4*B(29)+C5*B(30))+2*C3*C4*B(33)
      CC=C1**3*C(1)+CN2**3*C(28)+CO2**3*C(45)+C2**3*C(54)+H2**3*C(59)
     &+C3**3*C(60)
      CC=CC+CO**3*C(61)+HE**3*C(62)
      CC=CC+3*C1**2*(CN2*C(2)+CO2*C(3)+C2*C(4)+H2*C(5)+C3*C(6)+CO*C(7)
     &+C4*C(8))
      CC=CC+3*C1**2*(C5*C(9)+C6*C(10)+C7*C(11))
      CC=CC+3*CN2**2*(C1*C(12)+CO2*C(29)+C2*C(30)+H2*C(31)+C3*C(32)
     &+CO*C(33)+C4*C(34)
      CC=CC+3*CO2**2*(C1*C(17)+CN2*C(35)+C2*C(46)+H2*C(47)+C3*C(48))
      CC=CC+3*C2**2*(C1*C(20)+CN2*C(39)+CO2*C(49)+H2*C(55)+C3*C(56))
      CC=CC+3*H2**2*(C1*C(25)+CN2*C(42)+CO2*C(52)+C2*C(57))
      CC=CC+3*C3**2*(C1*C(26)+CN2*C(43)+CO2*C(53)+C2*C(58))
     &+3*CO**2*(C1*C(27)+CN2*C(44))
      CC=CC+6*C1*CN2*(CO2*C(13)+C2*C(14)+H2*C(15)+C3*C(16))
      CC=CC+6*C1*CO2*(C2*C(18)+C3*C(19))
      CC=CC+6*C1*C2*(H2*C(21)+C3*C(22)+C4*C(23)+C5*C(24))
      CC=CC+6*CN2*CO2*(C2*C(36)+H2*C(37)+C3*C(38))
      CC=CC+6*CN2*C2*(H2*C(40)+C3*C(41))
      CC=CC+6*CO2*C2*(H2*C(50)+C3*C(51))
C     ..... Iterations to find the Compressibility factor (CF)
      M=1
      ZF=1
200   DE=P/(ZF*R*T)
      Z=1+BB*DE+CC*DE*DE
      IF (ABS(ZF-Z).LT.0.000001) GOTO 240
      M=M+1
      ZF=Z
      GOTO 200
240   CF=Z
      END
```

```
      BLOCK DATA
      COMMON /COEFF/ X(111),Y(186)
      DATA (X(I),I=1,28)/               -0.298675E+0,  -0.213606E+0,  -0.356120E+0
     & -0.499337E+0,  -0.328913E-1,  -0.554110E+0,  -0.687290E-1,  -0.138708E+1
     & -0.771367E+0,  -0.775140E+0,  -0.209989E+1,  -0.239409E+1,  -0.144600E+0
     & -0.339693E+0,  -0.263553E+0,  +0.184506E-1,  -0.431268E+0,  -0.122189E+0,
     & +0.907220E-1,  -0.868340E+0,  -0.112000E+1,  -0.757226E-1,  -0.126372E+1
     & -0.188108E+1,  -0.202135E+1,  -0.107320E+1,  -0.174834E+0,  -0.142020E+1/
      DATA (X(I),I=29,56)/              -0.283644E+1,  -0.331426E+1,  -0.110596E-2
     & -0.259920E+1,  -0.435393E+1,  -0.130820E+0,  -0.708016E+1,  +0.206740E-1,
     & -0.111580E+2,  +0.133425E-2,  +0.104585E-2,  +0.144963E-2,  +0.205755E-2,
     & +0.158947E-3,  +0.185897E-2,  -0.239381E-6,  +0.689575E-2,  +0.143827E-2,
     & +0.184050E-2,  +0.979707E-2,  +0.112199E-1,  +0.740910E-3,  +0.161176E-2,
     & +0.106233E-2,  -0.111895E-3,  +0.184597E-2,  +0.521240E-3,  -0.198016E-2/
      DATA (X(I),I=57,84)/             +0.403760E-2,  +0.576913E-2,  +0.356052E-3,
     & +0.584661E-2,  +0.926431E-2,  +0.954081E-2,  +0.464810E-2,  +0.103728E-2,
     & -0.575230E-2,  +0.136800E-1,  +0.152019E-1,  +0.813385E-4,  +0.119650E-1,
     & +0.213786E-1,  +0.602540E-3,  +0.363100E-1,  -0.513060E-4,  +0.535740E-1,
     & +0.159761E-5,  -0.131159E-5,  -0.158061E-6,  -0.230786E-6,  -0.916521E-7,
     & -0.147793E-5,  +0.518195E-6,  -0.940171E-5,  +0.125744E-5,  -0.400000E-6/
      DATA (X(I),I=85,111)/            -0.125373E-4,  -0.143976E-4,  -0.911950E-6
     & -0.204429E-5,  -0.110155E-5,  +0.303122E-6,  -0.208060E-5,  +0.437181E-6,
     & +0.456313E-5,  -0.516570E-5,  -0.813744E-5,  -0.357678E-6,  -0.747500E-5,
     & -0.124525E-4,  -0.121639E-4,  -0.560520E-5,  -0.150332E-5,  +0.640000E-5,
     & -0.181244E-4,  -0.189007E-4,  -0.987220E-7,  -0.152910E-4,  -0.287483E-4
     & -0.644300E-6,  -0.503829E-4,  +0.724000E-7,  -0.684970E-4/
      DATA (Y(I),I=1,28)/              +0.927260E-2,  +0.889649E-2,  +0.117576E-1,
     & +0.160923E-1,  +0.233159E-3,  +0.892850E-2,  +0.736748E-2,  +0.227707E+0
     & +0.320344E-1,  +0.320344E-1,  +0.320344E-1,  +0.874115E-2,  +0.900338E-2,
     & +0.449508E-2,  +0.360589E-2,  -0.187767E-1,  +0.877002E-2,  -0.176574E-2,
     & -0.623055E-1,  -0.827905E-2,  -0.116095E-2,  -0.189480E+0,  -0.189480E+0,
     & -0.189480E+0,  -0.308914E-2,  -0.848395E-1,  +0.436399E-2,  +0.784980E-2/
      DATA (Y(I),I=29,56)/             +0.552066E-2,  -0.136584E-1,  -0.650954E-2
     & +0.674630E-2,  +0.615912E-2,  -0.101202E+0,  +0.358783E-2,  -0.768174E-5,
     & +0.269108E-2,  -0.301719E-1,  -0.369750E-1,  -0.297743E-3,  -0.506529E-1,
     & -0.877711E-2,  -0.859670E-1,  +0.418924E-2,  +0.205130E-2,  +0.154623E+0,
     & +0.843475E-2,  -0.594074E-1,  +0.133240E+0,  -0.516317E-2,  -0.944235E-1,
     & +0.810002E-2,  -0.149820E+0,  -0.621000E-1,  +0.843800E-2,  -0.144234E+0/
      DATA (Y(I),I=57,84)/             +0.294870E-1,  -0.216734E+0,  -0.104711E-2
     & -0.270290E+0,  +0.190870E-2,  0.00,          -0.376132E-4,  -0.410784E-4
     & -0.447615E-4,  -0.638586E-4,  +0.102923E-4,  +0.190637E-4,  -0.276578E-4
     & -0.139926E-2,  -0.691674E-4,  -0.691674E-4,  -0.691674E-4,  -0.443880E-4,
     & -0.365487E-4,  -0.530233E-6,  -0.145437E-4,  +0.142145E-3,  -0.167108E-4
     & +0.597684E-4,  -0.459485E-3,  +0.121913E-3,  +0.263812E-4,  +0.139040E-2/
      DATA (Y(I),I=85,112)/            +0.139040E-2,  +0.139040E-2,  -0.254973E-4
     & +0.605618E-3,  -0.100358E-4,  -0.398950E-4,  -0.168609E-4,  +0.106972E-3,
     & +0.519104E-4,  -0.365557E-5,  -0.272612E-4,  +0.723716E-5,  +0.806674E-5,
     & +0.385226E-4,  -0.596904E-5,  +0.222387E-3,  +0.289620E-3,  +0.168641E-4,
     & +0.363400E-3,  +0.639925E-4,  +0.579776E-5,  -0.126365E-4,  +0.348880E-4,
     & +0.965716E-3,  -0.355786E-4,  +0.430798E-3,  -0.807996E-2,  +0.579276E-4/
      DATA (Y(I),I=113,140)/           +0.671721E-3,  -0.484242E-4,  +0.994385E-3,
```

```
& +0.508050E-3,  -0.265166E-4,  +0.101275E-2,  -0.197083E-3,  +0.143604E-2,
& -0.364887E-5,  +0.171040E-2,  +0.420040E-5,  0.00,          +0.493066E-7,
& +0.603723E-7,  +0.505481E-7,  +0.779273E-7,  -0.227208E-7,  -0.979363E-7,
& +0.343051E-7,  +0.221564E-5,  0.00,          0.00,          0.00,
& +0.696584E-7,  +0.467680E-7,  -0.152576E-7,  +0.204722E-7,  -0.221855E-6/
  DATA (Y(I),I=141,168)/          -0.151577E-9,  -0.123785E-6,  -0.733765E-6,
& -0.239958E-6,  -0.512600E-7,  -0.238860E-5,  -0.238860E-5,  -0.238860E-5,
& -0.439625E-7,  -0.954102E-6,  +0.660155E-8,  +0.611870E-7,  +0.157169E-7,
& -0.179369E-6,  -0.887528E-7,  -0.234530E-7,  +0.391226E-7,  -0.121618E-5,
& -0.325798E-7,  -0.817506E-7,  +0.540142E-8,  -0.350693E-6,  -0.494852E-6,
& -0.330680E-7,  -0.577241E-6,  -0.107591E-6,  -0.877215E-6,  +0.136268E-7/
  DATA (Y(I),I=169,186)/          -0.837030E-7,  +0.156362E-5,  +0.450909E-7,
& -0.691157E-6,  +0.129643E-5,  -0.105347E-6,  -0.108080E-5,  +0.745489E-7,
& -0.152037E-5,  -0.885260E-6,  +0.381345E-7,  -0.163277E-5,  +0.334778E-6,
& -0.221182E-5,  +0.467095E-8,  -0.250010E-5,  -0.156800E-7,  0.00/
  END
```

EXAMPLE 4.7   *Calculate the Density of a Natural Gas Mixture at T = 280 K and P = 60 bars*

The composition is (mol %):

1)	Methane	73.5015
2)	Nitrogen	10.0214
3)	Hydrogen	9.4918
4)	Ethane	3.3152
5)	Carbon Dioxide+Ethene	1.6245
6)	Carbon Monoxide	0.9142
7)	Propane	0.7657
8)	Butanes	0.2458
9)	Pentanes+ Benzene	0.0740
10)	Hexanes+Ethylbenzene	0.0241
11)	Heptanes+Toluene	0.0148
12)	Octanes+Higher hydrocarbons	0.0070
13)	Helium	0.0
		100.00

Display 4.4 shows the calling program and the output.

DISPLAY 4.4   VGERG Sample Input/Output

```
PROGRAM EX0407
DATA T,P/260.,60./
DATA C1,C2,C3,C4,C5,C6,C7,C8/0.735015,0.033152,0.00765,
&0.002458,0.00074,0.000241,0.000148,0.000070/
DATA CN2,CO2,H2,CO,HE/0.100214,0.016245,0.094918,0.009142,0.0/
CALL VGERG(T,P,CF,DE,C1,CN2,CO2,C2,H2,C3,CO,C4,HE,C5,C6,C7,C8)
WRITE (*,1000) CF,DE
1000 FORMAT(3X,'COMPRESSION FACTOR ',F8.5,2X,'DENSITY (MOL/LT)'
&F8.5)
END
```

```
COMPRESSION FACTOR   0.90079
DENSITY (MOL/LT)      2.86108
```

❑

References

1. T.M. Reed and K.E. Gubbins, *Applied Statistical Mechanics* (McGraw-Hill, Kogakusha, 1973).
2. J.H. Dymond and E.B. Smith, *The Virial Coefficients of Pure Gases and Mixtures. A Critical Compilation* (Clarendon Press, Oxford, 1980).
3. G.C. Maitland, M. Rigby, E.B. Smith and W.A. Wakeham, *Intermolecular Forces. Their Origin and Determination* (Clarendon Press, Oxford, 1981).
4. H.W. Schamp Jr., E.A. Mason, A.C.B. Richardson and A. Altman, *Phys. Fluids* **1** (1958) 329, in [2].
5. H.J. Achtermann, J. Hong, W. Wagner and A. Pruss, *J. Chem. Eng. Data* **37** (1992) 414.
6. K.S. Pitzer and R.F. Curl, *Ind. Eng. Chem.* **50** (1958) 265.
7. C. Tsonopoulos and J. Prausnitz, *Cryogenics* October (1979) 315.
8. S.M. Walas, *Phase Equilibria in Chemical Engineering* (Butterworth Publishers, London, 1985).
9. M. Jaeschke, S. Audibert, P. van Caneghem, A.E. Humphreys, R. Janssen-van Rosmalen, Q. Pellei, J.P.J. Michels, J.A. Schouten and C.A. ten Seldam, *High Accuracy Compressibility Factor Calculation for Natural Gases and Similar Mixtures by Use of a Truncated Virial Equation* (GERG, Verlag des Vereins Deutscher Ingenieure, Dusseldorf, 1988).
10. M. Jaeschke, S. Audibert, P. van Caneghem, A.E. Humphreys, R. Janssen-van Rosmalen, Q. Pellei, J.A. Schouten and J.P Michels, *SPE Prod. Engng.* Aug. (1991) 343.

11. M. Jaeschke, S. Audibert, P. van Caneghem, A.E. Humphreys, R. Janssen-van Rosmalen, Q. Pellei, J.A. Schouten and J.P Michels, *SPE Prod. Engng.* Aug. (1991) 350.
12. A.E. Sherwood and J.M. Prausnitz, *J. Chem. Phys.* **41** (1964) 413.
13. R.C. Reid, J.M. Prausnitz and B.E. Poling, *The Properties of Gases and Liquids*, 4th Ed. (McGraw Hill, 1988).
14. B.I. Lee and M.G. Kesler, *A.I.Ch.E. J.* **21** (1975) 510.
15. M. Orbey and J.M. Vera, *A.I.Ch.E. J.* **29** (1983) 107.
16. P.L. Chueh and J.M. Prausnitz, *A.I.Ch.E. J.* **13** (1967) 896.

5
Corresponding States

We have already shown that the thermodynamic properties of real fluids may be separated into perfect-gas and residual parts, and that the former may be obtained either from the partition function or from empirical correlations of the perfect-gas heat capacity. Exact evaluation of residual thermodynamic properties from first principles would require calculation of the configuration integral and this is, in general, an impossible task. In the previous chapter, we discussed the exact expansion of the configuration integral as a power series in density leading to the virial equation of state. Unfortunately, the virial series fails to converge for dense fluids and another approach is required to cover the whole of the fluid region. The principle of corresponding states is a powerful predictive tool with just this kind of wide-ranging applicability. It underpins in some form the vast majority of thermodynamic models used in chemical engineering and, in this chapter, we describe both its scientific basis and its applications to pure fluids and mixtures.

The principle of corresponding states establishes a connection between the configuration integrals of different substances and thereby allows each of the configurational thermodynamic properties of one fluid to be expressed in terms of those of another, reference, fluid. Since configurational and residual thermodynamic properties are related in a very simple way, essentially the same results apply also to the latter. Unlike many other thermodynamic models, the principle of corresponding states is soundly based in statistical mechanics and can be a powerful predictive tool. The Principle will be discussed first for the case of spherical molecules which obey the elementary two-parameter corresponding states theory. Next, the method is extended by the introduction of a third parameter to the cases of non-polar and slightly-polar molecules of general shape. Finally, an empirical four-parameter modification of the method is discussed which is applicable to highly polar substances.

These methods are illustrated by the Lee and Kesler [1] implementation of the three-parameter corresponding-states scheme and by the Wu and Stiel [2] four-parameter modification for polar molecules. Computer programs implementing both of these schemes are included.

5.1. Two-parameter Corresponding States (Spherical Molecules)

The theoretical basis of the two-parameter corresponding states principle is the assumption that the intermolecular potentials of two substances may be rendered identical by the suitable choice of two scaling parameters, the one applied to the separation and the other to the energy.   Thus the intermolecular potential of a substance that conforms to the principle is taken to be

$$\varphi(r) = \varepsilon\, F(r/\sigma) , \qquad\qquad (5.1)$$

where ε and σ are respectively scaling parameters for energy and distance, and F is a *universal* function. Substances that obey Eq.(5.1) are said to be *conformal*. One of the great strengths of the method is that the function F need not be known. Instead, a reference substance is introduced, identified by the subscript 0, for which the thermodynamic properties of interest are known and this is used to eliminate F from the problem.   The configurational (and hence residual) properties of another conformal substance, identified by the subscript i, are thereby given in terms of those of the reference fluid. We shall also see that the parameters ε and σ may be eliminated in favour of measurable macroscopic quantities.

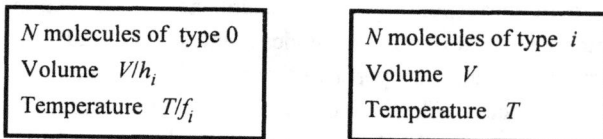

N molecules of type 0	N molecules of type i
Volume V/h_i	Volume V
Temperature T/f_i	Temperature T

Figure 5.1 Corresponding-states principle.

The consequences of conformality may be derived by means of the following thought experiment. Consider two conformal substances, one of which is designated as the reference fluid, contained in separate vessels of the same shape but different volumes as illustrated in Figure 5.1. Let there the N molecules of type i be contained in volume V at temperature T, while the N molecules of the reference fluid are contained in volume V/h_i at temperature T/f_i. Here, $h_i=(\sigma_i/\sigma_0)^3$ and $f_i=(\varepsilon_i/\varepsilon_0)$ are scaling ratios. We now suppose that the molecules are arranged in geometrical similar positions within their respective containers. Then, for each molecules in the system on the right with position vector $\mathbf{r}_i$ defined relative to the origin in that system, there is a corresponding molecules in the reference system with position vector $\mathbf{r}_0$ defined relative to the origin in that system and these position vectors are related by

$$\mathbf{r}_0 = \mathbf{r}_i / h_i^{1/3} . \tag{5.2}$$

Since it is assumed that the pair potentials are conformal and that either (i) the pair additivity approximation is obeyed, or (ii) the N-body potentials are also conformal, the configurational energies of the two systems are related by

$$\mathcal{U}_0(\mathbf{r}_{0,1}, \mathbf{r}_{0,2}, \cdots, \mathbf{r}_{0,N}) = \mathcal{U}_i(\mathbf{r}_{i,1}, \mathbf{r}_{i,2}, \cdots, \mathbf{r}_{i,N}) / f_i . \tag{5.3}$$

This relation must apply to any configuration because for each configuration of the reference system a geometrically similar one exists for the second system.

The configuration integral $\mathcal{L}_0$ for the reference system is given according to Eq.(1.31) by

$$\mathcal{L}_0(V/h_i, T/f_i) = \int \cdots \int_{V/h_i} \exp(-f_i \mathcal{U}_0 / k_B T) \, d\mathbf{r}_0^N \tag{5.4}$$

while that for the other system is

$$\mathcal{L}_i(V,T) = \int \cdots \int_V \exp(-\mathcal{U}_i / k_B T) \, d\mathbf{r}_i^N . \tag{5.5}$$

Upon changing the variables of integration from $\mathbf{r}_i$ to $\mathbf{r}_0$, in accordance with Eq.(5.2), and making use of Eq.(5.3), $\mathcal{L}_i$ becomes

$$\mathcal{L}_i(V,T) = \int \cdots \int_{V/h_i} \exp(-f_i \mathcal{U}_0 / k_B T) \, h_i^N \, d\mathbf{r}_0^N . \tag{5.6}$$

Then, comparing Eqs.(5.4) and (5.6), we see that the configuration integrals of the two systems are related by the simple equation:

$$\mathcal{L}_i(V,T) = h_i^N \mathcal{L}_0(V/h_i, T/f_i) . \tag{5.7}$$

Since the compression factor is a purely configurational property given by $Z = (V/N)(\partial \ln \mathcal{L}/\partial V)_T$, it follows from Eq.(5.7) that

$$Z_i(V,T) = Z_0(V/h_i, T/f_i) \ . \tag{5.8}$$

Thus the compression factor of one conformal substance may be equated with that of another at a scaled volume and a scaled temperature. As this relation must hold also at the critical point, it follows that the scaling parameters are related to the critical constants by

$$f_i = T_i^c/T_0^c \quad \text{and} \quad h_i = V_i^c/V_0^c \ , \tag{5.9}$$

and that the reduced pressure $P_r = P/P^c$ is the same function of the reduced volume $V_r = V/V^c$ and the reduced temperature $T_r = T/T^c$ in all conformal systems. Consequently, the compression factor is a universal function of T_r and V_r or, alternatively, of T_r and P_r.

The critical volume is rather difficult to measure accurately and it is often convenient to adopt T_r and P_r as the independent variables in a corresponding-states calculation. The scaling parameter h_i may be expressed in a form independent of V^c by noting that Z^c is the same for all conformal fluids so that

$$h_i = (P_0^c/P_i^c) f_i \ . \tag{5.10}$$

The residual properties of the system are obtained by noting that $Q^{\text{res}} = \mathcal{L}/V^N$ and hence

$$Q_i^{\text{res}}(V,T) = Q_0^{\text{res}}(V/h_i, T/f_i) \ . \tag{5.11}$$

It is then a simple matter to show (c.f. Table 1.1) that the residual Helmholtz free energy is given in terms of that of the reference system by

$$A_i^{\text{res}}(V,T) = f_i A_0^{\text{res}}(V/h_i, T/f_i) \ , \tag{5.12}$$

so that A^{res}/RT^c is a universal function of T_r and P_r in conformal systems. Eq.(5.12) may be used to obtain expressions for all of the residual thermodynamic properties.

Generalised charts are available giving the compression factor Z, as well as $(H_\mathrm{m}^\mathrm{res}/RT^c)$, $(S_\mathrm{m}^\mathrm{res}/R)$ and $\ln(f/P)$, in terms of reduced temperature and pressure. Of course, the principle of corresponding states tells us nothing whatever about the perfect-gas contributions to the enthalpy and entropy which must be evaluated by alternative means.

The simple corresponding-states treatment outlined above applies accurately only to a small number of substances. It is obeyed with great accuracy by the monatomic gases Ar, Ke and Xe but He and, to some extent, Ne deviate from the principle because of quantum effects at low temperatures. Several simple molecules, including N_2, CO and CH_4, deviate only slightly from the principle but most other molecules depart considerably. However, the extensions that will be described in the following section turn the principle into a powerful and wide-ranging predictive tool.

5.2. Three-parameter Model (Non-spherical Molecules)

In order to apply the principle of corresponding states with any accuracy to molecular fluids, it is necessary to take into account the non-spherical nature of the molecules. The anisotropic nature of the intermolecular potential φ in these cases has been described briefly in Chapter 2 and here we simply recall that φ is a function not only of the separation r but also of the relative orientation ϖ. Here the vector ϖ represents the full set of angular co-ordinates required to specify the relative orientation of two molecules. In this section we first derive the general consequences of anisotropic forces and then show how one may account for such forces by the inclusion of a third parameter in the principle of corresponding states.

The configuration integral of such a non-spherical system is given in the approximation of pair-wise additivity by

$$\mathscr{L} = \int \cdots \int \prod_{i=1}^{N-1} \prod_{j=i+1}^{N} \exp(-\phi_{ij}/k_\mathrm{B}T) \, dr^N d\varpi^N \,. \tag{5.13}$$

Consequently, the configurational properties depend upon the average over all possible orientations of $\exp(-\phi/k_\mathrm{B}T)$. To demonstrate the principal consequences of the presence of angular-dependent intermolecular forces, we shall derive an approximate expression for the effective isotropic intermolecular potential ϕ^* which satisfies the relation

$$\exp(-\phi^*/k_\mathrm{B}T) = \left\langle \exp(-\phi/k_\mathrm{B}T) \right\rangle_\varpi \,, \tag{5.14}$$

where $< \ >_\varpi$ denotes the unweighted average over all orientations of the two molecules. In terms of this effective potential, the configuration integral is simply

$$\mathscr{L} = \int \cdots \int \prod_{i=1}^{N-1} \prod_{j=i+1}^{N} \exp(-\phi_{ij}^*/k_\mathrm{B}T) \, d\mathbf{r}^N \ . \tag{5.15}$$

To show this, the intermolecular potential $\phi(r,\varpi)$ is written as the sum of isotropic and anisotropic terms:

$$\phi(\mathbf{r},\varpi) = \phi_0(\mathbf{r}) + \phi_1(\mathbf{r},\varpi) \ . \tag{5.16}$$

This separation is in general quite arbitrary but it can be made unique by requiring that $\phi_0(r)$ be the unweighted average of $\phi(r,\varpi)$ over all orientations:

$$\left\langle \phi(r,\varpi) \right\rangle_\varpi = \phi_0(r) \ . \tag{5.17}$$

We now expand $\exp(-\phi_1/k_\mathrm{B}T)$ as a power series,

$$\exp(-\phi_1/k_\mathrm{B}T) = 1 - (\phi_1/k_\mathrm{B}T) + \tfrac{1}{2}(\phi_1/k_\mathrm{B}T)^2 + \cdots \ , \tag{5.18}$$

take the average over all orientations of the two molecules,

$$\left\langle \exp(-\phi_1/k_\mathrm{B}T) \right\rangle_\varpi = 1 + \tfrac{1}{2}\left\langle (\phi_1/k_\mathrm{B}T)^2 \right\rangle_\varpi + \cdots \ , \tag{5.19}$$

and truncate the series after the term in ϕ_1^2. The angle average of $\exp(-\phi/k_\mathrm{B}T)$ is therefore given correct to second-order in the anisotropic terms by

$$\left\langle \exp(-\phi/k_\mathrm{B}T) \right\rangle_\varpi = \exp\{-(\phi_0 + \phi'/k_\mathrm{B}T)\} \ , \tag{5.20}$$

where

$$\phi'(r) = -\tfrac{1}{2}\left\langle \phi_1^2(r,\varpi) \right\rangle_\varpi \ . \tag{5.21}$$

We now have an effective, but temperature dependent, isotropic intermolecular potential given by

$$\phi^*(r) = \phi_0(r) + \phi'(r)/k_\mathrm{B}T \tag{5.22}$$

which determines the configurational thermodynamic properties through Eq.(5.15). According to Eq.(5.21), $\varphi'(r)$ is negative irrespective of the precise nature of the anisotropic terms in the intermolecular potential.

It is possible to develop these arguments into a perturbation expansion of the configuration integral applicable to specific examples of anisotropic forces. For example, the presence of a dipole moment μ may be accounted for by means of an expansion in powers of μ^2. However, to develop a more widely-applicable procedure it is necessary to have some general means of quantifying the non-sphericity of the system. A direct measure would be provided by the function $(\varphi'/k_B T)$ but this is of limited value as there is little precise information about the intermolecular potentials of real molecules and because $(\varphi'/k_B T)$ is a function both of a microscopic variable (r) and a macroscopic variable (T). Nevertheless, the preceding discussion suggests several semi-empirical extensions to the simple corresponding-states theory.

It is clear from Eq.(5.22) that the parameters ε^* and σ^* characterising the effective isotropic potential φ^* are slowly varying functions of T which approach those of the isotropic part of the potential φ_0 as $T \to \infty$. If we further assume that the effective isotropic potential of the substance in question is conformal with that of the reference fluid then the results of the previous section - Eqs.(5.8), (5.11), (5.12) and their consequences - remain formally valid but with scaling parameters f_i and h_i which are themselves slowly varying functions of temperature. In place of Eqs.(5.9), we therefore write

$$ f_i = (T_i^c / T_0^c)\, \Theta_i(T_r) \quad \text{and} \quad h_i = (V_i^c / V_0^c)\, \Phi_i(T_r) , \qquad (5.23) $$

where Θ_i and Φ_i are molecular shape factors for substance i with reference fluid 0. In fact, similar conclusions can be reached without recourse to molecular considerations by viewing h_i and f_i as the parameters which map the surfaces $Z_i(V,T)$ and $Z_0(V,T)$ on to each other through Eq.(5.8); in that case the shape factors may be functions of V_r as well as T_r [15].

It might be thought that the dependence of the shape factors on T_r and V_r would be entirely specific to the fluid and reference fluid under consideration so that, while formally true, Eqs.(5.23) would have little predictive capability. However, it turns out that Θ_i and Φ_i may be represented remarkably well by functions that incorporate only one or, at most, two substance-dependent parameters in addition to T^c and P^c [15,16]. By far the most common and successful choice for the third parameter is Pitzer's acentric factor ω which we discuss in the next section.

5.2.1. The Acentric Factor

In Figure 5.2 the reduced vapour pressure P^σ/P^c is plotted as a function of the reduced temperature T/T^c for selected substances. According to the principle of corresponding states, all spherical systems have the same reduced vapour pressure curve (P^σ/P^c vs T/T^c) and it can be seen in the figure that the data for Ar, Kr and Xe do indeed lie essentially on a single curve. It also happens that $\log(P^\sigma/P^c)$ is a nearly linear function of T/T^c. The results for non-spherical molecules also happen to lie almost on straight lines but with slopes that deviate from that of the simple fluids. This may be rationalised by considering the effective isotropic potential for a system of non-spherical molecules, Eq.(5.22), in which the effect of non-spherical terms is to make the effective potential always more attractive by an amount which increases with decreasing temperature. This has the effect of causing the vapour pressure to fall more rapidly as the temperature is reduced than it would in the absence of such anisotropic forces.

The observations above provide a basis for defining in terms of the slope of the vapour pressure curve a single measurable macroscopic parameter that correlates with the degree of anisotropy. Such a parameter was proposed by Pitzer [3,4] in 1955 who defined the *acentric factor* ω by means of the simple relation

$$\omega = -1 - \log_{10}\left[P^\sigma(T = 0.7T^c)\big/P^c\right]. \tag{5.24}$$

Although this is a somewhat arbitrary definition it has the following advantages:
1. For the spherical molecules Ar, Kr and Xe, which obey simple corresponding states with high accuracy, the reduced vapour pressure is close to 0.1 at $T = 0.7T^c$, so that $\omega \approx 0$ as required for any reasonable definition.
2. The condition $T = 0.7T^c$ is not far from the normal boiling temperature for many substances and vapour pressure data are usually readily available.
3. The acentric factor for non-spherical molecules correlates well with other indications of anisotropy such as molecular shape and dipole moments.
4. A three-parameter (T^c, P^c, ω) corresponding states treatment works extremely well for a wide range of substances.

The extended principle of corresponding states asserts that the same reduced equation of state, $P_r = \mathcal{F}_\omega(T_r,V_r)$, is obeyed by all fluids that have the same value of ω. Pitzer further postulated that the function $\mathcal{F}_\omega(T_r,V_r)$ is *linear* in ω so that a generalised property X is given as a function of reduced temperature and pressure by

$$X(T_r, P_r) = X_0(T_r, P_r) + \omega X_1(T_r, P_r) . \qquad (5.25)$$

Here, X_0 is known as the simple fluid term and X_1 as the correction term. Charts and equations representing the simple fluid and correction terms as functions of reduced temperature and pressure are available for $X = Z$, (H_m^{res}/RT^c), (S_m^{res}/R), and $\ln(f/P)$.

Although we shall not consider molecular shape factors further in this chapter, we note that an alternative implementation of the extended principle of corresponding states is possible in which Θ_i and Φ_i are considered as universal functions of T_r, V_r and ω.

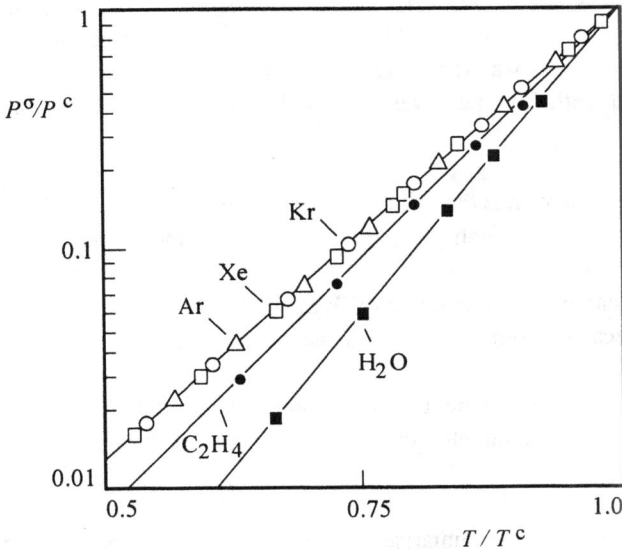

Figure 5.2 Reduced vapour pressure as a function of reduced temperature.

5.2.2. Pure Non-polar Fluids

Although the original three-parameter corresponding states charts and tables prepared by Pitzer and his collaborators [3,4] are still useful, they are limited to reduced temperatures above 0.8 and are unsuitable for use on a computer. The more recent correlation due to Lee and Kesler [1] covers an extended range and is easily incorporated into computer programs. In this scheme, two reference equations of state are used each similar in form to that of Benedict, Webb & Rubin (see Section 6.6.1). One of these equations represents the compression factor Z_0 of the simple fluids, while the other represents the compression factor Z_R of a complex reference fluid, namely *n*-octane. Expressions for the other configurational and residual properties of the reference fluids are determined by these equations of state and a generalised property X is obtained by interpolation (or extrapolation) with respect to ω between X_0 and X_R:

$$X = X_0 + (\omega/\omega_R)(X_R - X_0) .\qquad(5.26)$$

Here, $\omega_R = 0.3978$ is the acentric factor of *n*-octane and, by comparison with Eq.(5.25), one can identify $X_1 = (X_R - X_0)/\omega_R$ as the correction term. *n*-octane was chosen because it was the largest hydrocarbon for which reasonably accurate volumetric and enthalpy data were available over a sufficiently wide range of conditions.

The Lee-Kesler scheme has the following advantages:
1. It may be applied to both gas and liquid phases in the ranges $0.3 \le T_r \le 4.0$ and $0.01 \le P_r \le 10.0$.
2. It is applicable to a wide range of fluids including hydrocarbons from methane to hexadecane, aromatics, alkenes, acetylenes, inorganics, and slightly polar fluids.
3. It is easily extended to include multi-component mixtures.
4. The scheme is complete, consistent and easily incorporated in computer programs.

The Lee-Kesler scheme is summarised in Table 5.1. Both Z_0 and Z_R are represented by the same equation, Eq.(5.26), but with different values of the parameters. The dimensionless quantity v which appears here is defined by $v = (P^c/RT^c)V_m$, where V_m is the molar volume of the reference fluid in question at the specified temperature and pressure.

The calculation of Z for the fluid of interest at a given temperature and pressure may be accomplished in the following steps.
1. Calculate the reduced temperature $T_r = T/T^c$ and reduced pressure $P_r = P/P^c$.
2. Making use of the simple-fluid constants, solve Eq.(5.27) for the dimensionless volume $v = v_0$ and hence determine the compression factor of the simple reference fluid: $Z_0 = (P_r/T_r)v_0$.

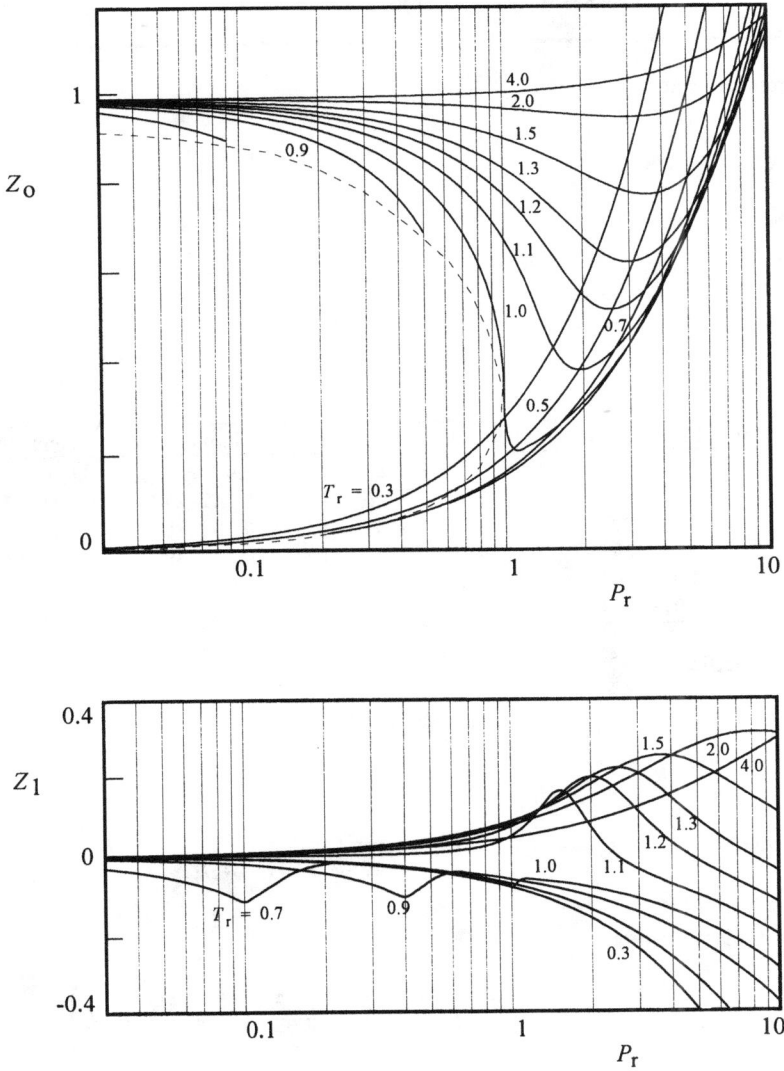

Figure 5.3 Plot of Z_0 and Z_1 as a function of reduced pressure P_r at selected reduced temperatures T_r.

Table 5.1 Lee-Kesler Equations, 1975 [1]

Compression Factor

$$Z = \left(\frac{P_r v}{T_r}\right) = 1 + \frac{B}{v} + \frac{C}{v^2} + \frac{D}{v^5} + \frac{c_4}{T_r^3 v^2}\left(\beta + \frac{\gamma}{v^2}\right)\exp\left(-\frac{\gamma}{v^2}\right) \quad (5.27)$$

where

$$B = b_1 - b_2/T_r - b_3/T_r^2 - b_4/T_r^3$$

$$C = c_1 - c_2/T_r + c_3/T_r^3$$

$$D = d_1 + d_2/T_r$$

Constant	Simple Fluid	Reference	Constant	Simple Fluid	Reference
b_1	0.1181193	0.2026579	c_3	0.0	0.016901
b_2	0.265728	0.331511	c_4	0.042724	0.041577
b_3	0.154790	0.027655	$d_1 \times 10^4$	0.155488	0.48736
b_4	0.030323	0.203488	$d_2 \times 10^4$	0.623689	0.0740336
c_1	0.0236744	0.0313385	β	0.65392	1.226
c_2	0.0186984	0.0503618	γ	0.060167	0.03754

Fugacity Coefficient

$$\ln\left(\frac{f}{P}\right) = Z - 1 - \ln(Z) + \frac{B}{v} + \frac{C}{2v^2} + \frac{D}{5v^5} + E \quad (5.28)$$

where

$$E = \frac{c_4}{2T_r^3 \gamma}\left\{\beta + 1 - \left(b + 1 + \frac{\gamma}{v^2}\right)\exp\left(-\frac{\gamma}{v^2}\right)\right\}. \quad (5.29)$$

Residual Molar Enthalpy

$$\frac{H_m^{res}}{RT^c} = T_r\left(Z - 1 - \frac{b_2 + 2b_3/T_r + 3b_4/T_r^2}{T_r v} - \frac{c_2 - 3c_3/T_r^2}{2T_r v^2} + \frac{d_2}{5T_r v^5} + 3E\right). \quad (5.30)$$

Residual Molar Entropy

$$\frac{S_m^{res}}{R} = \ln(Z) - \frac{b_1 + b_3/T_r^2 + 2b_4/T_r^3}{v} - \frac{c_1 - 2c_3/T_r^3}{2v^2} - \frac{d_1}{5v^5} + 2E. \quad (5.31)$$

Table 5.1(con/d) Lee-Kesler Equations, 1975 [1]

Residual Molar Isochoric Heat Capacity

$$\frac{C_{V,m}^{res}}{R} = \frac{2(b_3 + 3b_4/T_r)}{T_r^2 v} - \frac{3c_3}{T_r^3 v^2} - 6E \ . \tag{5.32}$$

Residual Molar Isobaric Heat Capacity

$$\frac{C_{P,m}^{res}}{R} = \frac{C_{V,m}^{res}}{R} - 1 - T_r \left(\frac{\partial P_r}{\partial T_r}\right)_v^2 \bigg/ \left(\frac{\partial P_r}{\partial v}\right)_{T_r} \tag{5.33}$$

where

$$\left(\frac{\partial P_r}{\partial T_r}\right)_v = \frac{1}{v}\left\{1 + \frac{b_1 + b_3/T_r^2 + 2b_4/T_r^3}{v} + \frac{c_1 - 2c_3/T_r^3}{v^2} + \frac{d_1}{v^5}\right.$$
$$\left. - \frac{2c_4}{T_r^3 v^2}\left(\beta + \frac{\gamma}{v^2}\right)\exp\left(-\frac{\gamma}{v^2}\right)\right\} \tag{5.34}$$

and

$$\left(\frac{\partial P_r}{\partial v}\right)_{T_r} = -\frac{T_r}{v^2}\left\{1 + \frac{2B}{v} + \frac{3C}{v^2} + \frac{6D}{v^5}\right.$$
$$\left. + \frac{c_4}{T_r^3 v^2}\left(3\beta + \frac{\gamma}{v^2}\left[5 - 2\beta - \frac{2\gamma}{v^2}\right]\right)\exp\left(-\frac{\gamma}{v^2}\right)\right\} \tag{5.35}$$

Vapour Pressure

$$\ln\left(\frac{P^\sigma}{P^c}\right) = 5.92714 - 6.09648/T_r - 1.28862\ln(T_r) + 0.169347T_r^6$$
$$+ \omega\{15.2518 - 15.6875/T_r - 13.4721\ln(T_r) + 0.43577T_r^6\} \tag{5.36}$$

Acentric Factor

$$\omega = \frac{\ln(0.101325\,\text{MPa}/P^c) - 5.92714 + 6.09648/T_r^b + 1.28862\ln T_r^b - 0.169347(T_r^b)^6}{15.2518 - 15.6875/T_r^b - 13.4721\ln T_r^b + 0.43577(T_r^b)^6} \tag{5.37}$$

3.   Then, making use of the constants for the complex reference fluid, solve
      Eq.(5.26) for $v = v_R$ and hence determine the compression factor
      $Z_R = (P_r/T_r)v_R$ of the complex reference fluid.
4.   Calculate the compression factor Z from Eq.(5.26).

Once v_0 and v_R have been evaluated, the fugacity and residual properties of each
reference fluid may be evaluated by means of the equations in Table 5.1; these are
then combined in accordance with Eq.(5.26) to obtain the properties of the fluid in
question.

      By analysis of a wide selection of experimental data for non-polar and slightly
polar fluids, Lee and Kesler were also able to develop a rather accurate correlation
of the vapour pressure in terms of T_r and ω. This correlation is given by Eq.(5.36).

      The acentric factor used within the Lee-Kesler scheme may be obtained from
experimental vapour pressure data in the usual way.   Alternatively, it may be
evaluated from the critical constants and the normal boiling temperature T^b by
application of Eq.(5.36) at $T = T^b$; this yields Eq.(5.37) for ω in terms of P^c and
$T_r^b = T^b/T^c$.

      The Lee-Kesler scheme was tested extensively against experimental values of
the compression factor and the enthalpy for a wide range of fluids and, for the
compression factor, the authors reported an average error of about 2 per cent and a
maximum of 4 per cent (for H_2S). The simple fluid correction factor Z_0 as well as
the correction term Z_1, both obtained from Eq.(5.27) on a number of isotherms are
illustrated in Figure 5.3, while detailed examples of the use of the scheme will be
presented in Section 5.5.1.

5.2.3. Non-polar Mixtures

      The most common method by which the principle of corresponding states may
be applied to a mixture is by application of *one-fluid* theories. In such theories, the
mixture is treated as a hypothetical pure substance that is conformal with a reference
fluid.   So-called *mixing-rules* are applied to relate the scaling parameters of the
mixture to those of the pure components. The simplest set of mixing rules are those
of Kay in which pseudo-critical constants are defined for the mixture as mole-
fraction-weighted averages of the pure-component values:

$$T_{mix}^c = \sum_i x_i T_i^c \quad \text{and} \quad P_{mix}^c = \sum_i x_i P_i^c . \tag{5.38}$$

In order to apply this scheme to more than just the simple fluids, one further defines an average acentric factor for the mixture:

$$\omega_{mix} = \sum_i x_i \, \omega_i \, . \tag{5.39}$$

Then, according to the extended principle of corresponding states, the compression factor of the mixture at reduced temperature $T_r = T/T^c_{mix}$ and reduced pressure $P_r = P/P^c_{mix}$ should equal that of a reference fluid, chosen such that $\omega = \omega_{mix}$, at the same reduced temperature and pressure.

Although simple to apply, Kay's rules are purely empirical and, except for mixtures of rather similar molecules, suffer from inferior accuracy. Much better results may be obtained by application of the so-called *van der Waals one-fluid theory* (VDW-1) in which the mixing rules take explicit account of the interactions between unlike molecules. The VDW-1 mixing rules may be expressed in terms of a pseudo-critical temperature and a pseudo-critical molar volume for the mixture defined in the following ways:

$$V^c_{mix} = \frac{1}{8} \sum_i \sum_j x_i \, x_j \left[\left(V^c_i\right)^{1/3} + \left(V^c_j\right)^{1/3} \right]^3 \tag{5.40}$$

$$T^c_{mix} = \frac{1}{8\,V^c_{mix}} \sum_i \sum_j x_i \, x_j \left[\left(V^c_i\right)^{1/3} + \left(V^c_j\right)^{1/3} \right]^3 \kappa_{ij} \sqrt{T^c_i \, T^c_j} \, . \tag{5.41}$$

We shall see in the next chapter (Section 6.4) that the quadratic dependence on mole fraction exhibited by these equations follows exactly from the theory of van der Waals equation of state. The specific form follows from the assumption of modified Lorentz-Bertholot combining rules for the unlike interactions: $\sigma_{ij} = \frac{1}{2}(\sigma_i + \sigma_j)$ and $\varepsilon_{ij} = \kappa_{ij} \sqrt{\varepsilon_i \, \varepsilon_j}$. The quantity κ_{ij} which appears in Eq.(5.41) is known as a binary interaction parameter and permits tailoring of the mixing rules to achieve improved agreement with experimental results for binary mixtures. Clearly, κ_{ij} can differ from unity only in the case of unlike interactions.

It is again convenient to eliminate the critical molar volumes of the pure components from Eq.(5.40) in favour of the critical pressures by means of the relation

$$V^c_i = Z^c_i \, RT^c_i / P^c_i \tag{5.42}$$

and to define a mean acentric factor for the mixture according to Eq.(5.39).

Used in combination with the Lee-Kesler equations for the properties of the reference fluid, the simple VDW-1 mixing rules given above have a powerful predictive capability. To simplify application of the method, Lee and Kesler [1] recommended use of the correlation

$$Z_i^c = 0.2905 - 0.085\omega_i \qquad (5.43)$$

for the critical compression factor. If the binary interaction parameters are each set equal to unity, it is then only necessary to know (T_i^c, P_i^c, ω_i) for each component in the mixture. Usually, the values of κ_{ij} are adjusted to optimise agreement between predictions and experimental results for the binary sub-systems formed from the most important components in the mixtures. Typically, some phase equilibrium properties (such as dew- or bubble-point pressures) are used for this purpose.

Although the Lee-Kesler scheme with VDW-1 mixing rules performs well for many different mixtures, rather large errors can arise when the mixture contains molecules of greatly different sizes (so-called asymmetric mixtures) even after optimisation of the κ_{ij}'s. Rather better results are obtained with the modified mixing rules of Plöcker *et al* [5] in which Eq.(5.41) for the pseudo-critical temperature is replaced by:

$$T_{\text{mix}}^c = \frac{1}{(8V_{\text{mix}}^c)^\eta} \sum_i \sum_j x_i\, x_j \left[\left(V_i^c\right)^{1/3} + \left(V_j^c\right)^{1/3} \right]^{3\eta} \kappa_{ij} \sqrt{T_i^c T_j^c} \ . \qquad (5.44)$$

Here, η is a parameter equal to unity in the VDW-1 mixing rules but to which, after an extensive comparison with experimental data, the value $\eta = 0.25$ was assigned. Values of the binary interaction parameters κ_{ij} in this mixing rule were determined by Plöcker *et al* [5] for about 150 non-polar or slightly-polar binary pairs. Correlations of the κ_{ij}'s were also presented for mixtures containing hydrocarbons, nitrogen, carbon dioxide and hydrogen.

As a demonstration of the application of the Lee-Kesler scheme for mixtures, in Section 5.6.2 the mixing rules of Plöcker *et al* are incorporated in a computer routine that will be employed for the estimation of mixture properties.

5.3. Four-parameter Model (Polar Molecules)

The scheme developed by Lee and Kesler [1] is very successful for many systems but its predictive capability for polar (and associating) molecules is generally poor. In an attempt to remedy this situation, Wu and Stiel [2] modified the Lee-Kesler scheme by incorporating the polar substance water as a third reference fluid. The compression factor is given in this extended scheme by

$$Z = Z_o + \omega Z_1 + Y Z_2 , \qquad (5.45)$$

where Y is a fourth parameter characteristic of polar molecules. Here, Z_o and Z_1 are given by the original Lee-Kesler scheme, while Z_2 was defined in terms of the compression factor Z_w of water by means of the equation

$$Z_2 = Z_w - (Z_o + \omega_w Z_1) \qquad (5.46)$$

in which $\omega_w = 0.344$ is the acentric factor of water. This definition ensures that for non-polar systems ($Y = 0$), the compression factor reverts to the Lee-Kesler formulation while, for $Y = 1$, Eq.(5.45) reduces to the compression factor Z_w of water. Wu and Stiel adopted the equation of state of water due to Keenan *et al.* [7]; this is given in Table 5.2. Although there is clearly no unique way of defining the parameter Y, Wu and Stiel argued that a definition based on the molar volume of the saturated liquid at a specified reduced temperature had the desired properties. Thus they defined Y by means of the equation

$$V_r^\sigma = 0.1326 - 0.0547\,\omega - 0.0222\,Y , \qquad (5.47)$$

where $V_r^\sigma = V_m^\sigma (P^c / RT^c)$ and V_m^σ is the molar volume of the saturated liquid at $T_r = 0.8$. It is therefore necessary to have an experimental value for this quantity in order to apply the Wu-Stiel method. Table A3 of Appendix A gives values of the parameter Y for selected compounds. Eq.(5.47) has the desired properties of giving $Y \approx 0$ for non-polar substances and $Y = 1$ for water.

For mixtures, the pseudo-critical constants and acentric factor are obtained from mixing rules such as the VDW-1 rules or, preferably, the mixing rules of Plöcker *et al* [5].

In Section 5.6.3 a FORTRAN subroutine will be presented which, combined with the Lee-Kesler subroutine, enables evaluation of the compression factor, enthalpy and entropy according to the Wu-Stiel equations. The modified scheme

Table 5.2   Equation of State for Water [3]

$$Z_w = \frac{MP}{\rho RT} = 1 + \rho_w Q + \rho_w^2 \left(\frac{\partial Q}{\partial \rho_w} \right)_\tau \tag{5.48}$$

where M is the molar mass, ρ is the mass density, $\rho_w = \rho/(\text{kg/m}^3)$, and for $\tau = 1000$ K/T,

$$Q = (\tau - \tau_{a1}) \sum_{j=1}^{7} \left\{ (\tau - \tau_{aj})^{j-2} \left[\sum_{i=1}^{8} A_{ij}(\rho_w - \rho_{aj})^{i-1} + e^{-4.8\rho_w}(A_{9j} + A_{10j}\rho_w) \right] \right\} \tag{5.49}$$

Also, for $j = 1$ $\tau_{aj} = 1.544912$ and $\rho_{aj} = 0.634$, while for $j > 1$ $\tau_{aj} = 2.5$ and $\rho_{aj} = 1.0$

i	A_{i1}	A_{i2}	A_{i3}	A_{i4}	A_{i5}	A_{i6}	A_{i7}
1	29.492937	-5.198586	6.833535	-0.1564104	-6.397241	-3.966140	-0.6904855
2	-132.13917	7.777918	-26.149751	-0.7254611	26.409282	15.453061	2.7407416
3	274.64632	-33.301902	65.326396	-9.2734289	-47.740374	-29.14247	-5.1028070
4	-360.93828	-16.254622	-26.181978	4.3125840	56.323130	29.568796	3.9636085
5	342.18431	-177.31074	0	0	0	0	0
6	-244.50042	127.48742	0	0	0	0	0
7	155.18535	137.46153	0	0	0	0	0
8	5.972849	155.97836	0	0	0	0	0
9	-410.30848	337.31180	-137.46618	6.7874983	136.87317	79.847970	13.041253
10	-416.05860	-209.88866	-733.96848	10.401717	645.81880	399.17570	71.531353

Residual enthalpy

$$\frac{H_w^{res}}{RT} = \rho_w \tau \left(\frac{\partial Q}{\partial \tau} \right)_{\rho_w} + \rho_w Q + \rho_w^2 \left(\frac{\partial Q}{\partial \rho_w} \right)_\tau \tag{5.50}$$

Residual entropy

$$\frac{S_w^{res}}{R} = \ln Z_w - \rho_w Q + \rho_w \tau \left(\frac{\partial Q}{\partial \tau} \right)_{\rho_w} \tag{5.51}$$

Fugacity coefficient

$$\ln(f/P)_w = (H_w^{res}/RT) - (S_w^{res}/R) \tag{5.52}$$

Table 5.2 (con/d) Equation of State for Water.

Residual isochoric heat capacity

$$\frac{C_{V,\mathrm{w}}^{\mathrm{res}}}{R} = -\rho_{\mathrm{w}}\tau^2 \left(\frac{\partial^2 Q}{\partial \tau^2}\right)_{\rho_{\mathrm{w}}} \tag{5.53}$$

Residual isobaric heat capacity

$$\frac{C_{P,\mathrm{w}}^{\mathrm{res}}}{R} = \frac{C_{V,\mathrm{w}}^{\mathrm{res}}}{R} - 1 + \frac{T}{\rho_{\mathrm{w}}^2}\left(\frac{\partial P}{\partial T}\right)_{\rho_{\mathrm{w}}}^2 \Bigg/ \left(\frac{\partial P}{\partial \rho_{\mathrm{w}}}\right)_T \tag{5.54}$$

---

gives significantly improved results for polar substances and this is illustrated in Figure 5.4 where the compression factor of liquid ammonia [9] is compared with predictions based on both the original Lee-Kesler scheme and the Wu-Stiel modification.

Alternative corresponding-states schemes that can be applied to polar substances have been devised by Wilding and Rowley [7] and by Wilding *et al* [8].

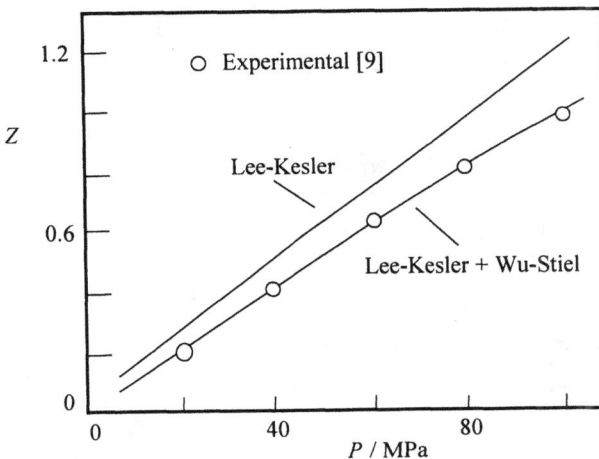

Figure 5.4 A comparison of the compressibility factor Z of liquid ammonia at 310 K as a function of pressure P.

In these methods, ω is replaced by a microscopic parameter, based on the radius of gyration, which reflects the actual non-spherical shape of the molecules. Polar and associating substances are characterised by two additional parameters, the one for the compression factor and the other for the enthalpic properties. Values for these parameters were given for 52 molecules [8].

5.4. Partial Fugacity Coefficient

The methods of the preceding sections enable calculation of the mixture fugacity coefficient $\varphi_{mix} = f/P$ according to the extended principle of corresponding states. Generally, this will be obtained in the form

$$\ln \varphi_{mix} = \ln \varphi_0 + \omega \ln \varphi_1 + Y \ln \varphi_2 , \qquad (5.55)$$

where φ_0 is the simple-fluid term, φ_1 is the usual non-polar correction term and φ_2 is the polarity term. In order to perform phase equilibrium calculations for mixtures, it is necessary to obtain the partial fugacity coefficient $\varphi_i = f_i/P$ for each component i in each phase. This quantity may be obtained from the mixture fugacity coefficient by means of the thermodynamic relation

$$\ln \varphi_i = \ln \varphi_{mix} - \sum_{k \neq i}^{n} x_k \left(\frac{\partial \ln \varphi_{mix}}{\partial x_k} \right)_{T,P,x_{j \neq i}} , \qquad (5.56)$$

where n is the number of components and x denotes mole fraction. Notice that the partial derivatives with respect to x_k are taken with all other mole fractions except x_i held constant. Since $\ln \varphi_{mix}$ is a function of the four variables $T_r = T/T^c_{mix}$, $P_r = P/P^c_{mix}$, ω_{mix} and Y_{mix}, the required isothermal-isobaric composition derivative of the mixture fugacity may be expanded as

$$\left(\frac{\partial \ln \varphi_{mix}}{\partial x_k} \right) = \left(\frac{\partial \ln \varphi_{mix}}{\partial T_r} \right)\left(\frac{\partial T_r}{\partial x_k} \right) + \left(\frac{\partial \ln \varphi_{mix}}{\partial P_r} \right)\left(\frac{\partial P_r}{\partial x_k} \right)$$
$$+ \left(\frac{\partial \ln \varphi_{mix}}{\partial \omega_{mix}} \right)\left(\frac{\partial \omega_{mix}}{\partial x_k} \right) + \left(\frac{\partial \ln \varphi_{mix}}{\partial Y_{mix}} \right)\left(\frac{\partial Y_{mix}}{\partial x_k} \right) \qquad (5.57)$$

Table 5.3 Derivatives of the Mixing Rules for the Wu-Stiel Corresponding-states Scheme.

$$\left(\frac{\partial T_{\text{mix}}^c}{\partial x_k}\right)_{x_{j\neq i}} = \frac{1}{(V_{\text{mix}}^c)^{\eta}}\left\{\sum_{l=1}^{n}\left[x_l\left\{(V_{lk}^c)^{\eta}T_{lk}^c - (V_{li}^c)^{\eta}T_{li}^c\right\}\right] - \eta(V_{\text{mix}}^c)^{\eta-1}\left(\frac{\partial V_{\text{mix}}^c}{\partial x_k}\right)T_{\text{mix}}^c\right\} \quad (5.58)$$

$$\left(\frac{\partial P_{\text{mix}}^c}{\partial x_k}\right)_{x_{j\neq i}} = \frac{P_{\text{mix}}^c}{Z_{\text{mix}}^c}\left(\frac{\partial Z_{\text{mix}}^c}{\partial x_k}\right)_{x_{j\neq i}} + \frac{P_{\text{mix}}^c}{T_{\text{mix}}^c}\left(\frac{\partial T_{\text{mix}}^c}{\partial x_k}\right)_{x_{j\neq i}} - \frac{P_{\text{mix}}^c}{V_{\text{mix}}^c}\left(\frac{\partial V_{\text{mix}}^c}{\partial x_k}\right)_{x_{j\neq i}} \quad (5.59)$$

$$\left(\frac{\partial \omega_{\text{mix}}}{\partial x_k}\right)_{x_{j\neq l}} = \omega_k - \omega_i, \qquad \left(\frac{\partial Y_{\text{mix}}}{\partial x_k}\right)_{x_{j\neq i}} = Y_k - Y_i \quad (5.60)$$

where

$$\left(\frac{\partial V_{\text{mix}}^c}{\partial x_k}\right)_{x_{j\neq i}} = 2\sum_{l=1}^{n}x_l\left(V_{lk}^c - V_{li}^c\right) \quad (5.61)$$

$$\left(\frac{\partial Z_{\text{mix}}^c}{\partial x_k}\right)_{x_{j\neq i}} = -0.085\left(\frac{\partial \omega_{\text{mix}}}{\partial x_k}\right)_{x_{j\neq l}} \quad (5.62)$$

$$T_{ij}^c = k_{ij}\sqrt{T_i^c T_j^c} \qquad \text{and} \qquad V_{ij}^c = \frac{1}{8}\left[\left(V_i^c\right)^{1/3} + \left(V_j^c\right)^{1/3}\right]^3 \quad (5.63)$$

Then, making use of the relations

$$\left(\frac{\partial \ln \phi_{\text{mix}}}{\partial T}\right)_{P,x_j} = -\frac{H_m^{\text{res}}}{RT^2} \qquad \text{and} \qquad \left(\frac{\partial \ln \phi_{\text{mix}}}{\partial P}\right)_{T,x_j} = -\frac{V_m^{\text{res}}}{RT} = \frac{1-Z}{P}, \quad (5.64)$$

it is easy to show that the required derivative may be written explicitly in terms of mixture departure functions and derivatives of the four mixing rules with respect to composition:

$$\left(\frac{\partial \ln \phi_{\mathrm{mix}}}{\partial x_k}\right)_{T,P,x_{j\neq i}} = \frac{1}{T}\frac{H^{\mathrm{res}}}{RT^{\mathrm{c}}_{\mathrm{mix}}}\left(\frac{\partial T^{\mathrm{c}}_{\mathrm{mix}}}{\partial x_k}\right)_{x_{j\neq i}} + \frac{1-Z}{P^{\mathrm{c}}_{\mathrm{mix}}}\left(\frac{\partial P^{\mathrm{c}}_{\mathrm{mix}}}{\partial x_k}\right)_{x_{j\neq i}}$$

$$+ \ln \phi_1 \left(\frac{\partial \omega_{\mathrm{mix}}}{\partial x_k}\right)_{x_{j\neq i}} + \ln \phi_2 \left(\frac{\partial Y_{\mathrm{mix}}}{\partial x_k}\right)_{x_{j\neq i}}.$$

(5.65)

It is clear from this equation that the partial fugacity coefficient is dependent upon the choice of mixing rules. In Table 5.3 we give the composition derivatives of $T^{\mathrm{c}}_{\mathrm{mix}}$, $P^{\mathrm{c}}_{\mathrm{mix}}$ and ω_{mix} corresponding to the mixing rules proposed by Plöcker *et al* [5] together with the composition derivative of Y_{mix} obtained from the linear combining rule

$$Y_{\mathrm{mix}} = \sum_{i=1}^{n} x_i Y_i$$

(5.66)

The mixing rules for the mixture pseudo-critical temperature and pseudo-critical pressure of course reduce to VDW-1 results if $\eta = 1$. In any case, to obtain accurate results for mixtures containing dissimilar molecules, it is usually necessary to adjust the binary interaction parameters κ_{ij} which modify the unlike terms in the mixture pseudo-critical temperature.

In Section 5.6.4 a subroutine that implements these equations will be described.

5.5. Summary

In this chapter the elementary principle of corresponding states was introduced. It was shown to have a sound theoretical foundation but to apply accurately only to a small number of so-called simple fluids. The extended principle of corresponding states including the acentric factor of Pitzer was then introduced together with the implementation due to Lee and Kesler. Mixing rules which permit the application of the method to multi-component mixtures were also described. The principle in this form is applicable with good accuracy to a wide range of pure substances and mixtures. Its main failings are in respect of highly asymmetric mixtures and polar or associating substances. The first of these limitations is best alleviated by a careful selection of mixing rules and the optimisation of binary interaction parameters while the second requires the incorporation of a fourth parameter. The Wu-Stiel method for polar fluids was described as an example.

5.6. Computational Implementation

In this section the following four FORTRAN subroutines will be presented:

1. **Lee-Kesler scheme for pure non-polar fluids** (Section 5.6.1)
   This subroutine evaluates thermodynamic properties of pure non-polar fluids according to the Lee-Kesler scheme supplemented by an empirical correlation of the perfect-gas isobaric heat capacity. The following properties are calculated for both liquid and vapour phases: acentric factor, compression factor, density, fugacity, enthalpy, entropy, and heat capacities.
2. **Lee-Kesler scheme for non-polar mixtures** (Section 5.6.2)
   This subroutine, used in combination with the previous one, permits the calculation of thermodynamic properties of non-polar (or slightly polar) fluid mixtures. The Plöcker mixing rules are used together with Lee's and Kesler's correlation of the critical compression factor.
3. **Wu-Stiel scheme for polar fluids** (Section 5.6.3)
   This subroutine, used in combination with the basic Lee-Kesler routine, permits the calculation of thermodynamic properties of pure polar fluids.
4. **Partial fugacity coefficient in a mixture** (Section 5.6.4)
   This subroutine returns the partial fugacity coefficient of a component in a mixture. The VDW-1 mixing rules are employed with Lee's and Kesler's correlation of Z^c.

5.6.1. Non-polar Fluids

The FORTRAN subroutine LEEKESL presented here and listed in Display 5.1 is based on the Lee-Kesler equations presented in Table 5.1 of Section 5.2.2. This subroutine calculates the acentric factor, vapour pressure, compression factor, density, fugacity, enthalpy, entropy, and the constant-volume and constant-pressure heat capacities. The following points should be noted:
- The prefect-gas contributions to the enthalpy, entropy and heat capacities are calculated from the heat-capacity correlation described in Section 2.10.1.
- Throughout this book, it is arbitrarily assumed that the enthalpy and entropy are equal to zero at $T = 273.15$ K and $P = 0.10133$ MPa.
- The subroutine's range of applicability is restricted (by convergence) to $0.4 < T_r < 4$ and $0.01 < P_r < 10$, and application near the critical point should be avoided.
- Switch (IS) refers to the number of components in a mixture and must be set to unity for pure fluids. Its function will be further discussed in the next section.

```
      SUBROUTINE LEEKESL
     &(T,P,TB,TC,PC,IS,IP,W1,W2,W3,W4,A,PS,Z,D,F,H,S,CV,CP)
C
C
C     The Subroutine calculates the following thermodynamic properties of non-polar
C     compounds, as a function of Temperature, T in [K], and Pressure, P in [Pa]:
C           1) Acentric Factor                    (A)        in [-]
C           2) Saturation Vapour Pressure         (PS)       in [Pa]
C           3) Compression Factor                 (Z)        in [-]
C           4) Molar Density                      (D)   in [MOL/M3]
C           5) Fugacity                           (F)        in [Pa]
C           6) Molar Enthalpy                     (H)     in [J/mol]
C           7) Molar Entropy                      (S)    in [J/mol/K]
C           8) Molar Isochoric Heat Capacity      (CV)   in [J/mol/K]
C           9) Molar Isobaric Heat Capacity       (CP)   in [J/mol/K]
C     based on the Corresponding States Equations developed in
C     B.I. Lee & M.G. Kesler, AIChE J. 21 510 (1975)
C     - Enthalpy and Entropy are arbitrarily assumed to be zero at 273.15 K, 0.1013 MPa.
C     - Perfect-gas properties are calculated from perfect-gas Heat Capacities as:
C        CP=W1+W2*T+W3*(T**2)+W4*(T**3) in [J/mol/K]
C     - Two Switches are required from the Subroutine:
C        IP = Phase specification Switch (Liquid: IP=1, Vapour: IP=2)
C        IS = Number of components in mixture (Pure fluid: IS=1)
C     - Other data required are: normal boiling temperature, TB in [K], critical
C        temperature, TC in [K], and critical pressure, PC in [Pa].
C
      COMMON /EQCON/ B1(2),B2(2),B3(2),B4(2),C1(2),C2(2),C3(2),C4(2),
     &D1(2),D2(2),BI(2),GA(2)
      DIMENSION ZF(2),FUG(2),ENTH(2),SS(2),CCV(2),CCP(2)
C.....Perfect-Gas Properties
      T0=273.15
      P0=0.10133E+6
      R=8.31451
      PGCP=W1+W2*T+W3*T*T+W4*(T**3)
      PGCV=PGCP-R
      PGH=W1*(T-T0)+W2*((T**2)-(T0**2))/2.+W3*((T**3)-(T0**3))/3.+
     &W4*((T**4)-(T0**4))/4.
      PGS=W1*LOG(T/T0)+W2*(T-T0)+W3*((T**2)-(T0**2))/2.+
     &W4*((T**3)-(T0**3))/3.-R*LOG(P/P0)
C.....Acentric Factor
      TR=T/TC
      PR=P/PC
      TBR=TB/TC
C.....Check for mixture (IS>1), or pure fluid (IS=1)
      IF (IS.GT.1) GOTO 10
      A1=LOG(0.10133E+6/PC)-5.92714+6.09648/TBR+1.28862*LOG(TBR)-
     &0.169347*TBR**6
      A2=15.2518-15.6875/TBR-13.4721*LOG(TBR)+0.43577*TBR**6
      A=A1/A2
```

```
C        Saturation Vapour Pressure
    10  PS=PC*EXP(5.92714-6.09648/TR-1.28862*LOG(TR)+0.169347*TR**6 +
        &A*(15.2518-15.6875/TR-13.4721*LOG(TR)+0.43577*TR**6))
C        Compression Factor
        DO 50 I=1,2
        IF (IP.EQ.1) THEN
        Z1=0.001
        Z2=0.010
    20 IF ((COMP(Z1,I,TR,PR)*COMP(Z2,I,TR,PR)).LT.0.0) GOTO 40
        Z1=Z2
        Z2=Z2+0.01
        GOTO 20
        ELSE
        Z1=1.1
        Z2=1.2
    30 IF ((COMP(Z1,I,TR,PR)*COMP(Z2,I,TR,PR)).LT.0.0) GOTO 40
        Z2=Z1
        Z1=Z1-0.05
        GOTO 30
        ENDIF
    40 Z3=(Z1+Z2)/2.
        IF (ABS(Z1-Z2).LT.0.000001) GOTO 50
        IF ((COMP(Z3,I,TR,PR)*COMP(Z1,I,TR,PR)).LT.0.0) Z2=Z3
        IF ((COMP(Z3,I,TR,PR)*COMP(Z1,I,TR,PR)).GT.0.0) Z1=Z3
        GOTO 40
    50 ZF(I)=Z3
        DO 100 I=1,2
        B=B1(I)-B2(I)/TR-B3(I)/(TR**2)-B4(I)/(TR**3)
        C=C1(I)-C2(I)/TR+C3(I)/(TR**3)
        D=D1(I)+D2(I)/TR
        VR=ZF(I)*TR/PR
        EE=(BI(I)+1.+GA(I)/(VR**2))*EXP(-GA(I)/(VR**2))
        E=C4(I)*(BI(I)+1.-EE)/(2.*(TR**3)*GA(I))
C        Fugacity
        FU=EXP(ZF(I)-1.-LOG(ZF(I))+B/VR+C/(2.*(VR**2))+D/(5.*(VR**5))+E)
        FUG(I)=LOG10(FU)
C        Residual Enthalpy
        ENT1=(B2(I)+2.*B3(I)/TR+3.*B4(I)/(TR**2))/(TR*VR)
        ENT2=(C2(I)-3.*C3(I)/(TR**2))/(2.*TR*VR*VR)
        ENT3=D2(I)/(5.*TR*(VR**5))
        ENTH(I)=TR*(ZF(I)-1.-ENT1-ENT2+ENT3+3.*E)
C        Residual Entropy
        S1=(B1(I)+B3(I)/(TR**2)+2.*B4(I)/(TR**3))/VR
        S2=(C1(I)-2.*C3(I)/(TR**3))/(2.*VR*VR)
        S3=D1(I)/(5.*(VR**5))
        SS(I)=LOG(ZF(I))-S1-S2-S3+2.*E
C        Residual Isochoric Heat Capacity
```

```
      CV1=2.*(B3(I)+3.*B4(I)/TR)/(VR*TR*TR)
      CV2=3.*C3(I)/(VR*VR*(TR**3))
      CCV(I)=CV1-CV2-6.*E
C.... Residual Isobaric Heat Capacity
      CP11=(B1(I)+B3(I)/(TR**2)+2.*B4(I)/(TR**3))/VR
      CP12=((C1(I)-2.*C3(I)/(TR**3))/(VR*VR))+(D1(I)/(VR**5))
      CP13=2.*C4(I)*(BI(I)+GA(I)/(VR*VR))*EXP(-GA(I)/(VR**2))
      CP13=CP13/(VR*VR*(TR**3))
      DPDT=(1.+CP11+CP12-CP13)/VR
      CP21=1.+2.*B/VR+3.*C/(VR*VR)+6.*D/(VR**5)
      CP22=3.*BI(I)+(GA(I)/(VR*VR))*(5.-2.*(BI(I)+GA(I)/(VR*VR)))
      CP22=(C4(I)/(VR*VR*(TR**3)))*CP22*EXP(-GA(I)/(VR*VR))
      DPDV=-(TR/(VR*VR))*(CP21+CP22)
  100 CCP(I)=CCV(I)-1.-TR*DPDT*DPDT/DPDV
C........Total Properties (& Compression Factor, Density, Fugacity)
      Z=ZF(1)+(A/0.3978)*(ZF(2)-ZF(1))
      D=P/(Z*R*T)
      F=P*10.**(FUG(1)+(A/0.3978)*(FUG(2)-FUG(1)))
      H=PGH+R*TC*(ENTH(1)+(A/0.3978)*(ENTH(2)-ENTH(1)))
      S=PGS+R*(SS(1)+(A/0.3978)*(SS(2)-SS(1)))
      CV=PGCV+R*(CCV(1)+(A/0.3978)*(CCV(2)-CCV(1)))
      CP=PGCP+R*(CCP(1)+(A/0.3978)*(CCP(2)-CCP(1)))
      IF (IS.GT.1) PS=99999999.9
      RETURN
      END

      FUNCTION COMP(Z,I,TR,PR)
      COMMON /EQCON/ B1(2),B2(2),B3(2),B4(2),C1(2),C2(2),C3(2),C4(2),
     &D1(2),D2(2),BI(2),GA(2),ZF(2),FUG(2),ENTH(2),SS(2),CCV(2),CCP(2)
      B=B1(I)-B2(I)/TR-B3(I)/(TR**2)-B4(I)/(TR**3)
      C=C1(I)-C2(I)/TR+C3(I)/(TR**3)
      D=D1(I)+D2(I)/TR
      VR=Z*TR/PR
      ZZ=1.+B/VR+C/(VR**2)+D/(VR**5)
      COMP=Z-ZZ-((C4(I)/((TR**3)*(VR**2)))*(BI(I)+(GA(I)/(VR**2)))*
     &EXP(-GA(I)/(VR**2)))
      RETURN
      END

      BLOCK DATA
      COMMON /EQCON/ B1(2),B2(2),B3(2),B4(2),C1(2),C2(2),C3(2),C4(2),
     &D1(2),D2(2),BI(2),GA(2)
      DATA B1,B2/0.1181193,0.2026579,0.2657280,0.3315110/
      DATA B3,B4/0.1547900,0.0276550,0.0303230,0.2034880/
      DATA C1,C2/0.0236744,0.0313385,0.0186984,0.0503618/
      DATA C3,C4/0.0000000,0.0169010,0.0427240,0.0415770/
      DATA D1,D2/0.155488E-4,0.48736E-4,0.623689E-4,0.0740336E-4/
      DATA BI,GA/0.65392,1.226,0.060167,0.03754/
      END
```

EXAMPLE 5.1	*Evaluation of the Density and Fugacity of n-Dodecane at* *a) T = 298.15 K, P = 0.1 MPa, and* *b) T = 358.16 K, P = 13.8 MPa,* *and of the Enthalpy Difference Between the two States.*

Subroutine LEEKESL is employed. Critical constants, boiling temperature, and heat-capacity coefficients are obtained from Table A1 in Appendix A.

The calling program is shown in Display 5.2. The state point is defined in the first DATA statement and is shown set to the first set of conditions. Switch IP, which specifies the phase, is set equal to 1 for the liquid phase, while switch IS is also set equal to unity becuase we are dealing with a pure fluid.

Running the program and collecting the results in a table, we obtain:

T / K	P / MPa	ρ_n / (mol·m^{-3})		f / (10^4 MPa)		ΔH_m / (kJ·mol^{-1})		
		L-K	[10]	L-K	[10]	L-K	L-K	[10]
298.15	0.10	4446	4375	0.20	0.25	-53.8		
							25.5	25.1
358.15	13.8	4277	4193	30.2	32.3	-28.3		

In the table, the experimental values of Snyder and Winnick [10] are included for purposes of comparison. In this particular example it can be seen that, despite the wide range of conditions and the fact that the fluid is in the liquid phase, the density agrees with experiment to within 2 per cent, while the fugacity and the enthalpy differences are also quite satisfactory.

It is also worth noticing that the value of the acentric factor obtained from the critical constants and the normal boiling temperature through Eq.(5.37) is in excellent agreement with the value given in Table A1 of Appendix A.

```
                              DISPLAY 5.2  LEEKESL Sample Input/Output
PROGRAM EX0501
DATA T,P,TB,TC,PC,IP,IS/298.15,0.101E+6,489.5,658.65,1.83E+6,1,1/
DATA W1,W2,W3,W4/-9.328,1,149,-6.347E-4,1.359E-7/
CALL LEEKESL(T,P,TB,TC,PC,IS,IP,W1,W2,W3,W4,A,PS,Z,D,F,H,S,CV,CP)
WRITE (*,1040) A,Z,D,F,H,S,CV,CP
1040  FORMAT(6X,'ACENTRIC FACTOR                       (-) =',F13.4,
     &/,5X,'COMPRESSION FACTOR                     (-) =',F13.4,
     &/,5X,'MOLAR DENSITY                    (MOL/M3) =',F10.1,
     &/,5X,'FUGACITY                             (PA) =',F10.1,
     &/,5X,'MOLAR ENTHALPY                   (J/MOL) =',F10.1,
     &/,5X,'MOLAR ENTROPY                  (J/MOL/K) =',F10.1,
     &/,5X,'MOLAR ISOCHORIC HEAT CAPACITY  (J/MOL/K) =',F10.1,
```

```
47,5X MOLAR ISOBARIC HEAT CAPACITY        (J/MOL/K) = ,F10.1,/
    END


ACENTRIC FACTOR                      (-)       0.5764
COMPRESSION FACTOR                   (-)       0.0092
MOLAR DENSITY                    (MOL/M3)     4446.0
FUGACITY                            (PA)        20.0
MOLAR ENTHALPY                    (J/MOL)    -53858.5
MOLAR ENTROPY                   (J/MOL/K)      -108.7
MOLAR ISOCHORIC HEAT CAPACITY   (J/MOL/K)       329.3
MOLAR ISOBARIC HEAT CAPACITY    (J/MOL/K)       356.0
```

❏

5.6.2. Non-polar Mixtures

As already discussed in Section 5.2.3, in the case of non-polar mixtures, the mixing rules of Plöcker *et al* [5] are widely used. In using them, it is convenient to employ Lee and Kesler's correlation of Z^c so that critical molar volumes are not required. These equations are incorporated in subroutine PLOCKER shown in Display 5.3. Note that PLOCKER, in its present form, requires binary interaction parameters and that values for 150 pairs exist in reference [5].

- This subroutine calculates the acentric factor, the pseudo-critical constants and the four perfect-gas heat capacity coefficients of the mixture and must be called from the main program before using LEEKESL to determine other mixture properties. Pure component parameters are passed to the subroutine as arguments.
- The switch IS must be set equal to the number of components.
- Subsequent use of the LEEKESL subroutine may be made as already described.

```
                                        Display 5.3  PLOCKER List
    SUBROUTINE PLOCKER
    &(IS,T,P,XX,TTB,TTC,PPC,PLK,Y1,Y2,Y3,Y4,A,TC,PC,W1,W2,W3,W4)
C
C   The Subroutine is to be used before Subroutine LEEKESL, in case of
C   multicomponent mixtures, to apply mixing rules to the Critical
C   Parameters and the Acentric Factor, based on the Corresponding
C   States Equations developed in
C   B.I. Lee & M.G. Kesler, AIChE J. 21:510 (1975)
C   - Since the Acentric Factor is calculated here, the corresponding part in the
C     LEEKESL Subroutine is not used (via switch IS)
C   - Perfect-gas Heat Capacity coefficients are calculated as mole fraction averages
C     of the pure components coefficients given as
C       CP(I)=Y1(I)+Y2(I)*T+Y3(I)*(T**2)+Y4(I)*(T**3)
```

```
C         - The number of components is set at parameter IS.
C         - Data required for every fluid are: mole fraction XX(I), normal boiling
C           temperature, TTB(I) in [K], critical temperature, TTC(I) in [K],
C           critical pressure, PPC(I) in [Pa], and the binary interaction
C           parameters PLK(I,J).
C
          DIMENSION XX(IS),TTB(IS),TTC(IS),PPC(IS)
          DIMENSION PLK(IS,IS),Y1(IS),Y2(IS),Y3(IS),Y4(IS),VC(10)
C.........Acentric Factor & Critical Volume of Component I
          A=0.0
          R=8.31451
          DO 200 I=1,IS
          TR=T/TTC(I)
          PR=P/PPC(I)
          TBR=TTB(I)/TTC(I)
          A1=LOG(0.10133E+6/PPC(I))-5.92714+6.09648/TBR+1.28862*LOG(TBR)-
         &0.169347*TBR**6
          A2=15.2518-15.6875/TBR-13.4721*LOG(TBR)+0.43577*TBR**6
          AM=A1/A2
          VC(I)=(0.2905-0.085*AM)*R*TTC(I)/PPC(I)
     200  A=A+XX(I)*AM
C.........Mixture Critical Volume, Critical Temperature and Critical Pressure
          VVC=0.0
          DO 220 J=1,IS
          DO 220 K=1,IS
          VV=((VC(J)**(1./3.))+(VC(K)**(1./3.)))**3
     220  VVC=VVC+XX(J)*XX(K)*VV
          VVC=VVC/8.
          TC=0.0
          DO 240 J=1,IS
          DO 240 K=1,IS
          VV=((VC(J)**(1./3.))+(VC(K)**(1./3.)))**(3.*0.25)
     240  TC=TC+XX(J)*XX(K)*VV*((TTC(J)*TTC(K))**0.5)
          TC=TC/((8.*VVC)**0.25)
          PC=(0.2905-0.085*A)*R*TC/VVC
C.........Mixture Heat Capacity Coefficients
          W1=0.0
          W2=0.0
          W3=0.0
          W4=0.0
          DO 260 I=1,IS
          W1=W1+XX(I)*Y1(I)
          W2=W2+XX(I)*Y2(I)
          W3=W3+XX(I)*Y3(I)
     260  W4=W4+XX(I)*Y4(I)
          RETURN
          END
```

EXAMPLE 5.2	*Evaluation of the Isobaric Specific Heat Capacity of the Mixture (0.766 C$_3$H$_8$ + 0.234 CH$_4$) at* *a) T = 310 K, P = 1.7 MPa and at* *b) T = 420 K, P = 13.7 MPa.*

The calling program is shown in Display 5.4 for the first set of conditions. Values of the normal boiling temperature, critical constants and heat-capacity coefficients are given in Appendix A. The binary interaction coefficient is 1.113 [5]. The mixture molar mass is 0.037531 kg·mol^{-1}.

At $T = 310$ K, $P = 1.7$ MPa the program gives $C_p = 2.20$ kJ·kg^{-1}·K^{-1}, while at $T = 420$ K and $P = 13.7$ MPa, the value obtained is 3.60 kJ·kg^{-1}·K^{-1}. Experimental values quoted [11] are 2.5 kJ·kg^{-1}·K^{-1} and 3.4 kJ·kg^{-1}·K^{-1} respectively. The agreement is reasonable.

Note that if the binary interaction coefficient was unity, the equivalent values would have been 2.15 kJ·kg^{-1}·K^{-1} and 3.46 kJ·kg^{-1}·K^{-1}. In this case the difference is not large.

```
                                    DISPLAY 5.4   PLOCKER Sample Input/Output
      PROGRAM EX0502
      PARAMETER (IS=2)
      DIMENSION XX(IS),TTB(IS),TTC(IS),PPC(IS)
      DIMENSION Y1(IS),Y2(IS),Y3(IS),Y4(IS), PLK(IS,IS)
      DATA T,P,IP/310.,1.7E+6,2/
      DATA XX(1),TTB(1),TTC(1),PPC(1)/0.234,111.6,190.55,4.599E+6/
      DATA Y1(1),Y2(1),Y3(1),Y4(1)/19.25,5.213E-2,1.197E-5,-1.132E-8/
      DATA XX(2),TTB(2),TTC(2),PPC(2)/.766,231.1,369.85,4.247E+6/
      DATA Y1(2),Y2(2),Y3(2),Y4(2)/-4.224,3.063E-1,-1.586E-4,3.215E-8/
      DATA PLK(1,1),PLK(1,2),PLK(2,1),PLK(2,2)/1.,1.113,1.113,1./
      CALL PLOCKER
     &(IS,T,P,XX,TTB,TTC,PPC,PLK,Y1,Y2,Y3,Y4,A,TC,PC,W1,W2,W3,W4)
      CALL LEEKESL
     &(T,P,TB,TC,PC,IS,IP,W1,W2,W3,W4,A,PS,Z,D,F,H,S,CV,CP)
      WRITE (*,1040) (CP/37.531)
1040  FORMAT(3X,'ISOBARIC SPECIFIC HEAT CAPACITY (KJ/KG/K) =',F5.2)
      END



 ISOBARIC SPECIFIC HEAT CAPACITY (KJ/KG/K) = 2.20
```

☐

| **EXAMPLE 5.3** | *Evaluation the Specific Enthalpy Difference h(T = 590K, P = 9.7MPa) - h(T = 470K, P = 1.4 MPa) for the Mixture (0.324 n-C$_8$H$_{18}$ + 0.676 C$_6$H$_6$).* |

The calling program is shown in Display 5.5. The binary interaction parameter is 0.987 [5]. All the routines are called twice, once for each set of conditions, and the specific enthalpy difference obtained is 332.9 kJ·kg^{-1}. The experimental value [12] is 315 kJ·kg^{-1} and the agreement is reasonable.

```
                    DISPLAY 5.5  PLOCKER Sample Input/Output
   PROGRAM EX0503
   PARAMETER (IS=2)
   DIMENSION XX(IS),TTB(IS),TTC(IS),PPC(IS)
   DIMENSION PLK(IS),Y1(IS),Y2(IS),Y3(IS),Y4(IS)
   DATA IP,XX(1),TTB(1),TTC(1),PPC(1)/1,0.676,353.2,562.2,4.895E+6/
   DATA Y1(1),Y2(1),Y3(1),Y4(1)/-33.92,0.4739,-3.017E-4,7.13E-8/
   DATA XX(2),TTB(2),TTC(2),PPC(2)/0.324,398.8,568.95,2.49E+6/
   DATA Y1(2),Y2(2),Y3(2),Y4(2)/-6.096,0.7712,-4.195E-4,8.855E-8/
   DATA PLK(1,1),PLK(1,2),PLK(2,1),PLK(2,2)/1.,0.987,0.987,1./
   DO 300 I=1,2
   IF (I.EQ.1) T=590
   IF (I.EQ.1) P=9.7E+6
   IF (I.EQ.2) T=470
   IF (I.EQ.2) P=1.4E+6
   CALL PLOCKER
  &(IS,T,P,XX,TTB,TTC,PPC,PLK,Y1,Y2,Y3,Y4,A,TC,PC,W1,W2,W3,W4)
   CALL LEEKESL
  &(T,P,TB,TC,PC,IS,IP,W1,W2,W3,W4,A,PS,Z,D,F,H,S,CV,CP)
   IF (I.EQ.1) H1=H
300 IF (I.EQ.2) HT=(H1-H)/89.815
   WRITE (*,1040) HT
1040 FORMAT(3X,'SPECIFIC ENTHALPY DIFFERENCE (KJ/KG) =',F5.1)
   END


SPECIFIC ENTHALPY DIFFERENCE (KJ/KG) = 332.9
```

❑

These examples demonstrate the predictive power of the three-parameter corresponding states scheme for non-polar fluids. The examples were chosen such that experimental values were available with which to compare the results but, at the same time, wide ranges of temperature and pressure were examined. The agreement in all cases was relatively good.

5.6.3. *Polar Fluids*

The procedure developed by Wu and Stiel is used here. Subroutine WUSTIEL shown in Display 5.6 calculates the polar correction term in the four-parameter model for polar fluids while subroutine LEEKESL is used as before to obtain the other terms. The following points should be noted:
- Subroutine WUSTIEL calculates only corrections to the compression factor, enthalpy and entropy differences and fugacity. It should be called after subroutine LEEKESL and the results combined to obtain the final property values.
- The value of the polarity parameter Y must be passed to subroutine LEEKESL by the main program as shown in the listing. All other variables required by the WUSTIEL subroutine are obtained from the LEEKESL subroutine, through its arguments.
- The switch (IS) should be set to unity for pure fluids.
- For very polar fluids, the acentric factor must be entered instead of using Eq.(5.37).

```
                                                    Display 5.6  WUSTIEL List
      SUBROUTINE LEEKESL
      &(T,P,TB,TC,PC,Y,IS,IP,W1,W2,W3,W4,A,PS,Z,D,F,H,S,CV,CP)
C
C       In case of Polar fluids two 'additions' must be made to the LEEKESL soubroutine
C       presented in Display 5.1. These are
C         1)  In the LEEKESL subroutine's argument the variable Y (the polarity factor,
C             entered from main calling program) must be included as shown above.
C         2)  The following lines related to subroutine WUSTIEL must be included
C             before the RETURN statement.
C
C..... Corrections for Polar fluids
      CALL WUSTIEL(IP,TR,PR,ZW,HW,SW)
      Z2=ZW-(ZF(1)+0.344*(ZF(2)-ZF(1))/0.3978)
      Z=Z+Y*Z2
      D=P/(Z*R*T)
      ENTH2=HW*TR-(ENTH(1)+0.344*(ENTH(2)-ENTH(1))/0.3978)
      H=H+(Y*ENTH2*R*TC)
      S2=SW-(SS(1)+0.344*(SS(2)-SS(1))/0.3978)
      S=S+(Y*S2*R)
      FW=LOG10(EXP(HW-SW))
      FUG2=FW-(FUG(1)+0.344*(FUG(2)-FUG(1))/0.3978)
      F=P*10**((FUG(1)+(A/0.3978)*(FUG(2)-FUG(1)))+Y*FUG2)
      IF (ABS(S2).GT.2.0.OR.ABS(ENTH2).GT.1.) WRITE (*,1050)
1050  FORMAT(12X,'>>> TR OR PR OUT OF RANGE, NO CONVERGENCE <<<')
      RETURN
      END
```

```
      SUBROUTINE WUSTIEL (IP,TR,PR,ZW,HW,SW)
C
C     The Subroutine calculates the following corrections to the thermodynamic
C     properties of polar compounds, as a function of Reduced Temperature,
C     TR and Pressure, PR.
C     1) Compression Factor Correction      (ZW)    in [-]
C     2) Molar Enthalpy Correction          (HW)    in [J/mol/K]
C     3) Molar Entropy Correction           (SW)    in [J/mol/K]
C     (while Fugacity correction can be obtained from ln(f/P)w=HW-SW)
C     based on the Corresponding States Equations developed in
C     G.Z.A. Wu and L.I.Stiel, AIChE J. 31:1632 (1985)
C     - The Subroutine must be called from subroutine LEEKESL as
C       described above.
C     - Phase of fluid (IP) is specified in LEEKESL routine as described before.
C     - Parameter Y (polarity factor) must be entered from the calling program.
C
      COMMON /EQCONZW/ A(10,7),TAJ(7),RAJ(7)
      T=647.29*TR
      P=22.088*PR
C........ Compression Factor Correction
      IF (IP.EQ.1) THEN
      Z1=0.001
      Z2=0.010
   21 IF ((COMPZW(Z1,T,P)*COMPZW(Z2,T,P)).LT.0.0) GOTO 41
      Z1=Z2
      Z2=Z2+0.01
      GOTO 21
      ELSE
      Z1=1.1
      Z2=1.2
   31 IF ((COMPZW(Z1,T,P)*COMPZW(Z2,T,P)).LT.0.0) GOTO 41
      Z2=Z1
      Z1=Z1-0.05
      GOTO 31
      ENDIF
   41 Z3=(Z1+Z2)/2.
      IF (ABS(Z1-Z2).LT.0.000001) GOTO 51
      IF ((COMPZW(Z3,T,P)*COMPZW(Z1,T,P)).LT.0.0) Z2=Z3
      IF ((COMPZW(Z3,T,P)*COMPZW(Z1,T,P)).GT.0.0) Z1=Z3
      GOTO 41
   51 ZW=Z3
C........ Molar Enthalpy and Entropy Correction Term
      TA=1000./T
      RW=P/(0.46151*ZW*T)
      Q=0.
      DQT=0.
      DO 71 J=1,7
      EX=(EXP(-4.8*RW))*(A(9,J)+A(10,J)*RW)
      QS=0.
      DO 61 I=1,8
```

```
 61   QS=QS+A(I,J)*((RW-RAJ(J))**(I-1))
      Q=Q+((TA-TAJ(J))**(J-2))*(QS+EX)*(TA-TAJ(1))
 71   DQT=DQT+((TA-TAJ(J))**(J-2))*(QS+EX)+
     &(FLOAT(J-2)*(TA-TAJ(J))**(J-3))*(QS+EX)*(TA-TAJ(1))
      HW=(ZW-1.)+(RW*TA*DQT)
      SW=LOG(ZW)-(RW*Q)+(RW*TA*DQT)
      RETURN
      END

      FUNCTION COMPZW(Z,T,P)
      COMMON /EQCONZW/ A(10,7),TAJ(7),RAJ(7)
      TA=1000./T
      RW=P/(0.46151*Z*T)
      Q=0.
      DQ=0.
      DO 111 J=1,7
      EX=(EXP(-4.8*RW))*(A(9,J)+A(10,J)*RW)
      DEX=(EXP(-4.8*RW))*(-4.8*A(9,J)-4.8*A(10,J)*RW+A(10,J))
      QS=0.
      DQS=0.
      DO 101 I=1,8
      QS=QS+A(I,J)*((RW-RAJ(J))**(I-1))
101   DQS=DQS+(FLOAT(I-1)*A(I,J)*((RW-RAJ(J))**(I-2)))
      Q=Q+((TA-TAJ(J))**(J-2))*(QS+EX)*(TA-TAJ(1))
111   DQ=DQ+((TA-TAJ(J))**(J-2))*(DQS+DEX)*(TA-TAJ(1))
      COMPZW=Z-1.-(RW*Q)-(RW*RW*DQ)
      RETURN
      END

      BLOCK DATA
      COMMON /EQCONZW/ A(10,7),TAJ(7),RAJ(7)
      DATA TAJ/1.544912,2.5,2.5,2.5,2.5,2.5,2.5/
      DATA RAJ/0.634,1.,1.,1.,1.,1.,1./
      DATA (A(1,K),K=1,7)/29.492937,-5.198586,6.8335354,-0.1564104,
     &-6.3972405,-3.9661401,-0.69048554/
      DATA (A(2,K),K=1,7)/-132.13917,7.7779182,-26.149751,-0.72546108,
     &26.409282,15.453061,2.7407416/
      DATA (A(3,K),K=1,7)/274.64632,-33.301902,65.326396,-9.2734289,
     &-47.740374,-29.142470,-5.1028070/
      DATA (A(4,K),K=1,7)/-360.93828,-16.254622,-26.181978,4.3125840,
     &56.323130,29.568796,3.9636085/
      DATA (A(5,K),K=1,7)/342.18431,-177.31074,0.,0.,0.,0.,0./
      DATA (A(6,K),K=1,7)/-244.50042,127.48742,0.,0.,0.,0.,0./
      DATA (A(7,K),K=1,7)/155.18535,137.46153,0.,0.,0.,0.,0./
      DATA (A(8,K),K=1,7)/5.9723487,155.97836,0.,0.,0.,0.,0./
      DATA (A(9,K),K=1,7)/-410.30848,337.31180,-137.46618,6.7874983,
     &136.87317,79.847970,13.041253/
      DATA (A(10,K),K=1,7)/-416.05860,-209.88866,-733.96848,10.401717,
     &645.81880,399.17570,71.531353/
      END
```

EXAMPLE 5.4	*Evaluation of the Specific Enthalpy, Specific Entropy and Density of 1,1,1,2-Tetrafluoroethane (R134a) at*

a) $T = 250$ *K,* $P = 2$ *MPa, and*
b) $T = 450$ *K,* $P = 10$ *MPa.*

To evaluate the properties of R134a according to the Wu-Stiel procedure we require the polarity parameter Y. This can be obtained through Eq.(5.47) if the molar volume of the saturated liquid at $T_r = 0.8$ is known. For R134a the molar volume under these conditions is 84.89 cm³·mol⁻¹ [17]. Running the program with $Y = 0$ the acentric factor is calculated to be 0.3226. Hence $Y = 0.195$. Note that, in Table A3 in Appendix A, the value quoted for Y parameter is 0.180, which is near the aforementioned value.

In Display 5.7 the calling program that calculates the required input parameters is shown. The program calls LEEKESL subroutine twice, once for each set of conditions. Note that in the first state, R134a is liquid while in the second case it is compressed gas.

T / K	P / MPa	ρ / (kg·m⁻³)			Δh / (kJ·kg⁻¹)			Δs / (kJ·kg⁻¹·K⁻¹)		
		[14]	W-S	L-K	[14]	W-S	L-K	[14	W-S	L-K
250	2	1371	1371	1322	324	318	319	0.91	0.89	0.89
450	10	475	473	464						

$\Delta h = h(T = 450$ K, $P = 10$ MPa$) - h(T = 250$ K, $P = 2$ MPa$)$
$\Delta s = S(T = 450$ K, $P = 10$ MPa$) - S(T = 250$ K, $P = 2$ MPa$)$

In the table, a comparison is shown between (a) the results obtained from the Lee-Kesler scheme incorporating the Wu-Stiel procedure (W-S), (b) the results obtained from the original Lee-Kesler scheme (L-K), and (c) the experimental values [13]. It can be seen that the agreement between the values obtained with the Wu-Stiel procedure and the experimental values is excellent and that the simpler Lee-Kesler scheme also gives quite reasonable results.

```
                            DISPLAY 5.7  WUSTIEL Sample Input/Ouput
PROGRAM EX0504
DATA T,P,TB,TC,PC,Y,IS/250.,2.E+6,247.,374.21,4.056E+6,0.195,1/
DATA W1,W2,W3,W4/16.7803,0.28634,-2.2732E-4,1.13305E-7/
DO 100 I=1,2
IP=I
T=T+FLOAT(I-1)*200
P=P+FLOAT(I-1)*8.E+6
```

```
       H1=H
       S1=S
       PS1=PS
       D1=D
100    CALL LEEKESL(T,P,TB,TC,PC,Y,IS,IP,W1,W2,W3,W4,A,PS,Z,D,F,H,S,CV,CP)
       WMOL=102.0314
       WRITE (-,1040) A,PS1*1 E-6,PS*1 E-6,(D1*WMOL/1000.),(D*WMOL/1000)
     & ,(H1/WMOL),(H/WMOL),(S1/WMOL),(S/WMOL)
1040   FORMAT(5X,'> TEMPERATURE, PRESSURE           250K, 2MPA',
     &   450K, 10MPA',/,5X,'ACENTRIC FACTOR                (-) =',F17.3,
     &/,5X,'SATURATION VAPOUR PRESSURE    (MPA) =',2F11.1,
     &/,5X,'DENSITY                       (KG/M3) =',2F11.1,
     &/,5X,'ENTHALPY                      (KJ/KG) =',2F11.1,
     &/,5X,'ENTROPY                       (KJ/KG/K) =',2F12.2)
       WRITE (-,1060) (H-H1)/WMOL, (S-S1)/WMOL
1060   FORMAT(1X,/,5X,'ENTHALPY DIFFERENCE            (KJ/KG) =',F17.1,
     &/,5X,'ENTROPY DIFFERENCE            (KJ/KG/K) =', F18.2)
       END
```

TEMPERATURE, PRESSURE		250K, 2MPA	450K, 10MPA
ACENTRIC FACTOR	(-) =		0.323
SATURATION VAPOUR PRESSURE	(MPA) =	0.1	17.6
DENSITY	(KG/M3) =	1371.2	472.5
ENTHALPY	(KJ/KG) =	-233.6	84.1
ENTROPY	(KJ/KG/K) =	-0.94	-0.06
ENTHALPY DIFFERENCE	[KJ/KG] =		317.7
ENTROPY DIFFERENCE	[KJ/KG/K] =		0.89

❑

5.6.4. Partial Fugacity Coefficient

In Section 5.4 a procedure was described in order to calculate the partial fugacity coefficient of component i in the mixture, from the mixture fugacity, based on the Plöcker's mixing rules shown in Table 5.3.

Subroutine NPFUGI, shown in Display 5.8, performs this calculation for the specific case of non-polar mixtures (its extension to polar mixtures implies the use of an additional term). Plöcker's parameter η, as well as the binary interaction parameters κ_{ij} must be specified.

In the beginning of the subroutine, the mixture critical parameters are calculated, then the derivatives shown in Table 5.3 are calculated and from them the partial fugacity coefficient is obtained.

```
      SUBROUTINE NPFUGI
      &(IS,T,X,TC,PC,OMEG,HORTCM,PK,HTA,PHIMIX,PHIMIX1,ZMIX,PHI)
C
C          The Subroutine calculates the partial fugacity coefficient (PHI)
C          of component i in a mixture, from the mixture fugacity. The
C          Plocker mixing rules (Table 5.3) are employed.
C          Data required are:
C          - Temperature, T in [K], No of components, IS,
C          - For each component:  mole fractions, X in [-], critical
C            temperature, TC in [K], critical pressure, PC in [Pa],
C            acentric Factor, OMEG in [-], and
C          - For the mixture: residual enthalpy term, HORTCM in [-],
C            fugacity, PHIMIX in [-], 1st correction to fugacity, PHIMIX1 in
C            [-] and compression factor, Zmix in [-].
C          - Plocker's parameter HTA and binary interaction parameters
C            PK(I,J) must also be specified.
C          Mixture critical parameters are calculated in the beginning.
C
      DIMENSION X(IS),TC(IS),PC(IS),OMEG(IS),PHI(IS),PK(IS,IS),VC(10)
      R=8.31451
C..........Mixture Acentric Factor (AFMIX)
      AFMIX=0.
      DO 100 I=1,IS
  100 AFMIX=AFMIX+X(I)*OMEG(I)
C..........Component Critical Volume (VC(I)) & Mixture's (VCMIX)
      DO 120 I=1,IS
  120 VC(I)=(0.2905-0.085*OMEG(I))*R*TC(I)/PC(I)
      VVC=0.0
      DO 140 J=1,IS
      DO 140 K=1,IS
      VV=((VC(J)**(1./3.))+(VC(K)**(1./3.)))**3
  140 VVC=VVC+X(J)*X(K)*VV
      VCMIX=VVC/8.
C..........Mixture Critical Parameters (TCMIX,ZCMIX,PCMIX)
      TCC=0.0
      DO 160 J=1,IS
      DO 160 K=1,IS
      VV=((VC(J)**(1./3.))+(VC(K)**(1./3.)))**(3.*HTA)
  160 TCC=TCC+X(J)*X(K)*VV*PK(J,K)*((TC(J)*TC(K))**0.5)
      TCMIX=TCC/((8.*VCMIX)**HTA)
      ZCMIX=(0.2905-0.085*AFMIX)
      PCMIX=ZCMIX*R*TCMIX/VCMIX
C..........Calculation of Derivatives
      DO 600 IC=1,IS
      DPHI=0.
      DO 500 K=1,IS
      IF (K.EQ.IC) GO TO 500
```

```
      SUMVC=0.
      SUMTC=0.
      DO 400 L=1,IS
      TCLK=PK(L,K)*((TC(L)*TC(K))**0.5)
      TCLI=PK(L,IC)*((TC(L)*TC(IC))**0.5)
      VCLK=(1./8.)*((VC(L)**3.+VC(K)**3.)**(1./3.))
      VCLI=(1./8.)*((VC(L)**3.+VC(IC)**3.)**(1./3.))
      SUMVC=SUMVC+X(L)*(VCLK-VCLI)
400   SUMTC=SUMTC+X(L)*((VCLK**HTA)*TCLK-(VCLI**HTA)*TCLI)
      DVCDX=2.*SUMVC
      DTCDX=(1./(VCMIX**HTA))*(SUMVC-HTA*(VCMIX**(HTA-1.))
     &*DVDX*TCMIX)
      DOMDX=OMEG(K)-OMEG(IC)
      DZCDX=-0.85*DOMDX
      DPCDX=PCMIX*((ZDCDX/ZCMIX)+(DTCDX/TCMIX)-(DVCDX/VCMIX))
      DLNFDX=(HORTCM*DTCDX/T)+(DPCDX*(1.-ZMIX)/PCMIX)
     &+(DOMDX*LOG(PHIMIX1))
      DPHI=DPHI+X(K)*DLNFDX
500   CONTINUE
600   PHI(IC)=EXP(LOG(PHIMIX)-DPHI)
      RETURN
      END
```

EXAMPLE 5.5   *Calculation of the Partial Fugacity Coefficients for an Equimolar Mixture of Methyl Ethyl Ketone and Toluene at T = 323.15 K and P = 25 kPa.*

The critical parameters and acentric factors required for the pure components are obtained from Appendix A, while the residual enthalpy term, mixture fugacity, first correction to the mixture fugacity and mixture compression factor are obtained by employing the LEEKESL and PLOCKER subroutines as described in Section 5.6.2.

The calling program and its output are shown in Display 5.9. Component 1 is methyl ethyl ketone. In this example the value of the parameter η, as well as the binary interaction parameters, were all taken equal to unity.

As shown in Display 5.9, the partial fugacity coefficients calculated by employing the subroutine NPFUGI are 0.986 and 0.987. Values for the partial fugacity coefficients for the two components are quoted in literature (page 343 of [14]) as 0.987 and 0.983. In this case, the agreement is very good (see also Example 6.4).

```
                                   DISPLAY 5.9  NPFUGI Sample Input/Output
      PROGRAM EX0505
      PARAMETER (IS=2)
      DIMENSION X(IS),TC(IS),PC(IS),OMEG(IS),PHI(IS),PK(IS,IS)
      DATA T/323.15/
      DATA X(1),TC(1),PC(1),OMEG(1)/0.5,535.6,4.15E+6,0.329/
      DATA X(2),TC(2),PC(2),OMEG(2)/0.5,563.0,4.11E+6,0.263/
      DATA PK(1,1),PK(1,2),PK(2,1),PK(2,2)/1.,1.,1.,1./
      DATA HORTCM,PHIMIX,PHIMIX1,ZMIX/-0.02835,0.9864,0.98673,0.9862/
      HTA=1.
      CALL NPFUGI
     &(IS,T,X,TC,PC,OMEG,HORTCM,PK,HTA,PHIMIX,PHIMIX1,ZMIX,PHI)
      WRITE (*,1000) PHI(1),PHI(2)
 1000 FORMAT(1X,'FUGACITY COEFFICIENTS IN MIXTURE OF ',
     &/,3X,'COMPONENT 1 : ',F5.3,/,3X,'COMPONENT 2 : ',F5.3)
      END


FUGACITY COEFFICIENTS IN MIXTURE OF
  COMPONENT 1 : 0.986
  COMPONENT 2 : 0.987
```

□

References

1. B.I. Lee and M.G. Kesler, *AIChE J.* **21** (1975) 510.
2. G.Z.A. Wu and L.I. Stiel, *AIChE J.* **31** (1985)1632.
3. K.S. Pitzer, *J. Am. Chem. Soc.* **77** (1955) 3427.
4. K.S. Pitzer, D.Z. Lippman, R.F. Curl, C.M. Huggins and D.E. Petersen, *J. Am. Chem. Soc.* **77** (1955) 3433.
5. U.Plöcker, H. Knapp and J. Prausnitz, *Ind. Eng. Chem. Process Des. Dev.* **17** (1978) 324.
6. J.H. Keenan, F.G. Keyes, P.G. Hill and J.G. Moore, *Steam Tables* (Wiley, New York, 1969).
7. W.V. Wilding and R.L. Rowley, *Int. J. Thermophys.* **7** (1986) 525.
8. W.V. Wilding, J.K. Johnson and R.L. Rowley, *Int. J. Thermophys.* **8** (1987) 717.
9. L. Haar and J. Gallagher, *J. Phys. Chem. Ref. Data* **7** (1978) 635.
10. P.S. Snyder and J. Winnick, *Proceedings of 5th Symp. Thermophys. Prop.*, ASME, Boston, (1970) 115.
11. V.F. Yesavage, D.L. Katz and J.E. Powers, *J. Chem. Eng. Data* **14** (1969) 139.
12. J.M. Lenoir and K.E. Hayworth, *J. Chem. Eng. Data* **16** (1971) 280.
13. *Thermophysical Properties of Environmentally Acceptable Refrigerants* (Japanese Association of Refrigeration, Tokyo, 1991).

14. J.M. Smith and H.C. van Ness, *Introduction to Chemical Engineering Thermodynamics* 4th Ed. (McGraw-Hill, 1987).
15. J.S. Rowlinson and J.R. Watson, *Chem. Eng. Sci.* **24** (1969) 1565.
16. J.W. Leach, P.S. Chappelear and T.W. Leland, *AIChE J.* **14** (1968) 568.
17. R. Tillner-Roth and H.D. Baehr, *J. Chem. Thermodyn.* **25** (1993) 277.

6

Equations of State

Equations of state are widely used in the prediction of thermodynamic properties of pure fluids and fluid mixtures, in part because they provide a thermodynamically consistent route to the configurational properties of both gaseous and liquid phases. Consequently, equation-of-state methods may be used to determine phase-equilibrium conditions as well as other properties. Indeed, the most well-known application of such methods in chemical engineering lies in the field of high-pressure vapour-liquid equilibria (VLE) where the equation-of-state approach is the method of choice for the vast majority of systems. The main exceptions are systems containing associating substances for which activity coefficient models, which will be described in Chapter 7, may provide a more accurate route to the fugacity of the liquid phase. The application of both methods to phase equilibrium problems will be discussed in detail in Chapter 8.

The main objective of this chapter is to provide an introduction to the predictive capabilities of the most commonly encountered equations of state. Following a few historical notes and a classification of the various kinds of equations of state, the most widely used equations will be reviewed. In each case applications to both pure substances and multicomponent mixtures will be discussed. Finally the ranges of applicability will be outlined and some examples given which will illustrate the capabilities of the approach. More advanced examples involving phase equilibria will be presented in Chapter 8.

6.1. Introduction and Historical Notes

The term *equation of state* is used to describe an empirically-derived function which provides a relation between pressure, density, temperature and (for a mixture) composition; such a relation provides a prescription for the calculation of all of the configurational and residual thermodynamic properties of the system within some domain of applicability. Sometimes, the term *equation of state* is attached to a more fundamental relation which may be used to obtain both perfect-gas and residual properties. Many equations of state can represent adequately the properties of the gas phase, some are applied only to the liquid, but the most important category of equation-of-state models contains those that may applied in the same form to both gaseous and liquid phases. There are no equations of state applicable simultaneously to solid, liquid and gas.

The functional form of most equations of state has been arrived at by combining some semi-rigorous theoretical analysis of molecular interactions with a parameterisation that permits satisfactory reproduction of certain experimental data such as vapour pressures and compression factors.

The history of the equation of state dates back to at least 1662 when Boyle conducted his experiments on air and deduced that, at a given temperature, the volume of a fixed mass of gas is inversely proportional to its pressure (PV = constant). The effect of temperature was observed by Charles in 1787 and later by Gay-Lussac who, in 1802, found the dependence of volume on temperature to be linear at constant pressure. Together with Dalton's law of partial pressures, postulated in 1801, these observations suggest the relation

$$P = (\sum_i n_i) RT / V \qquad (6.1)$$

which we would now recognise as the equation of state of a perfect-gas mixture.

The transition between vapour and liquid phases received systematic attention in 1823 from Faraday but it was not until the work of Andrews on carbon dioxide (1869) that the volumetric and phase behaviour of a pure fluid was established over appreciable ranges of temperature and density. This behaviour is illustrated by the three-dimensional phase diagram shown, together with its projections on to the *P-V* and *P-T* planes, in Figure 6.1. Also shown for purposes of comparison is the behaviour of the perfect-gas.

The pioneering experimental work of Andrews and others paved the way towards the modern view of the equation of state and led van der Waals to postulate in his dissertation of 1873 "On the continuity of the gas and liquid states" the famous equation which now bears his name:

$$P = \frac{RT}{V_m - b} - \frac{a}{V_m^2} . \qquad (6.2)$$

This equation is cubic in the molar volume and contains two parameters a and b which may be adjusted to reproduce the critical temperature and pressure for a pure

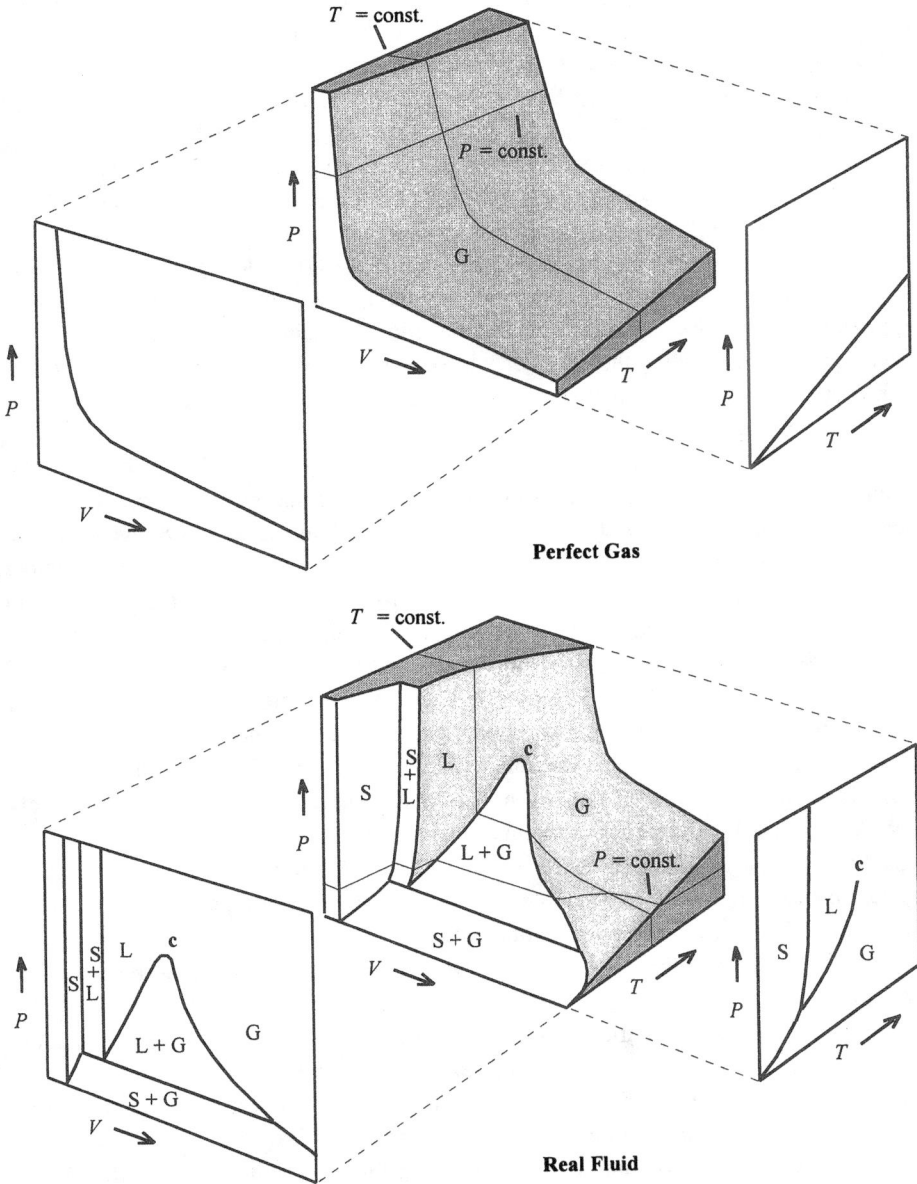

Figure 6.1 *P-V-T* behaviour of a real fluid and a perfect gas.

fluid. van der Waals' equation was the first equation of state capable of predicting both gaseous and liquid phases and the majority of equation-of-state models in use today are simple empirical modifications which retain its basic cubic form.

The equation of van der Waals has a semi-theoretical basis which we will review briefly below. More rigorous statistical-mechanical analysis led Ursell in 1912 to a derivation of the virial equation of state which had itself been proposed on purely empirical grounds by Thiesen in 1885. The virial equation has been emulated in part by many empirical equations of state which express the pressure in terms of a finite polynomial in $1/V_m$ augmented by empirical terms designed to extend the range of densities to which the equation may be applied. Notable examples of this approach include the equations of Beattie and Bridgeman (1927) and Benedict, Webb and Rubin (1940), the latter of which is applicable to gas and liquid phases. Other attempts to build on theoretical results include the work of Carnahan and Starling (1972) [6].

Meanwhile, scores of empirical modifications of van der Waals' equation had been proposed. The modern development of this field started with the equation of Redlich and Kwong (1949) [3], and led to such workhorses as the Soave equation (1972) [4] and the equation of Peng and Robinson (1976) [5]. These are cubic equations whose parameters are determined for a pure fluid from the critical constants and the acentric factor while, for a mixture, combining rules are used to express the parameters in terms of pure-component values. Vapour properties and vapour-liquid equilibrium conditions predicted by these models are usually in fair agreement with experimental observations, although the prediction of liquid density is generally rather poor. The combination of simplicity of form with reasonably reliable predictions of VLE conditions has made cubic equations the most widely used in chemical engineering problems. More accurate results can be obtained but usually only by recourse to much more complicated equations of state, such as generalisations of the Benedict-Webb-Rubin equation, with parameters fitted to wide-ranging and accurate experimental data.

Today the development of equations of state remains an active field of research, primarily in areas of:

1. Highly-accurate equations, often with many constants, for important pure substances such as water, ammonia, carbon dioxide etc.

2. Accurate equation-of-state models for specific mixtures such as those encountered in the natural gas and petroleum industries.

3. Simple equations of state that combine satisfactory predictive capabilities with the computational efficiency required for detailed simulations of chemical processing operations such as multistage separation processes for mixtures.

4. Models for complex systems such as electrolytes, polymers, coal liquids, and highly polar substances.

The following equations of state will be described further in this chapter:
- van der Waals equation
- The cubic equations of Redlich and Kwong, Soave, and Peng and Robinson
- A modern variant of the Benedict-Webb-Rubin (BWR) equation.

Together with the discussion of the virial equation (Chapter 4) and that connected with corresponding-states equations of state (Chapter 5), the discussion should provide the reader with an insight into the nature, applicability and use of equations of state.

6.2. Classification of Equations of State

The need for accurate predictions of the thermodynamic properties of many fluids and mixtures has led to the development of a rich diversity of equations of state with differing degrees of empiricism, predictive capability and mathematical form. Before proceeding with the discussion of specific equations of state it is useful to make some general classifications into which they may fall.

The main types of equation of state may be classified conveniently according to their mathematical form as follows:

1. **Standard *P-V-T* forms.** The type of equation of state may be written for a pure fluid as

$$P = P(T, V_m) \qquad \text{or} \qquad Z = Z(T, V_m) \qquad (6.3)$$

while, for a mixture of n components, there are a further $n-1$ independent composition variables. Sub-classifications may be introduced according to the structure of the function P or Z:
- **Truncated virial equations** in which P is given by a polynomial in $1/V_m$ with temperature and composition dependent coefficients.
- **Cubic equations** in which P is given by a cubic function of V_m containing two parameters which are functions of composition and possibly also of temperature.
- **Complex empirical equations** which represent P by some combination of polynomial and other terms (*eg* BWR and related equations).

It is also possible to invert the functional relationships to give V_m or Z in terms of T, P and the composition variables. However, although this choice of independent variables may be convenient in the analysis of experimental data for a single fluid phase, it is almost never used in thermodynamic modelling.

2.   **Fundamental form.** A fundamental equation gives one of the state functions in terms of its natural independent variables. By far the most common choice is the molar Helmholtz energy A_m as a function of temperature and molar volume:

$$A_m = A_m(T, V_m) \qquad (6.4)$$

In order to achieve a precise representation of the experimental data over a wide range of conditions, the structure of the functional relationship is often very complicated. Invariably, A_m is split into perfect-gas and residual parts which are parameterised separately. Although accurate fundamental equations of state exist for only about twenty of the most important pure fluids, one or more of these may form the basis of a corresponding-states treatment of the residual properties of a wide variety of other fluids including mixtures.

Table 6.1   Properties derived from equations of state

Residual Enthalpy, Entropy and Heat Capacity

$$H_m^{\text{res}} = \int\limits_{\infty}^{V_m} \left[V_m \left(\frac{\partial P}{\partial V_m} \right)_T + T \left(\frac{\partial P}{\partial T} \right)_{V_m} \right] dV_m = RT^2 \int\limits_{\infty}^{V_m} \left(\frac{\partial Z}{\partial T} \right)_{V_m} \frac{dV_m}{V_m} + RT - PV_m \qquad (6.5)$$

$$S_m^{\text{res}} = \int\limits_{\infty}^{V_m} \left[\left(\frac{\partial P}{\partial T} \right)_{V_m} - \frac{R}{V_m} \right] dV_m + R \ln Z = \int\limits_{\infty}^{V_m} \left[Z - 1 + T \left(\frac{\partial Z}{\partial T} \right)_{V_m} \right] \frac{dV_m}{V_m} + R \ln Z \qquad (6.6)$$

Fugacity Coefficient

$$\ln \phi_i = \frac{1}{RT} \int\limits_{V}^{\infty} \left[\left(\frac{\partial P}{\partial n_i} \right)_{T, V, n_{j \neq i}} - \frac{RT}{V} \right] dV - \ln Z \qquad (6.7)$$

6.3. Properties from Equations of State

A fundamental equation of state may be used to obtain all of the thermodynamic properties of the fluid; however, we shall not consider such formulations further in this chapter. The more usual *P-V-T* equation of state permits the calculation of all configurational and residual thermodynamic properties of the fluid within that equation's domain of applicability. Explicit expressions for these properties may be obtained by means of the appropriate operations on the function $P(T,V_m)$; the cases of the residual enthalpy and entropy and the fugacity are delineated in Table 6.1 (see also Appendix B).

6.4. van der Waals Equation of State

The simplest cubic equation of state is that of van der Waals. Although this equation is never very accurate for real fluids, it is based on sound theoretical approximations and predicts behaviour which is, physically, essentially correct. For these reasons it is appropriate to discuss briefly the basis of the equation for pure fluids and mixtures.

In deriving his equation, van der Waals assumed that in a fluid of interacting molecules, each molecule moves independently in a uniform potential field provided by the other molecules. The only constraint considered on the motion of the molecules was that they could not penetrate the cores of other molecules. These assumptions are consistent with a model in which the molecules are treated as rigid spheres between which there is an infinitesimal attractive interaction of infinitely long range such that the inter-molecular potential energy is:

$$\lim_{\delta \to \infty} \begin{cases} \varphi(r) = +\infty & r < \sigma \\ \varphi(r) = -(\varepsilon/\delta^3) & \sigma \leq r < \delta\sigma \\ \varphi(r) = 0 & r \geq \delta\sigma \end{cases} . \tag{6.8}$$

van der Waals further assumed that the distribution of molecules around any chosen molecule is random outside the excluded region $r < \sigma$. Consequently, the number of molecules within a spherical shell of thickness dr and radius r centred on a reference molecule is $(N/V)(4\pi r^2 dr)$, where (N/V) is the mean number density. Hence the potential energy of interaction between a given molecule at $r = 0$ and all other molecules in the system is

$$\Phi = \underset{\delta \to \infty}{\mathrm{Lim}} \left[\int_{\sigma}^{\delta\sigma} (-\varepsilon/\delta^3)(N/V) 4\pi r^2 \, dr \right] = -\frac{2Na}{N_A^2 V} \quad \text{where} \quad a = \tfrac{2}{3}\pi N_A^2 \sigma^3 \varepsilon \,. \quad (6.9)$$

This 'mean-field' energy is simply proportional to the density of the system and is independent of both the position of the other molecules and the temperature. The total configurational potential energy of the system of N molecules in any allowed configuration is then simply

$$\mathcal{U} = \tfrac{1}{2}(N\Phi) = -(N/N_A)^2(a/V) \,, \quad (6.10)$$

where the factor (1/2) arises from the requirement that each interaction is counted only once. It is important to recognise that this expression applies only for configurations of the system in which no two molecules overlap; configurations in which overlap would exist correspond to infinite configurational potential energy.

Having obtained this particularly simple form for $\mathcal{U}$, it is straightforward to determine the configuration integral $\mathcal{L}$ and hence the pressure P. The configuration integral is obtained by combining equations (1.31) and (6.10) with the result

$$\mathcal{L} = \int \dots \int_{V_f} \exp(-\Phi N/2k_B T) \, d\mathbf{r}^N = V_f^N \exp(-\Phi N/2k_B T) \,. \quad (6.11)$$

Here, V_f, the so-called free volume, is the volume accessible to a given molecule and is equal to the system volume less the volume excluded by the other molecules. van der Waals employed a particularly simple argument to evaluate the free volume in which only binary interactions were considered. According to this argument a given molecule excludes a spherical volume of radius σ and this excluded volume is shared by two molecules giving an excluded volume of $(2/3)\pi\sigma^3$ per molecule. The free volume in a system of N molecules is therefore

$$V_f = V - (N/N_A)b \,, \quad (6.12)$$

where

$$b = \tfrac{2}{3}\pi N_A \sigma^3 \quad (6.13)$$

is the molar excluded volume. In terms of the parameters a and b, the logarithm of the configuration integral is then given by

$$\ln \mathcal{L} = N \ln \left[V - (N/N_A)b \right] + (N/N_A)^2 \left[a/Vk_B T \right] \,. \quad (6.14)$$

and, since $P = k_B T(\partial \ln \mathcal{L}/\partial V)_T$, we obtain finally

$$P = \frac{RT}{V_m - b} - \frac{a}{V_m^2} \quad \text{or} \quad \left(P + \frac{a}{V_m^2}\right)(V_m - b) = RT . \qquad (6.15)$$

At a fixed temperature this equation is cubic in the molar volume and it predicts the existence of both gas and liquid phases below a critical temperature. In this region there are three real roots for $V_m(T,P)$ the largest of which corresponds to the gas phase, the smallest to the liquid phase, and the intermediate one (which corresponds to a mechanically-unstable state of negative compressibility) has no physical significance. As T^c is approached from below, these roots converge on the critical molar volume V_m^c and above T^c there is only one real root for $V_m(T,P)$.

The critical point is defined by the conditions

$$\left(\frac{\partial P}{\partial V}\right)_T = 0 \quad \text{and} \quad \left(\frac{\partial^2 P}{\partial V^2}\right)_T = 0 \qquad (6.16)$$

which may be solved to obtain expressions for a and b in terms of the experimentally determined critical constants:

$$a = \frac{3P^c}{(V_m^c)^2} = \frac{27(RT^c)^2}{64\,P^c} \quad \text{and} \quad b = \frac{V_m^c}{3} = \frac{RT^c}{8P^c} . \qquad (6.17)$$

Eliminating a and b in favour of the above expressions, van der Waals' equation may be written in reduced form as

$$P_r = \frac{8T_r}{(3V_r - 1)} - \frac{3}{V_r^2} . \qquad (6.18)$$

Isotherms calculated from this equation are shown in Figure 6.2 together with the vapour-liquid co-existence curve. Solution of the phase-equilibrium problem will be discussed in detail in Chapter 8.

When compared with experimental data, the predictions of van der Waals' equation are found to be qualitatively correct but quantitatively rather poor. One of its main quantitative failures is that it predicts a fixed value for the critical compression factor, $Z^c = 0.375$, which is always larger than experimental values (usually less than 0.3). Furthermore the vapour-liquid coexistence curve is found to differ considerably from experimental data. Nevertheless the equation is based on

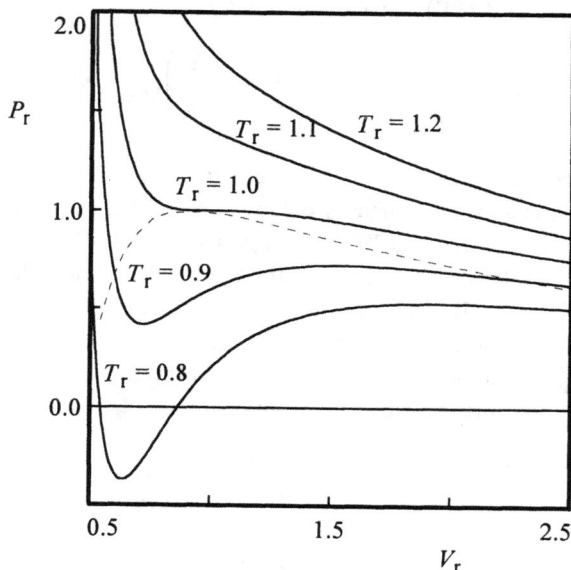

Figure 6.2 Isotherms calculated from van der Waals' equation of state (——);
the vapour-liquid co-existence curve (- - -) is also shown.

sound physics and it never predicts nonsense. Its quantitative failings can be at least
partially rectified rather easily, as we shall see, by fairly simple empirical
modifications which retain the cubic form.

Before proceeding to consider modern cubic equations, it is appropriate to
consider the application of van der Waals' model to the case of a multi-component
mixture. The results of this analysis constitute the van der Waals One-Fluid mixing
rules which were stated in Chapter 5 in connection with the principle of corresponding states. They also form the basis of the usual combining rules for modern cubic
equations of state.

It is actually rather easy to apply the assumptions adopted by van der Waals to
the case of a mixture of n components in which the interaction potential between
molecules i and j (which may be like or unlike) is given by

$$\lim_{\delta \to \infty} \begin{cases} \Phi_{ij}(r) = +\infty & r < \sigma_{ij} \\ \Phi_{ij}(r) = -(\varepsilon_{ij}/\delta^3) & \sigma_{ij} \leq r < \delta\sigma_{ij} \\ \Phi_{ij}(r) = 0 & r \geq \delta\sigma_{ij} \end{cases} . \tag{6.19}$$

The mean field potential at a given molecule now has contributions from both like and unlike interactions and one then obtains an expression for the total configurational potential energy which is identical in form to Eq. (6.9) but with a composition-dependent parameter a given by

$$a = \sum_{i=1}^{n} \sum_{j=1}^{n} x_i x_j a_{ij} , \qquad (6.20)$$

where

$$a_{ij} = \frac{2}{3} \pi N_A^2 \sigma_{ij}^3 \varepsilon_{ij} . \qquad (6.21)$$

Similarly, the free volume has contributions from like and unlike interactions and, again considering only binary interactions, one arrives at Eq.(6.12) but with the molar-excluded volume b dependent on composition through the relation

$$b = \sum_{i=1}^{n} \sum_{j=1}^{n} x_i x_j b_{ij} \qquad (6.22)$$

with

$$b_{ij} = \frac{2}{3} \pi N_A \sigma_{ij}^3 \qquad (6.23)$$

The remainder of the derivation is unchanged and one obtains again Eq.(6.15) but with the composition-dependent parameters defined above. These arguments lead directly to the van der Waals one-fluid mixing rules discussed in Chapter 5 when one introduces scaling parameters h_{ij} and f_{ij} to relate the distance and energy parameters, σ_{ij} and ε_{ij}, to those of a reference fluid.

Although the combining rules given above for the mixture parameters a and b are correct subject to the assumptions adopted by van der Waals, only Eq.(6.20) is usually applied without further approximation. Instead, the almost universal practice is to express the molar excluded volume in a mixture as a simple mole-fraction-weighted average of pure component values:

$$b = \sum_{i=1}^{n} x_i b_i . \qquad (6.24)$$

Furthermore, in order to apply the combining rule for the a parameter one must have an expression for the energy parameter ε_{ij} in the unlike interaction potential. This is

Table 6.2   van der Waals equation of state, 1873 [1]

Pressure

$$P = \frac{RT}{V_m - b} - \frac{a}{V_m^2}$$

(6.25)

Compression factor

$$Z = \frac{V_m}{V_m - b} - \frac{a}{RTV_m}$$

(6.26)

Reduced form

$$P_r = \frac{8T_r}{(3V_r - 1)} - \frac{3}{V_r^2}$$

(6.27)

where

$$a = \left(\sum_{i=1}^{n} x_i \sqrt{a_i} \right)^2 \qquad a_i = \frac{27\left(RT_i^c\right)^2}{64 P_i^c}$$

(6.28)

$$b = \sum_{i=1}^{n} x_i b_i \qquad b_i = \frac{RT_i^c}{8 P_i^c}$$

(6.29)

Residual Enthalpy, Entropy and Fugacity Coefficient

$$\frac{H_m^{res}}{RT} = \frac{b}{V_m - b} - \frac{2a}{RTV_m}$$

(6.30)

$$\frac{S_m^{res}}{R} = \ln\left[Z\left(1 - \frac{b}{V_m} \right) \right]$$

(6.31)

$$\ln \phi_i = \frac{b_i}{V_m - b} - \frac{2\sqrt{aa_i}}{RTV_m} - \ln\left[Z\left(1 - \frac{b}{V_m} \right) \right]$$

(6.32)

---

usually assumed to be given by the geometric mean of the pure-component values ε_{ii} and ε_{jj} so that the parameter a for the mixture reduces to the particularly simple form

$$a = \left(\sum_{i=1}^{n} x_i \sqrt{a_i} \right)^2 . \qquad (6.33)$$

With many cubic equations of state, it is useful to introduce binary interaction parameters in the combining rule for a_{ij}; however, there is little advantage in the present case.

We conclude this section with Table 6.2 which summarised the principal results and gives formulae for the residual enthalpy and entropy and for the fugacity coefficient according to van der Waals' equation of state with 'conventional' combining rules. Exampe 6.1 describes the derivation of the residual molar entropy from van der Waals' equation.

EXAMPLE 6.1   *Derivation of the Residual Molar Entropy for the van der Waals' Equation of State.*

The residual molar entropy is given by Eq.(6.6),

$$S_m^{res} = \int_{\infty}^{V_m} \left[\left(\frac{\partial P}{\partial T} \right)_{V_m} - \frac{R}{V_m} \right] dV_m + R \ln Z$$

and for van der Waals' equation

$$\left(\frac{\partial P}{\partial T} \right)_{V_m} = \frac{R}{\left(V_m - b \right)}$$

Substituting above, we obtain

$$\frac{S_m^{res}}{R} = \ln\left(1 - \frac{b}{V_m} \right) + \ln Z$$

and hence Eq.(6.31).

❑

6.5. Generalised Cubic Equations of State

van der Waals' equation represents the pressure by the sum of two terms, the one containing the parameter b arising from repulsive interactions and the other containing the attractive parameter a. Although the equation is not particularly accurate there are some obvious ways in which it can be improved:

1. The simple expression for the free volume, which neglects many body encounters, is crude. This may be corrected by replacing the repulsive term in the pressure by the correct one for rigid spheres.

2. The attractive term is derived on the basis of a uniform distribution of molecules, independent of T and V_m. This is clearly a fairly crude approximation and the situation can be improved by empirical modifications to the term a/V_m^2.

The equation of state of hard spheres is known essentially exactly and may be represented rather accurately by rational polynomials in the density. An equation of state combining such a function with the simple mean-field attractive term of van der Waals' has been proposed by Carnahan and Starling [6] and certainly leads to greatly improved accuracy. Nevertheless, most applications of equations of state in chemical engineering refer to conditions of sub-critical pressures where the failure of the simple vdW repulsive term is thought to be less important than the shortcomings of the attractive term. Consequently, attempts at improvement have usually concentrated on the attractive term and some of these will be discussed here. The most general cubic equation possible which retains the vdW repulsive term may be written

$$P = \frac{RT}{V_m - b} - \frac{a(T)}{(V_m + c_1 b)(V_m + c_2 b)} . \tag{6.34}$$

All of the modified cubic equations which are considered here conform to this general type and are characterised by integer values of the parameters c_1 and c_2. All except van der Waals equation have a as a function of temperature containing fixed parameters, in addition to the critical constants, which were determined by fitting experimental vapour pressure data for a range of substances. These modified equations, in particular those shown in Table 6.3 - Redlich-Kwong (RK), Soave (RKS), and Peng-Robinson (PR) - do not represent any great scientific advance but are nevertheless very important as engineering modelling tools. Their development was driven by the need to deal with the phase equilibria of complex multicomponent mixtures. In the following sections, these three cubic equations of state will be discussed with particular emphasis on their application to mixtures.

Table 6.3   Cubic Equations of State

van der Waals (vdW), 1873	$$P = \dfrac{RT}{V_m - b} - \dfrac{a}{V_m^2}$$	(6.35)
Redlich - Kwong (RK), 1949	$$P = \dfrac{RT}{V_m - b} - \dfrac{a}{V_m(V_m + b)\sqrt{T}}$$	(6.36)
Soave (RKS), 1972	$$P = \dfrac{RT}{V_m - b} - \dfrac{a\,\alpha(T)}{V_m(V_m + b)}$$	(6.37)
Peng - Robinson (PR), 1976	$$P = \dfrac{RT}{V_m - b} - \dfrac{a\,\alpha(T)}{V_m(V_m + b) + b(V_m - b)}$$	(6.38)

6.5.1. Redlich - Kwong Equation of State

At the time of its introduction, 1949 [3], this equation was considered to be a considerable improvement over other available equations of relatively simple forms. In his book published in 1978 [8], Redlich states that there was no particular theoretical basis for the equation, rather it is to be regarded as an effective empirical modification of its predecessors.

The full form of the Redlich-Kwong equation is given in Table 6.4 together with relations for all the parameters involved. The mixture combining rules shown here are the conventional ones discussed above in connection with van der Waals' equation but now incorporating binary interaction parameters in the attractive term.

About 110 modifications to the Redlich-Kwong equation have been proposed to date including the widely used version due to Soave which will be discussed in the next section. Although a great improvement over van der Waals' equation, the basic Redlich-Kwong equation gives useful results for only a few rather simple fluids. This is because the parameters of the equation are based entirely on the two critical constants T^c and P^c and do not incorporate the acentric factor.

Table 6.4   Redlich-Kwong Equation of State, 1949 [3]

$$P = \frac{RT}{V_\text{m} - b} - \frac{a}{V_\text{m}(V_\text{m} + b)\sqrt{T}} \tag{6.39}$$

or

$$Z^3 - Z^2 + \left(A - B^2 - B\right)Z - AB = 0 \tag{6.40}$$

where

$$A = \sum_{i=1}^{n} \sum_{j=1}^{n} x_i\, x_j\, (1 - k_{ij})\sqrt{A_i A_j} \tag{6.41}$$

$$B = \sum_{i=1}^{n} x_i\, B_i \tag{6.42}$$

$$a = \sum_{i=1}^{n} \sum_{j=1}^{n} x_i\, x_j\, (1 - k_{ij})\sqrt{a_i\, a_j} \tag{6.43}$$

$$b = \sum_{i=1}^{n} x_i b_i \tag{6.44}$$

and

$$A_i = 0.42748\, \frac{P_{\text{r},i}}{T_{\text{r},i}^{2.5}} \qquad B_i = 0.08664\, \frac{P_{\text{r},i}}{T_{\text{r},i}} \tag{6.45}$$

$$a_i = 0.42748\, \frac{R^2 (T_i^\text{c})^{2.5}}{P_i^\text{c}} \qquad b_i = 0.08664\, \frac{R T_i^\text{c}}{P_i^\text{c}} \tag{6.46}$$

Note: $k_{ii} = 0$.

6.5.2. Soave Equation of State

The Soave modification of the Redlich-Kwong equation involved replacing the term $(a/T^{0.5})$ by a more complicated function of the temperature, $a\alpha(T)$, incorporating the acentric factor. The form of this function was devised primarily to obtain a good representation of the vapour pressure curve for a number of hydrocarbons. Subsequently, this term was modified further by Graboski and Daubert [9] and it is their form of the equation that we give in Table 6.5.

The Soave equation together with that of Peng and Robinson are today probably the two most widely used equations of state. Because of the way in which the attractive terms have been tailored to achieve a fit to vapour pressure data, and the incorporation of the acentric factor, these equations usually permit VLE calculations to be made with acceptable accuracy. Single phase properties such as enthalpy and entropy may also be acceptable but liquid densities are usually rather poor. This poor performance is essentially due to the fact that the equations are really oversimplifications of the *P-V-T* surface which have been forced to fit vapour pressure data without any constraints on the densities of the co-existing phases. These equations perform best for light hydrocarbons and other small non-polar molecules but reasonable results may be obtained for more complicated substances including polar molecules. Reliable results should not be expected for systems with hydrogen bonding or other forms of molecular association.

A comprehensive study of the predictive capabilities of various thermo-dynamic models, sponsored by the API and reported by Daubert *et al.* [10], concluded that the Soave and Peng-Robinson equations were of roughly equal reliability for VLE calculations although the representation of *PVT* data in the viscinity of the critical point was better with the Peng-Robinson equation. Generally, the more complicated Lee-Kesler corresponding-states model gives results superior to both RKS and PR equations but the latter require only about one-tenth of the computer time of the former; even with modern computers, this can still be important in some complicated flowsheeting problems.

Table 6.5 Soave Equation of State, 1972 [4]

$$P = \frac{RT}{V_m - b} - \frac{a\,\alpha}{V_m(V_m + b)} \tag{6.47}$$

or

$$Z^3 - (1 - B)Z^2 + (A - B^2 - B)Z - AB = 0 \tag{6.48}$$

where $\quad A = \sum_{i=1}^{n} \sum_{j=1}^{n} x_i x_j (1 - k_{ij})\sqrt{A_i A_j} \tag{6.49}$

$$B = \sum_{i=1}^{n} x_i B_i \tag{6.50}$$

Table 6.5 (con/d)   Soave Equation of State, 1972 [4]

$$a\alpha = \sum_{i=1}^{n}\sum_{j=1}^{n} x_i x_j (1 - k_{ij})\sqrt{(a_i \alpha_i)(a_j \alpha_j)} \tag{6.51}$$

$$b = \sum_{i=1}^{n} x_i b_i \tag{6.52}$$

and $\quad A_i = 0.42747\, \alpha_i\, \dfrac{P_{r,i}}{T_{r,i}^2} \qquad\qquad B_i = 0.08664\, \dfrac{P_{r,i}}{T_{r,i}} \tag{6.53}$

$$a_i = 0.42747\, \frac{(RT_i^c)^2}{P_i^c} \qquad\qquad b_i = 0.08664\, \frac{RT_i^c}{P_i^c} \tag{6.54}$$

$$\alpha_i = \left[1 + n_i\left(1 - \sqrt{T_{r,i}}\right)\right]^2 \qquad n_i = 0.48508 + 1.55171\omega_i - 0.15613\omega_i^2 \tag{6.55}$$

Residual Molar Enthalpy, Entropy and Partial Fugacity Coefficient

$$\frac{H_m^{res}}{RT} = Z - 1 - \frac{A}{B}\left[1 + \frac{D}{a\alpha}\right]\ln\!\left(1 + \frac{B}{Z}\right) \tag{6.56}$$

$$\frac{S_m^{res}}{R} = \ln[Z - B] - \frac{BD}{a\alpha A}\ln\!\left(1 + \frac{B}{Z}\right) \tag{6.57}$$

$$\ln\varphi_i = \frac{B_i}{B}(Z-1) - \ln(Z - B) + \frac{A}{B}\left[\frac{B_i}{B} - \frac{2}{a\alpha}\sum_{j=1}^{N} x_j(1 - k_{ij})\sqrt{(a_i\alpha_i)(a_j\alpha_j)}\right]\ln\!\left(1 + \frac{B}{Z}\right) \tag{6.58}$$

where

$$D = \sum_{i=1}^{n}\sum_{j=1}^{n} x_i x_j\, n_j(1 - k_{ij})\sqrt{(a_i\alpha_i)a_j\, T_{rj}} \tag{6.59}$$

Note: $k_{ii} = 0$.

6.5.3. *Peng - Robinson Equation of State*

The equation of Peng and Robinson [5] is structurally rather similar to the RKS equation and, like the RKS, requires only the critical constants and the acentric factor for its application to a pure fluid. The different choice of the integer parameters c_1 and c_2 (Eq. 6.41) results in an improvement in the predicted critical compression factor to $Z^c = 0.307$ and this leads to generally improved predictions of liquid density. The equation is summarised in Table 6.6 where expressions for the residual molar enthalpy and entropy and for the fugacity coefficients are also given.

Table 6.6 Peng - Robinson Equation of State, 1976 [5]

$$P = \frac{RT}{V_m - b} - \frac{a\,\alpha}{V_m(V_m + b) + b(V_m - b)} \tag{6.60}$$

or

$$Z^3 - (1 - B)Z^2 + \left(A - 3B^2 - 2B\right)Z - \left(AB - B^2 - B^3\right) = 0 \tag{6.61}$$

where

$$A = \sum_{i=1}^{n} \sum_{j=1}^{n} x_i\, x_j\, (1 - k_{ij})\sqrt{A_i A_j} \tag{6.62}$$

$$B = \sum_{i=1}^{n} x_i\, B_i \tag{6.63}$$

$$a\alpha = \sum_{i=1}^{n} \sum_{j=1}^{n} x_i\, x_j\, (1 - k_{ij})\sqrt{(a_i\, \alpha_i)(a_j\, \alpha_j)} \tag{6.64}$$

$$b = \sum_{i=1}^{n} x_i\, b_i \tag{6.65}$$

and

$$A_i = 0.45724\, \alpha_i\, \frac{P_{r,i}}{T_{r,i}^2} \qquad B_i = 0.07780\, \frac{P_{r,i}}{T_{r,i}} \tag{6.66}$$

$$a_i = 0.45724\, \frac{(RT_i^c)^2}{P_i^c} \qquad b_i = 0.07780\, \frac{RT_i^c}{P_i^c} \tag{6.67}$$

Note: $k_{ii} = 0$.

Table 6.6 (con/d)   Peng - Robinson Equation of State, 1976 [5]

$$a_i = \left[1 + n_i \left(1 - \sqrt{T_{r,i}}\right)\right]^2 \qquad n_i = 0.37464 + 1.54226\omega_i - 0.26992\omega_i^2 \qquad (6.68)$$

Residual Molar Enthalpy, Entropy and Partial Fugacity Coefficient

$$\frac{H_m^{res}}{RT} = Z - 1 - \frac{A}{2.828B}\left[1 + \frac{D}{a\alpha}\right]\ln\left(\frac{Z + 2.414B}{Z - 0.414B}\right) \qquad (6.69)$$

$$\frac{S_m^{res}}{R} = \ln[Z - B] - \frac{BD}{2.828\,a\alpha\,A}\ln\left(\frac{Z + 2.414B}{Z - 0.414B}\right) \qquad (6.70)$$

$$\ln\varphi_i = \frac{B_i}{B}(Z - 1) - \ln(Z - B)$$

$$+ \frac{A}{2.828B}\left[\frac{B_i}{B} - \frac{2}{a\alpha}\sum_{j=1}^{N} x_j(1 - k_{ij})\sqrt{(a_i\,\alpha_i)(a_j\,\alpha_j)}\right]\ln\left(\frac{Z + 2.414B}{Z - 0.414B}\right) \qquad (6.71)$$

where

$$D = \sum_{i=1}^{n}\sum_{j=1}^{n} x_i x_j\, n_j(1 - k_{ij})\sqrt{(a_i\,\alpha_i)a_j\,T_{r,j}} \qquad (6.72)$$

Note: $k_{ii} = 0$.

6.5.4. Volume Translation Methods for Cubic Equation of States

As we have indicated, the major failing of cubic equations of state is that they provide only very rough predictions of liquid density. It turns out that a simple empirical correction, known as volume translation, can be devised which does much to improve this situation without affecting at all (P, T, x, y) calculations in vapour-liquid equilibrium problems [22]. The essential feature of this method for a pure fluid is that the molar volume V_m in the chosen cubic equation of state is replaced by a translated molar volume V_m' defined as

$$V_m' = V_m + c(T) \qquad (6.73)$$

Here, $c(T)$ is a correction which depends only upon the fluid and the temperature. To calculate the molar volume at a specified temperature and pressure, one first solves the equation of state for V_m and then obtains the corrected molar volume from Eq.(6.73). Since volume translation does not alter the shape of the P-V isotherms, it has no effect on the predicted of vapour pressures and $c(T)$ may therefore be adjusted to obtain agreement with experimental molar volumes of the saturated liquid. For example, Magoulas and Tassios [23] have proposed a volume-translated version of the Peng - Robinson equation of state in which $c(T)$ is represented by a universal function of T_r and ω with parameters that were fitted to the molar volumes of a number of saturated liquid alkanes.

Figure 6.3 illustrates the results for the case of ethane. The unmodified Peng - Robinson equation does provide a reasonable description of the saturated vapour density but the predicted critical density is nearly 10 per cent too low (corresponding to $Z^c = 0.307$ for the Peng - Robinson equation compared with 0.279 for ethane [24]) and the predicted liquid density is nearly 10 per cent too high at the normal boiling temperature. The volume-translated Peng - Robinson equation [23] provides

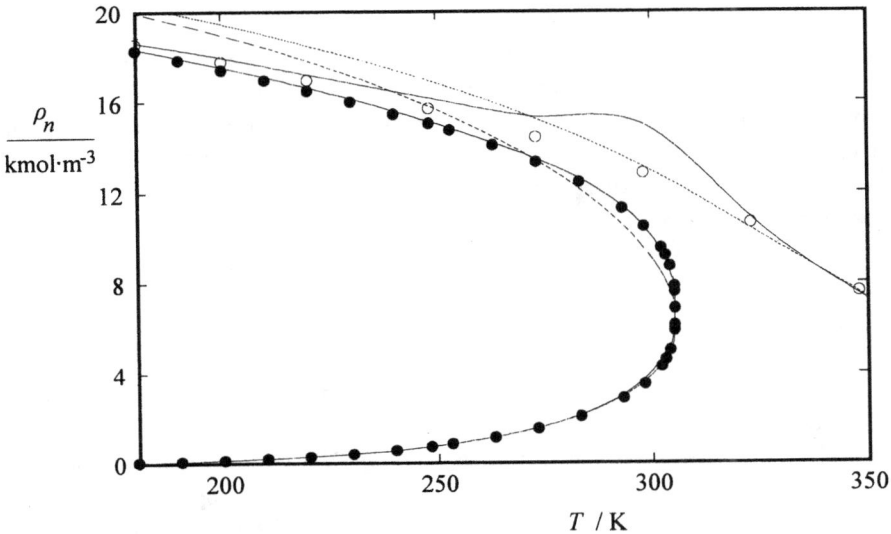

Figure 6.3 Comparison of experimental densities for fluid ethane with the predictions of the Peng-Robinson equation with and without volume translation. ●, experimental results for the saturated vapour and liquid [24,25]; O, experimental results for the isobar at $P=10$ MPa. Dashed lines are calculated from the Peng-Robinson equation and solid lines from the Peng-Robinson equation with volume translation [23].

a much improved representation of the density of the saturated liquid while the saturated vapour densities are almost unaffected. Densities of the compressed liquid at $T_r \leq 0.8$ are also greatly improved by the volume translation but, as the results on the 10 MPa isobar illustrate, large errors appear at high pressures when the temperature is near T^c. These errors arise because c is constant along an isotherm and $c(T_r = 1)$ must be large to obtain the correct critical density.

Volume translation can be applied to mixtures provided that the mixture volume correction c_{mix} is a linear function of the component mole fractions x_i:

$$c_{mix}(T, x_1, x_2 \cdots x_n) = \sum_{i=1}^{n} x_i \, c_i(T) \,. \tag{6.74}$$

This form of c_{mix} ensures that the predicted (P, T, x, y) phase behaviour remains unchanged [22].

The method of volume translation is very easy to apply in problems where predictions of phase behaviour and fluid density are required as one then simply corrects the molar volumes obtained from the unmodified equation of state by subtraction of c or c_{mix}. Caloric properties, such as residual molar enthalpy and entropy, which involve derivatives of the equation of state with respect to temperature, will differ from those given by the unmodified equation of state in a more complicated way and formulae that are specific to a particular functional form for $c(T)$ are required. The field of volume-translated cubic equations of state is presently an active area of research and it seems likely that the method will find wider application in the future.

6.6. Complex Equations of State

Every equation of state that has been proposed has more or less severe limitations with regard to the kinds of substances and range of conditions to which it may be applied. Some equations are better for predicting *PVT* properties while others are superior for phase equilibrium calculations. There is little hope that a universal equation of state of moderate complexity ever will be discovered. Nevertheless, there are a number of equations, more complex in form than the cubic equations, which do offer improved accuracy under most circumstances. One of the best examples, widely used today, is the Benedict-Webb-Rubin equation. This was developed originally in 1940 but has since been modified by several workers. We shall discuss only the Han-Starling modification.

6.6.1. Benedict, Webb & Rubin Equation of State

The celebrated equation of Benedict, Webb & Rubin [11-14] was devised as an improvement on the Beattie - Bridgeman equation of 1927 which was itself a truncated virial equation. The BWR equation gives the pressure as a polynomial in the density, with temperature-dependent coefficients, augmented by an exponential term intended to compensate for the truncation of the virial series. The defect of the Beattie-Bridgeman equation that the BWR equation rectified was the inability to represent behaviour of fluids at supercritical densities.

Experimental *PVT* data on gases, critical properties, and vapour pressures were employed in the evaluation of the eight parameters in the original BWR equation. Each substance required its own set of parameters and this was a limitation as coefficients were obtained for fewer than fifty substances, mostly hydrocarbons [15]. Nevertheless, for those systems the BWR equation became something of an industry standard, offering superior performance in all thermodynamic properties compared with cubic equations.

In order to permit wider application of the equation, it was desirable to develop generalised coefficients in terms of the critical constants and acentric factors of pure substances together with combining rules which permit application to mixtures. Such a generalisation was indeed developed in 1972 by Han and Starling [16]. Their full version of the equation for pure substances and mixtures is given in Table 6.7 together with the resulting relations for the residual enthalpy and entropy and for the fugacity coefficients.

Because of its high degree of non-linearity, the BWR equation is more difficult to use than a cubic equation (for which analytical solution for the roots is possible); however, the Newton-Raphson method is generally satisfactory (see Section 6.8.3).

A comparison of the BWR(HS) equation with the RKS, PR, Lee-Kesler and virial equations will be given in Section 6.8.4. Further examples of the BWR equation will be presented in Chapter 8 where phase-equilibrium calculations will be discussed.

Table 6.7 Benedict-Webb-Rubin (Han-Starling) Equation of State, 1972 [16]

$$P = \rho_n RT + \left(B_0 RT - A_0 - \frac{C_0}{T^2} + \frac{D_0}{T^3} - \frac{E_0}{T^4}\right)\rho_n^2 + \left(bRT - a - \frac{d}{T}\right)\rho_n^3$$

$$+ a\left(a + \frac{d}{T}\right)\rho_n^6 + \frac{c\rho_n^3}{T^2}\left(1 + \gamma\rho_n^2\right)\exp\left(-\gamma\rho_n^2\right)$$

(6.75)

where $\displaystyle A_0 = \sum_{i=1}^{n}\sum_{j=1}^{n} x_i x_j (1 - k_{ij})\sqrt{A_{0i} A_{0j}}$ $\displaystyle B_0 = \sum_{i=1}^{n} x_i B_{0i}$ (6.76)

$$C_0 = \sum_{i=1}^{n}\sum_{j=1}^{n} x_i x_j (1 - k_{ij})^3 \sqrt{C_{0i} C_{0j}} \qquad D_0 = \sum_{i=1}^{n}\sum_{j=1}^{n} x_i x_j (1 - k_{ij})^4 \sqrt{D_{0i} D_{0j}}$$

(6.77)

$$E_0 = \sum_{i=1}^{n}\sum_{j=1}^{n} x_i x_j (1 - k_{ij})^5 \sqrt{E_{0i} E_{0j}}$$

(6.78)

and $\displaystyle a = \left[\sum_{i=1}^{n} x_i a_i^{1/3}\right]^3$ $\displaystyle b = \left[\sum_{i=1}^{n} x_i b_i^{1/3}\right]^3$ $\displaystyle c = \left[\sum_{i=1}^{n} x_i c_i^{1/3}\right]^3$ (6.79)

$\displaystyle d = \left[\sum_{i=1}^{n} x_i d_i^{1/3}\right]^3$ $\displaystyle a = \left[\sum_{i=1}^{n} x_i a_i^{1/3}\right]^3$ $\displaystyle \gamma = \left[\sum_{i=1}^{n} x_i \gamma_i^{1/2}\right]^2$ (6.80)

while

$$A_{0i} = \frac{RT_i^c}{\rho_{ni}^c}\left(1.28438 - 0.920731\omega_i\right) \qquad B_{0i} = \frac{1}{\rho_{ni}^c}\left(0.443690 + 0.115449\omega_i\right)$$

(6.81)

$$C_{0i} = \frac{R(T_i^c)^3}{\rho_{ni}^c}\left(0.356306 + 1.70871\omega_i\right) \qquad D_{0i} = \frac{R(T_i^c)^4}{\rho_{ni}^c}\left(0.0307452 + 0.179433\omega_i\right)$$

(6.82)

$$E_{0i} = \frac{R(T_i^c)^5}{\rho_{ni}^c}\left(0.006450 - 0.022143\omega_i\right)\exp(-3.8\omega_i)$$

(6.83)

Note: $k_{ii} = 0$.

Table 6.7(con/d) Benedict-Webb-Rubin (Han-Starling) Equation of State [16]

Parameters a_i, b_i, c_i, d_i and α_i and γ_i are given by

$$a_i = \frac{RT_i^c}{(\rho_{ni}^c)^2}\left(0.484011 + 0.754130\omega_i\right) \qquad b_i = \frac{1}{(\rho_{ni}^c)^2}\left(0.528629 + 0.349261\omega_i\right) \qquad (6.84)$$

$$c_i = \frac{R(T_i^c)^3}{(\rho_{ni}^c)^2}\left(0.504087 + 1.32245\omega_i\right) \qquad d_i = \frac{R(T_i^c)^2}{(\rho_{ni}^c)^2}\left(0.0732828 + 0.463492\omega_i\right) \quad (6.85)$$

$$\alpha_i = \frac{1}{(\rho_{ni}^c)^2}\left(0.0705233 - 0.044448\omega_i\right) \qquad \gamma_i = \frac{1}{(\rho_{ni}^c)^2}\left(0.544979 - 0.270896\omega_i\right) \qquad (6.86)$$

Residual Molar Enthalpy

$$H_m^{res} = \rho_n\left(B_0 RT - 2A_0 - \frac{4C_0}{T^2} + \frac{5D_0}{T^3} - \frac{6E_0}{T^4}\right) + \frac{\rho_n^2}{2}\left(2bRT - 3a - \frac{4d}{T}\right)$$

$$+ \frac{\rho_n^5 a}{5}\left(6a + \frac{7d}{T}\right) + \frac{c}{\gamma T^2}\left[3 - \left(3 + \frac{\gamma\rho_n^2}{2} - \gamma^2\rho_n^4\right)\right]\exp\left(-\gamma\rho_n^2\right) \qquad (6.87)$$

Residual Molar Entropy

$$S_m^{res} = \sum_{i=1}^{N} x_i R\ln\left(\frac{Z}{x_i}\right) - \rho_n\left(B_0 R + \frac{2C_0}{T^3} - \frac{3D_0}{T^4} + \frac{4E_0}{T^5}\right) - \frac{\rho_n^2}{2}\left(bR + \frac{d}{T^2}\right)$$

$$+ \frac{\rho_n^5 a d}{5T^2} + \frac{2c}{\gamma T^3}\left[1 - \left(1 + \frac{\gamma\rho_n^2}{2}\right)\exp\left(-\gamma\rho_n^2\right)\right] \qquad (6.88)$$

Table 6.7(con/d)   Benedict-Webb-Rubin (Han-Starling) Equation of State [16]

Partial Fugacity Coefficient

$$RT \ln(\phi_i P) = RT \ln(\rho_n RT) + \rho_n(B_0 + B_{0i})RT + 2\rho_n \sum_{j=1}^{N} x_j \left(-(1-k_{ij}) \sqrt{A_{0j} A_{0i}} \right.$$

$$\left. -(1-k_{ij})^3 \frac{\sqrt{C_{0j} C_{0i}}}{T^2} + (1-k_{ij})^4 \frac{\sqrt{D_{0j} D_{0i}}}{T^3} - (1-k_{ij})^5 \frac{\sqrt{E_{0j} E_{0i}}}{T^4} \right)$$

$$+ \frac{\rho_n^2}{2} \left[3RT(b^2 b_i)^{1/3} - 3(a^2 a_i)^{1/3} - \frac{3}{T}(d^2 d_i)^{1/3} \right] + \frac{\alpha \rho_n^5}{5} \left[3(a^2 a_i)^{1/3} + \frac{3}{T}(d^2 d_i)^{1/3} \right]$$

$$+ \frac{3\rho_n^5}{5}\left(a + \frac{d}{T}\right)(a^2 a_i)^{1/3} + \frac{3\rho_n^2}{T^2}(c^2 c_i)^{1/3}\left[\frac{1-\exp(-\gamma\rho_n^2)}{\gamma\rho_n^2} - \frac{\exp(-\gamma\rho_n^2)}{2} \right]$$

$$- \frac{2c}{\gamma T^2} \sqrt{\frac{\gamma_i}{\gamma}} \left[1 - \left(1 + \gamma\rho_n^2 + \frac{\gamma^2 \rho_n^4}{2} \right) \exp(-\gamma\rho_n^2) \right] \tag{6.89}$$

6.7. The Critical Region

In the approach to the gas-liquid critical point of a pure fluid many thermodynamic properties change drastically, apparently approaching either zero or infinity, and at the critical point itself it is no longer possible to distinguish gas from liquid. It would, perhaps, be too much to expect the largely empirical equations of state considered in this book to reproduce all of this behaviour faultlessly. Indeed, while some of the correct features are predicted, serious failures that cannot easily be rectified are shared by almost all thermodynamic models in use today. It is therefore of some importance to be aware of these failings and of their consequences; that is the purpose of the present section

6.7.1. The Critical Point of a Pure Fluid

The equations of state considered in this book are analytical functions in the sense that the pressure may be expanded in a Taylor series about any given state point. In particular, for a pure fluid, these equations of state may be written as a Taylor expansion in T and ρ_n about the critical point. Usually, the equation of state is required only to satisfy the conditions[1]

$$\left(\frac{\partial P}{\partial \rho_n}\right)_T = 0 \quad \text{and} \quad \left(\frac{\partial^2 P}{\partial \rho_n^2}\right)_T = 0 \tag{6.90}$$

at the critical point so that the leading non-trivial terms in the Taylor expansion give [26]

$$\Delta P = P_{30}(\Delta \rho_n)^3 + P_{10}(\Delta T) + P_{11}(\Delta T)(\Delta \rho_n) + \cdots . \tag{6.91}$$

Here, $\Delta P = (P - P^c)$, $\Delta T = (T - T^c)$, $\Delta \rho_n = (\rho_n - \rho_n^c)$ and P_{ij} are constants related to the derivatives of P taken at the critical point. It follows from this expansion that the asymptotic form of the critical isotherm ($\Delta T = 0$) should obey the power law

$$|\Delta P| \sim |\Delta \rho_n|^\delta \tag{6.92}$$

with $\delta = 3$. Similarly, on the critical isochore ($\Delta \rho_n = 0$) the asymptotic form of the isothermal compressibility κ_T should follow the power law[2]

$$\kappa_T \sim |\Delta T|^{-\gamma} \tag{6.93}$$

with $\gamma = 1$, and the isochoric heat capacity in the homogeneous fluid ($\Delta T > 0$) is predicted to remain finite so that

$$C_V \sim |\Delta T|^{-\alpha} \tag{6.94}$$

with $\alpha = 0$. Finally, one can show that the Eq.(6.91) implies that the chemical potential should be constant along a path with $\Delta \rho_n \propto (\Delta T)^{1/2}$ so that along the vapour-liquid coexistence curve,

---

[1] These conditions are equivalent to Eqs.(6.16).
[2] $\kappa_T^{-1} = \rho_n (\partial P / \partial \rho_n)_T$

$$\Delta \rho_n \sim (\Delta T)^{\beta} \tag{6.95}$$

with $\beta = 1/2$.

All equations of state which, at the critical point, are both analytic and constrained only to satisfy Eqs.(6.90) lead to the same power laws and to the same values of the critical exponents α, β, γ and δ. Such an equation of state is said to be classical.

Unfortunate, while the experimental evidence supports the asymptotic power laws it does not agree with the classical values of the critical exponents. In particular the shape of the critical isotherm is much flatter than predicted and the coexistence curve is approximately cubic rather than quadratic. It has also been found that, rather than remaining finite at the critical point, C_V in fact diverges weakly. Great efforts have been expended to gain our present understanding of the nature of the critical region. These efforts have yielded universal values of the critical exponents [27,28], which are summarised in Table 6.8, and led to a reasonably well-developed theory for the whole critical region that describes both the asymptotic behaviour and the transition to classical behaviour far from the critical point [29]. Unfortunate, the theory is very complicated, its predictive capability is not yet well-developed and the results are unlikely to find wide application soon. It should also be mentioned that some complex classical equations of state have been developed for specific pure fluids that give accurate results except in a very small region close to the critical point where, like any classical equation, they must fail (see, for example, [30] and [31]).

Table 6.8 Experimental and Theoretical Values of the Critical Exponents

Critical exponent	Experiment [26]	Non-classical theory [28]	Classical theory
α	≈ 0.1	0.11	0
β	≈ 0.35	0.326	0.5
γ	≈ 1.2	1.24	1
δ	≈ 4.4	4.82	3

6.7.2. Critical Behaviour in Mixtures

Although the critical behaviour of a pure fluid is interesting, it is the properties of mixtures that are usually of the greater importance. Thus the failings of classical thermodynamic models for mixtures in the critical region is a matter of more practical concern and we therefore consider here some aspects of the problem.

In general, when a single homogeneous fluid phase of a mixture is caused to cross a phase boundary it splits into two phases that differ in compositions and other properties. It is possible to construct a locus of (T, P) points which defines the phase boundary for a given mixture and to determine the properties of the second fluid phase which forms when it is crossed at any location from the single-phase region. A critical point, if it exists, is one at which the properties of that second fluid phase become identical to that of the parent phase. In contrast to the situation with pure fluids, a mixture is not usually mechanically unstable at its critical point so that κ_T remains finite. It should be noted that the two phases in question here may be either liquid and vapour or two liquid phases.

Classical thermodynamic models, such as most equations of state and the activity coefficient models that will be discussed in the next chapter, presume that the chemical potential is an analytic function of temperature, pressure and composition. All necessarily lead to the same kind of limiting behaviour in the approach to a critical point and, for binary mixtures, one finds that the compositions x' and x'' of the coexisting phases follow the asymptotic power laws

$$|x' - x''| \sim |\Delta T|^\beta \qquad \text{on} \quad \Delta P = 0 \qquad (6.96)$$

or

$$|x' - x''| \sim |\Delta P|^\beta \qquad \text{on} \quad \Delta T = 0, \qquad (6.97)$$

with $\beta = 1/2$. As with pure fluids, the power laws themselves are confirmed by experiment but not the classical value of the critical exponent β; instead, values closer to 1/3 are found. According to the modern theoretical view of critical phenomena [27], the value of β which appears here should be identical to that determined from the coexistence curve of a pure fluid through Eq.(6.95) and this is in agreement with experiment.

The consequences of adopting a classical thermodynamic model are that predicted phase compositions near to a critical point itself are likely to exhibit quite large errors. The predicted location of the critical point is to a considerable extent dependent upon the binary interaction parameters in a particular thermodynamic model and it may be possible to force agreement with experimental data in that respect. When this is not done, there may be appreciable errors associated with the location of the critical point possibly leading to entirely incorrect predictions of the phase behaviour in the critical region. As with pure fluids, non-classical equations of state can be constructed that give essentially correct critical behaviour; however, their complexity is such that they are unlikely to find wide application in the near future.

6.8. Summary

The following conclusions can be drawn (these points will be amplified in the comparative examples that follow):
a) from the cubic equations of state, the Peng - Robinson and the Soave equations are the ones in most common use today.
b) The Benedict, Webb & Rubin (Han-Starling) equation of state is generally more accurate than any cubic equation of state. In the specific case of the partial fugacity coefficient, its ease of use makes the BWR(HS) superior to the Lee-Kesler corresponding-states scheme.
c) For all other thermodynamic properties, the Lee-Kesler approach (combined with the Wu-Stiel terms for polar substances) is usually to be preferred unless computing time is an important parameter.
d) In cases of gas mixtures at low densities, the virial equation is probably the most accurate equation of state.
e) All of the above fail in the immediate neighbourhood of the critical point.

6.9. Computational Implementation

In this section three FORTRAN subroutines will be presented implementing in turn, the RKS, the PR and the BWR(HS) equations of state. With all three routines, the density, enthalpy and entropy of the mixture, as well as the fugacity coefficient of each component, may be calculated. Binary interaction parameters are entered at run time. Pure substances may be treated by setting the number of components equal to one. In Section 6.9.4 comparative examples will be shown.

For the solution of each equation of state the Newton-Raphson method is employed by which successive approximations to the solution x of an equation $f(x)=0$ are calculated as

$$x_{n+1} = x_n - \frac{f(x_n)}{\partial f(x_n)/\partial x_n} \tag{6.98}$$

and solution is found when

$$|x_{n+1} - x_n| < \varepsilon \tag{6.99}$$

for some suitably small value of ε.

The perfect-gas enthalpy and entropy contributions are calculated from the isobaric perfect-gas heat capacity which is represented, as elsewhere in the book, by a polynomial in temperature containing four coefficients (see Section 2.10.1).

6.9.1. Soave Equation

The routine RKS shown in Display 6.1, calculates according to the equations presented in Table 6.5 the density, enthalpy, entropy and fugacity coefficient(s) for a pure substance or mixture at a specified temperature, pressure and composition. Additional inputs to the routine are:

- the critical temperature and pressure, and acentric factor for each component.
- four coefficients for the isobaric perfect-gas heat capacity for each component
- binary interaction parameters (set to zero, if not known).
- the number of components (IS), and the phase (Liquid, IP=1; Gas, IP=2).

Routine RKS is employed in the comparative examples shown in Section 6.9.4.

```
                                                     DISPLAY 6.1   RKS List
         SUBROUTINE RKS
      &(T,P,IP,IS,X,TC,PC,OMEG,Y1,Y2,Y3,Y4,BK,D,H,S,FUG)
C
C        The subroutine calculates the following thermodynamic properties
C        of mixtures, as a function of temperature, T in [K], and pressure, P in [PA]:
C            1) Molar Density                   (D)  in [MOL/M3]
C            2) Molar Enthalpy                  (H)  in [J/MOL]
C            3) Molar Entropy                   (S)  in [J/MOL/K]
C            4) Partial Fugacity Coefficient    (FUG)  in [-]
C        based on the Soave Equation of State, shown in Table 6.5
C
C        - Enthalpy and Entropy are arbitrarily assumed to be equal to zero
C          at 273.15 K and 0.1013 MPA.
C        - Perfect-gas properties are calculated from perfect-gas Heat
C          Capacities as: CP=W1+W2*T+W3*(T**2)+W4*(T**3) in [J/MOL/K]
C          where W1=Sum(Xi*Y1i) etc.
C        - Two Switches are required from the subroutine:
C          IP = Phase specification switch (Liquid=1, Vapour=2)
C          IS = Number of components in mixture (Pure fluid=1)
C        - Other data required for each component (I) are: mole fraction X (-), critical
C          temperature, TC in [K], critical pressure in [PA], and acentric factor, OMEG.
C        - Binary interaction parameters BK(I,J) must also be specified.
C
         DIMENSION X(IS),TC(IS),PC(IS),OMEG(IS),Y1(IS),Y2(IS),Y3(IS),Y4(IS)
         DIMENSION BK(IS,IS),FUG(IS)
         DIMENSION A(10),B(10),FM(10),ALP(10),AS(10)
         DATA R,T0,P0/8.314,273.15,0.101325E+6/
C.......Parameters calculation AT,BT,CT,DT & AST,ALP
         BT=0.
         DO 100 I=1,IS
         AS(I)=0.42747*R*R*TC(I)*TC(I)/PC(I)
         FM(I)=0.48508+1.55171*OMEG(I)-0.15613*(OMEG(I)**2)
         ALP(I)=(1.+FM(I)*(1.-(T/TC(I))**.5))**2
         A(I)=0.42747*ALP(I)*(P/PC(I))/((T/TC(I))**2)
```

```
        B(I)=0.08664*(P/PC(I))/(T/TC(I))
100   BT=BT+X(I)*B(I)
        AAT=0.
        AT=0.
        DT=0.
        DO 200 I=1,IS
        DO 200 J=1,IS
        BIP=(1.-BK(I,J))
        AAT=AAT+X(I)*X(J)*BIP*((AS(I)*AS(J)*ALP(I)*ALP(J))**.5)
        AT=AT+X(I)*X(J)*BIP*((A(I)*A(J))**.5)
200   DT=DT+X(I)*X(J)*FM(J)*BIP*((AS(I)*ALP(I)*AS(J)*(T/TC(J)))**.5)
C.......Density Calculation
        IF (IP.EQ.1) Z=0.01
        IF (IP.EQ.2) Z=0.8
300   F=(Z**3)-Z*Z+(AT-BT**2-BT)*Z-(AT*BT)
        F1=(3.*Z*Z)-(2.*Z)+(AT-(BT**2)-BT)
        H=F/F1
        Z=Z-H
        IF (ABS(H/Z).LE.0.0001) GOTO 400
        GOTO 300
400   D=P/(Z*R*T)
C.......Perfect-gas Enthalpy & Entropy
        W1=0.
        W2=0.
        W3=0.
        W4=0.
        DO 500 I=1,IS
        W1=W1+X(I)*Y1(I)
        W2=W2+X(I)*Y2(I)
        W3=W3+X(I)*Y3(I)
500   W4=W4+X(I)*Y4(I)
        PGH=W1*(T-T0)+W2*((T**2)-(T0**2))/2.+W3*((T**3)-(T0**3))/3.
      &+W4*((T**4)-(T0**4))/4.
        PGS=W1*LOG(T/T0)+W2*(T-T0)+W3*((T**2)-(T0**2))/2.
      &+W4*((T**3)-(T0**3))/3.-R*LOG(P/P0)
C.......Molar Enthalpy, Entropy & Partial Fugacities
        DH=-1.+Z-(AT/BT)*(1.+(DT/AAT))*LOG(1.+(BT*D*R*T/P))
        H=(PGH+R*T*DH)
        DS=LOG(Z-BT)+(BT*DT/(AT*AAT))*LOG(1.+(BT*D*R*T/P))
        S=PGS+(R*DS)
        DO 700 I=1,IS
        XAAT=0.
        DO 800 J=1,IS
800   XAAT=XAAT+X(J)*(1.-BK(I,J))*((AS(I)*AS(J)*ALP(I)*ALP(J))**.5)
        FILN=(B(I)*(Z-1.)/BT)-LOG(Z-BT)+(AT/BT)*((B(I)/BT)-(2.*XAAT/AAT))
      &*LOG(1.+(BT*D*R*T/P))
700   FUG(I)=EXP(FILN)
        RETURN
        END
```

6.9.2. *Peng - Robinson Equation*

The routine PR shown in Display 6.2 performs identical tasks to the routine RKS described above and is called in the same way; it employs the Peng - Robinson equation of state. Routine PR will also be employed in the comparative examples presented in Section 6.9.4. and again in the phase-equilibrium calculations presented in Chapter 8.

```
      SUBROUTINE PR
     &(T,P,IP,IS,X,TC,PC,OMEG,Y1,Y2,Y3,Y4,BK,D,H,S,FUG)
C
C     The subroutine calculates the following thermodynamic properties
C     of mixtures, as a function of temperature, T in [K], and pressure, P in [PA]:
C          1) Molar Density              (D)   in [MOL/M3]
C          2) Molar Enthalpy             (H)   in [J/MOL]
C          3) Molar Entropy              (S)   in [J/MOL/K]
C          4) Partial Fugacity Coefficient   (FUG)  in [-]
C     based on the Peng & Robinson Equation of State, shown in Table 6.6
C
C     - Enthalpy and Entropy are arbitrarily assumed to be equal to zero
C       at 273.15 K and 0.1013 MPA.
C     - Perfect-gas properties are calculated from perfect-gas Heat
C       Capacities as: CP=W1+W2*T+W3*(T**2)+W4*(T**3) in [J/MOL/K]
C       where W1=Sum(Xi*Y1i) etc.
C     - Two Switches are required from the subroutine:
C       IP = Phase specification switch (Liquid=1, Vapour=2)
C       IS = Number of components in mixture (Pure fluid=1)
C     - Other data required for each component (I) are:
C       mole fraction X (-), critical temperature, TC in [K], critical
C       pressure in [PA], and acentric factor, OMEG.
C     - Binary interaction parameters BK(I,J) must also be specified.
C
      DIMENSION X(IS),TC(IS),PC(IS),OMEG(IS),Y1(IS),Y2(IS),Y3(IS),Y4(IS)
      DIMENSION BK(IS,IS),FUG(IS)
      DIMENSION A(10),B(10),FM(10),ALP(10),AS(10)
      DATA R,T0,P0/8.314,273.15,0.101325E+6/
C........Parameters calculation AT,BT,CT,DT & AST,ALP
      BT=0.
      DO 100 I=1,IS
      AS(I)=0.45724*R*R*TC(I)*TC(I)/PC(I)
      FM(I)=0.37464+1.54226*OMEG(I)-0.26992*(OMEG(I)**2)
      ALP(I)=(1.+FM(I)*(1.-(T/TC(I))**.5))**2
      A(I)=0.45724*ALP(I)*(P/PC(I))/((T/TC(I))**2)
      B(I)=0.0778*(P/PC(I))/(T/TC(I))
  100 BT=BT+X(I)*B(I)
      AAT=0.
      AT=0.
```

```
      DT=0.
      DO 200 I=1,IS
      DO 200 J=1,IS
      BIP=(1.-BK(I,J))
      AAT=AAT+X(I)*X(J)*BIP*((AS(I)*AS(J)*ALP(I)*ALP(J))**.5)
      AT=AT+X(I)*X(J)*BIP*((A(I)*A(J))**.5)
  200 DT=DT+X(I)*X(J)*FM(J)*BIP*((AS(I)*ALP(I)*AS(J)*(T/TC(J)))**.5)
C...... Density Calculation
      IF (IP.EQ.1) Z=0.01
      IF (IP.EQ.2) Z=0.8
  300 F=(Z**3)-(1.-BT)*Z*Z+(AT-3.*(BT**2)-2.*BT)*Z-(AT*BT-BT*BT-BT**3)
      F1=(3.*Z*Z)-(2.*Z*(1.-BT))+(AT-3.*(BT**2)-2.*BT)
      H=F/F1
      Z=Z-H
      IF (ABS(H/Z).LE.0.0001) GOTO 400
      GOTO 300
  400 D=P/(Z*R*T)
C...... Perfect-gas Enthalpy & Entropy
      W1=0.
      W2=0.
      W3=0.
      W4=0.
      DO 500 I=1,IS
      W1=W1+X(I)*Y1(I)
      W2=W2+X(I)*Y2(I)
      W3=W3+X(I)*Y3(I)
  500 W4=W4+X(I)*Y4(I)
      PGH=W1*(T-T0)+W2*((T**2)-(T0**2))/2.+W3*((T**3)-(T0**3))/3.
     &+W4*((T**4)-(T0**4))/4.
      PGS=W1*LOG(T/T0)+W2*(T-T0)+W3*((T**2)-(T0**2))/2.
     &+W4*((T**3)-(T0**3))/3.-R*LOG(P/P0)
C...... Molar Enthalpy & Entropy
      DH=1.-Z+(AT/(2.828*BT))*(1.+DT/AAT)*LOG((Z+2.414*BT)/(Z-0.414*BT))
      H=(PGH-R*T*DH)
      DS=LOG(Z-BT)-(BT-DT/(2.828*AT*AAT))*LOG((Z+2.414*BT)/
     &(Z-0.414*BT))
      S=PGS+(R*DS)
C...... Partial Fugacities
      DO 700 I=1,IS
      XAAT=0.
      DO 800 J=1,IS
      BIP=(1.-BK(I,J))
  800 XAAT=XAAT+X(J)*BIP*((AS(I)*AS(J)*ALP(I)*ALP(J))**.5)
      FILN=(B(I)*(Z-1.)/BT)-LOG(Z-BT)+ (AT/(2.828*BT))*((B(I)/BT)-
     &(2.*XAAT/AAT))*LOG((Z+2.414*BT)/(Z-0.414*BT))
  700 FUG(I)=EXP(FILN)
      RETURN
      END
```

6.9.3. *Benedict, Webb & Rubin (Han-Starling) Equation*

The routine BWR shown in Display 6.3 performs identical tasks to the routines RKS and PR described above and is called in the same way; it employs the BWR(HS) equation of state. Routine BWR will be employed in the comparative examples presented in Section 6.9.4 and in the phase-equilibrium examples in Chapter 8.

DISPLAY 6.3   BWR List

```
      SUBROUTINE BWR
     &(T,P,IP,IS,X,TC,DC,OMEG,Y1,Y2,Y3,Y4,BK,D,H,S,FUG)
C
C     The subroutine calculates the following thermodynamic properties of mixtures,
C     as a function of temperature, T in [K], and pressure, P in [PA]:
C        1) Molar Density                    (D)   in [MOL/M3]
C        2) Molar Enthalpy                   (H)   in [J/MOL]
C        3) Molar Entropy                    (S)   in [J/MOL/K]
C        4) Partial Fugacity Coefficient     (FUG) in [-]
C     based on the B-W-R Han-Starling EoS shown in Table 6.7.
C
C     - Enthalpy and Entropy are arbitrarily assumed to be equal to zero
C       at 273.15 K and 0.1013 MPA.
C     - Perfect-gas properties are calculated from perfect-gas Heat
C       Capacities as: CP=W1+W2*T+W3*(T**2)+W4*(T**3) in [J/MOL/K]
C       where W1=Sum(Xi*Y1i) etc.
C     - Two Switches are required from the subroutine:
C       IP = Phase specification switch (Liquid=1, Vapour=2)
C       IS = Number of components in mixture (Pure fluid=1)
C     - Other data required for each component (I) are:
C       mole fraction X (-), critical temperature, TC in [K], critical
C       density in [MOL/M3], and acentric factor, OMEG.
C     - Binary interaction parameters BK(I,J) must also be specified.
C
      DIMENSION X(IS),TC(IS),DC(IS),OMEG(IS),Y1(IS),Y2(IS),Y3(IS),Y4(IS)
      DIMENSION BK(IS,IS),FUG(IS)
      DIMENSION A0(10),B0(10),C0(10),D0(10),E0(10),AS(10),BS(10),CS(10),
     &DS(10),ALP(10),GAM(10),A(11),B(11)
      DATA A/0.443690,1.28438,0.356306,0.544979,0.528629,0.484011,
     &0.0705233,0.504087,0.0307452,0.0732828,0.006450/
      DATA B/0.115449,-0.920731,1.70871,-0.270896,0.349261,0.754130,
     &-0.044448,1.32245,0.179433,0.463492,-0.022143/
      DATA R,T0,P0/8.314,273.15,0.101325E+6/
C
C     Parameters A0(I),B0(I),C0(I),D0(I),E0(I),
C                AS(I),BS(I),CS(I),DS(I),    &   ALP(I),GAM(I)
C
      DO 100 I=1,IS
      A0(I)=(R*TC(I)/DC(I))*(A(2)+B(2)*OMEG(I))
      B0(I)=(1./DC(I))*(A(1)+B(1)*OMEG(I))
      C0(I)=(R*(TC(I)**3)/DC(I))*(A(3)+B(3)*OMEG(I))
```

```
      D0(I)=(R*(TC(I)**4)/DC(I))*(A(9)+B(9)*OMEG(I))
      E0(I)=(R*(TC(I)**5)/DC(I))*(A(11)+B(11)*OMEG(I)*EXP(-3.8*OMEG(I)))
      AS(I)=(R*TC(I)/(DC(I)**2))*(A(6)+B(6)*OMEG(I))
      BS(I)=(1./(DC(I)**2))*(A(5)+B(5)*OMEG(I))
      CS(I)=(R*(TC(I)**3)/(DC(I)**2))*(A(8)+B(8)*OMEG(I))
      DS(I)=(R*(TC(I)**2)/(DC(I)**2))*(A(10)+B(10)*OMEG(I))
      ALP(I)=(1./(DC(I)**3))*(A(7)+B(7)*OMEG(I))
  100 GAM(I)=(1./(DC(I)**2))*(A(4)+B(4)*OMEG(I))
C     Total parameters A0T,B0T,C0T,D0T,E0T,
C                         AST,BST,CST,DST,  & ALPT,GAMT & WMOLT
      A0T=0.
      B0T=0.
      C0T=0.
      D0T=0.
      E0T=0.
      AST=0.
      BST=0.
      CST=0.
      DST=0.
      ALPT=0.
      GAMT=0.
      WMOLT=0.
      DO 200 I=1,IS
      B0T=B0T+X(I)*B0(I)
      AST=AST+X(I)*(AS(I)**(1./3.))
      BST=BST+X(I)*(BS(I)**(1./3.))
      CST=CST+X(I)*(CS(I)**(1./3.))
      DST=DST+X(I)*(DS(I)**(1./3.))
      ALPT=ALPT+X(I)*(ALP(I)**(1./3.))
      GAMT=GAMT+X(I)*(GAM(I)**(1./2.))
      DO 200 J=1,IS
      BIP=(1.-BK(I,J))
      A0T=A0T+X(I)*X(J)*BIP*((A0(I)*A0(J))**0.5)
      C0T=C0T+X(I)*X(J)*(BIP**3.)*((C0(I)*C0(J))**0.5)
      D0T=D0T+X(I)*X(J)*(BIP**4.)*((D0(I)*D0(J))**0.5)
      E0T=E0T+X(I)*X(J)*(BIP**5.)*((E0(I)*E0(J))**0.5)
  200 CONTINUE
      AST=AST**3
      BST=BST**3
      CST=CST**3
      DST=DST**3
      ALPT=ALPT**3
      GAMT=GAMT**2
C     Density calculation
      IF (IP.EQ.1) D=100000.
      IF (IP.EQ.2) D=10.
  300 F=-P+(D*R*T)
      F=F+(B0T*R*T-A0T-(C0T/(T**2))+(D0T/(T**3))-(E0T/(T**4)))*D*D
      F=F+(BST*R*T-AST-(DST/T))*(D**3)+ALPT*(AST+(DST/T))*(D**6)
```

```
      F=F+(CST/(T**2))*(D**3)*(1.+GAMT*D*D)*EXP(-GAMT*D*D)
      DF=(R*T)+2.*D*(B0T*R*T-A0T-(C0T/(T**2))+(D0T/(T**3))-(E0T/(T**4)))
      DF=DF+3.*D*D*(BST*R*T-AST-(DST/T))
      DF=DF+6.*(D**5)*ALPT*(AST+(DST/T))
      DF=DF+(CST/(T**2))*3.*(D**2)*(1.+GAMT*D*D)*EXP(-GAMT*D*D)
      DF=DF+(CST/(T**2))*(D**3)*(2.*GAMT*D)*EXP(-GAMT*D*D)
      DF=DF-2.*GAMT*D*(CST/(T**2))*(D**3)*(1.+GAMT*D*D)*EXP(-GAMT*D*D)
      H=F/DF
      IF (ABS(H/D).LE.0.000001) GOTO 400
      D=D-H
      GOTO 300
C..........Perfect-gas Enthalpy & Entropy
  400 W1=0.
      W2=0.
      W3=0.
      W4=0.
      DO 500 I=1,IS
      W1=W1+X(I)*Y1(I)
      W2=W2+X(I)*Y2(I)
      W3=W3+X(I)*Y3(I)
  500 W4=W4+X(I)*Y4(I)
      PGH=W1*(T-T0)+W2*((T**2)-(T0**2))/2.+W3*((T**3)-(T0**3))/3.
     &+W4*((T**4)-(T0**4))/4.
      PGS=W1*LOG(T/T0)+W2*(T-T0)+W3*((T**2)-(T0**2))/2.
     &+W4*((T**3)-(T0**3))/3.-R*LOG(P/P0)
C..........Molar Enthalpy & Entropy
      GDD=GAMT*D*D
      H=PGH+D*((B0T*R*T)-(2.*A0T)-(4.*C0T/(T*T))+(5.*D0T/(T**3))-
     &(6.*E0T/(T**4))) + 0.5*(D**2)*(2.*BST*R*T-(3.*AST)-(4.*DST/T))
      H=H + 0.2*(D**5)*ALPT*((6.*AST)+(7.*DST/T))
      H=H + (CST/(GAMT*T*T))*(3.-(3.+0.5*GDD-GDD*GDD)*EXP(-GDD))
      SR=0.
      DO 600 I=1,IS
  600 SR=SR-X(I)*R*LOG(R*T*D*X(I)/P0)
      S=PGS+SR
      S=-D*((B0T*R)+(2.*C0T/(T**3))-(3.*D0T/(T**4))+(4.*E0T/(T**5)))
      S=S - 0.5*(D**2)*((BST*R)+(DST/(T**2)))
      S=S + 0.2*(D**5)*ALPT*DST/(T**2)
      S=S + (2.*CST/(GAMT*(T**3)))*(1.-(1.+0.5*GDD)*EXP(-GDD))
C..........Partial Fugacities
      DO 800 I=1,IS
      FU=(R*T*LOG(D*R*T))+(D*R*T*(B0T+B0(I)))
      FUS=0.
      DO 700 J=1,IS
      BIP=(1.-BK(I,J))
  700 FUS=FUS+X(J)*(-BIP*((A0(J)*A0(I))**0.5)-(BIP**3.)*(((C0(J)*C0(I))**0.5)/(T*T))
     &+(BIP**4.)*(((D0(J)*D0(I))**0.5)/(T**3))-(BIP**5.)*(((E0(J)*E0(I))**0.5)/(T**4)))
      AAI=(AST*AST*AS(I))**(1./3.)
```

```
       BBI=(BST*BST*BS(I))**(1./3.)
       CCI=(CST*CST*CS(I))**(1./3.)
       DDI=(DST*DST*DS(I))**(1./3.)
       ALPI=(ALPT*ALPT*ALP(I))**(1./3.)
       GDD=GAMT*D*D
       FU=FU+(2.*D*FUS)+(3.*D*D/2.)*((BBI*R*T)-AAI-(DDI/T))
       FU=FU+(3.*ALPT*(D**5)/5.)*(AAI+(DDI/T))
       FU=FU+(3.*(D**5)/5.)*(AST+(DST/T))*ALPI
       FU=FU+(3.*CCI*D*D/(T**2))*(((1.-EXP(-GDD))/GDD)-(0.5*EXP(-GDD)))
       FU=FU-(2.*CST/(GAMT*T*T))*((GAM(I)/GAMT)**.5)*
      &(1.-(1.+GDD+0.5*GDD*GDD)*EXP(-GDD))
  800  FUG(I)=(1./P)*EXP(FU/(R*T))
       RETURN
       END
```

6.9.4. Comparative Examples

In this section comparative examples employing the routines described above will be presented. Properties are calculated from:
1. the RKS equation,
2. the PR equation,
3. the BWR(HS) equation,
4. the Lee-Kesler corresponding-states method (augmented by the Wu-Stiel terms for polar molecules, and
5. the virial equation truncated after the third virial coefficient
Wherever possible, the results are compared with experimental values.

| EXAMPLE 6.2 | *Calculation of the Density and Fugacity Coefficient of R134a (1,1,1,2-tetrafluoroethane) at (a) T = 250 K, P = 2 MPa, and (b) T = 450 K, P = 10 MPa. Enthalpy and Entropy Differences Between the two States are also Calculated.* |

The calling program and a sample output are shown in Display 6.4 for the first set of conditions. Critical constants and the acentric factor were obtained from Appendix A. The calling program indicates a single component by setting switch IS = 1.

The results are shown in Table 6.9. In the same table values obtained with the Lee-Kesler and the Lee-Kesler-Wu-Stiel methods (see Example 5.4) are shown together with the experimental values [17]. It should be noted that R134a is a polar substance and as such it offers a fairly severe test of the methods.

Table 6.9 Results for Example 6.2

	ρ	h	s	ϕ
	kg·m^{-3}	kJ·kg^{-1}	kJ·kg^{-1}·K^{-1}	
	$T = 250$ K, $P = 2$ MPa			
RKS	1200	-238.5	-0.64	0.0601
PR	1356	-235.2	-0.66	0.0598
BWR	1364	-236.1	-0.43	0.0604
L-K	1322	-234.5	-0.95	0.0600
L-K (+ W-S)	1371	-233.6	-0.94	0.0600
Expt. [17]	1371			
	$T = 450$ K, $P = 10$ MPa			
RKS	417	88.5	0.03	0.6614
PR	444	87.1	0.02	0.6237
BWR	452	89.1	-0.09	0.6374
L-K	464	84.3	-0.06	0.6333
L-K (+ W-S)	473	84.1	-0.06	0.6333
Expt. [17]	475			
		Δh	Δs	
		kJ·kg^{-1}	kJ·kg^{-1}·K^{-1}	
RKS		327	0.67	
PR		321	0.68	
BWR		325	0.34	
L-K		319	0.89	
L-K (+ W-S)		318	0.88	
Expt. [17]		324	0.91	

$\Delta h = h(T = 450$ K, $P = 10$ MPa$) - h(T = 250$ K, $P = 2$ MPa$)$
$\Delta s = s(T = 450$ K, $P = 10$ MPa$) - s(T = 250$ K, $P = 2$ MPa$)$

Based on the above results, the following observations may be made:
-   The Lee-Kesler-Wu-Stiel method produces excellent results for the density. The BWR(HS) also does well. The PR equation is superior to the RKS.
-   All models produce comparable values for the fugacity coefficients.
-   All models produce comparable values for the enthalpy increment, in agreement with experiment, but there is some disagreement for the entropy difference.

DISPLAY 6.4 Sample Input/Output

```
PROGRAM EX0602
PARAMETER (IS=1)
DIMENSION X(IS),TC(IS),PC(IS),DC(IS),OMEG(IS)
DIMENSION Y1(IS),Y2(IS),Y3(IS),Y4(IS),BK(IS,IS),FUG(IS)
CHARACTER*4 EOS(3)
DATA EOS/'RKS','P&R','BWR'/
```

```
      DATA X(1),TC(1),PC(1),DC(1),OMEG(1),BK(1,1)/1.,374.21,4.056E+6,
     &5048.,0.325,0./
      DATA Y1(1),Y2(1),Y3(1),Y4(1)/16.7803,0.28634,-2.2732E-4,1.1330E-7/
      DATA T,P,IP/250.,2.E+6,1/
C     DATA T,P,IP/450.,10.E+6,1/
      WMOL=0.020314
      WRITE (*,1000) T,P*1.E-6
1000  FORMAT(5X,'@ ',F6.2,' K & ',F6.2,' MPa',/,
     &5X,' EOS   DENSITY ENTHALPY ENTROPY PART.FUG ',/,5X,
     &'      (KG/M3)  (KJ/KG) (KJ/KG/K)   (-)')
      DO 10 J=1,3
      IF (J.EQ.1) CALL RKS(T,P,IP,IS,X,TC,PC,OMEG,Y1,Y2,Y3,Y4,BK,D,H,S,FUG)
      IF (J.EQ.2) CALL PR (T,P,IP,IS,X,TC,PC,OMEG,Y1,Y2,Y3,Y4,BK,D,H,S,FUG)
      IF (J.EQ.3) CALL BWR(T,P,IP,IS,X,TC,DC,OMEG,Y1,Y2,Y3,Y4,BK,D,H,S,FUG)
100   WRITE (*,1200) EOS(J),D*WMOL,H/WMOL/1000.,S/WMOL/1000.,FUG(1)
1200  FORMAT(5X,A6,F9.1,F9.1,F10.2,F11.4)
      END


 @ 250.00 K & 2.00 MPa
 EOS   DENSITY ENTHALPY ENTROPY PART.FUG
       (KG/M3)  (KJ/KG) (KJ/KG/K)   (-)
 RKS    1199.9   -238.5     -0.64    0.0601
 PR     1255.5   -235.2     -0.66    0.0898
 BWR    1263.5   -236.      -0.43    0.0604
```

□

EXAMPLE 6.3   *Calculation of the Density of the Mixture (0.85 CH_4 + 0.15 C_2H_6) at T = 288.15 K and P = 6 MPa.*

The calling program and the output are shown in Display 6.5. Data required are obtained from Table A1 in Appendix A. The binary interaction parameter for this mixture for the Soave [4] and Peng-Robinson [5] equations of state is taken to be zero, while $k_{12} = 0.01$ in the case of the Benedict-Webb-Rubin (Han-Starling) equation of state [16].

```
                                            DISPLAY 6.5  Sample Input/Output
      PROGRAM EX0603
      PARAMETER (IS=2)
      DIMENSION X(IS),TC(IS),PC(IS),DC(IS),OMEG(IS)
      DIMENSION Y1(IS),Y2(IS),Y3(IS),Y4(IS),BK(IS,IS),FUG(IS)
      CHARACTER*4 EOS(3)
      DATA EOS/'RKS','P&R','BWR'/
      DATA X(1),TC(1),PC(1),DC(1),OMEG(1)/.85,190.55,4.599E+6,10080.6,0.01/
      DATA X(2),TC(2),PC(2),DC(2),OMEG(2)/.15,305.33,4.871E+6,6954.1,0.099/
      DATA Y1(1),Y2(1),Y3(1),Y4(1)/0.,0.,0.,0./
      DATA Y1(2),Y2(2),Y3(2),Y4(2)/0.,0.,0.,0./
```

```
      DATA BK(1,1),BK(2,2)/0.,0./
      DATA T,P,IP/288.15,6.E+6,2/
      WRITE (*,1000) T,P*1.E-6
1000  FORMAT(5X,'@ ',F6.2,' K & ',F6.2,' MPA',/, 5X,' EOS   DENSITY (KMOL/M3)')
C.....RKS
      BK(1,2)=0.
      BK(2,1)=0.
      CALL RKS(T,P,IP,IS,X,TC,PC,OMEG,Y1,Y2,Y3,Y4,BK,D,H,S,FUG)
      WRITE (*,1200) EOS(1),D/1000.
C.....PR
      BK(1,2)=0.
      BK(2,1)=0.
      CALL PR (T,P,IP,IS,X,TC,PC,OMEG,Y1,Y2,Y3,Y4,BK,D,H,S,FUG)
      WRITE (*,1200) EOS(2),D/1000.
C.....BWR
      BK(1,2)=0.01
      BK(2,1)=0.01
      CALL BWR(T,P,IP,IS,X,TC,DC,OMEG,Y1,Y2,Y3,Y4,BK,D,H,S,FUG)
      WRITE (*,1200) EOS(3),D/1000.
1200  FORMAT(5X,A6,F7.3)
      END

  @ 288.15 K & 6.00 MPa
  EOS        DENSITY (KMOL/M3)
  RKS           2.955
  PR            3.060
  BWR           2.977
```

Table 6.10 Density of $(0.85\ CH_4 + 0.15\ C_2H_6)$ at $T = 288.15$ K and $P = 6$ MPa

	RKS	PR	BWR	L-K	PITZCURL	VGERG
ρ_n / kmol·m^{-3}	2.955	3.060	2.977	2.967	2.973	2.978

The results are shown also in Table 6.10. In the same table densities calculated from the following models are given: (a) Lee-Kesler method, (b) the truncated virial equation with Pitzer-Curl coefficients and mixing rules from Table 4.3 (PITZCURL routine), and (c) the GERG equation (VGERG routine). These were obtained directly from Example 4.6.

The density obtained from the VGERG routine should be accurate to within 0.1 per cent. It can be seen that in this particular case the BWR(HS) equation is in excellent agreement with that value. It is also worthwhile noting, had k_{12} for the Benedict-Webb-Rubin (Han-Starling) equation of state been set equal to zero, then the value for the density would have been 2.984 kmol·m^{-3} (0.24 per cent higher).

❑

EXAMPLE 6.4   *Calculation of the Partial Fugacity Coefficient for an Equimolar Mixture of Methyl Ethyl Ketone and Toluene at T = 323.15 K and P = 25 kPa.*

The calling program and the output are shown in Display 6.6. Data required are obtained from Table A1 in Appendix A. In this case the binary interaction parameter is not known and is therefore taken to be zero. The results are also shown in Table 6.11, together with the fugacity coefficients obtained in Example 5.5 from the Lee-Kesler-Wu-Stiel method and the experimental values (page 343 of [18]).

In this case, all values obtained are near to the experimental ones. The corresponding-states method and the BWR(HS) equation produce the best values.

Table 6.11 Partial Fugacity Coefficients*

	φ_1	φ_2
RKS	0.991	0.989
PR	0.991	0.989
BWR	0.988	0.986
L-K (+W-S)	0.986	0.987
Expt. [18]	0.987	0.983

*Component 1 is methyl ethyl ketone.

DISPLAY 6.6 Sample Input/Output

```
PROGRAM EX0604
PARAMETER (IS=2)
DIMENSION X(IS),TC(IS),PC(IS),DC(IS),OMEG(IS)
DIMENSION Y1(IS),Y2(IS),Y3(IS),Y4(IS),BK(IS,IS),FUG(IS)
CHARACTER*4 EOS(3)
DATA EOS/'RKS','P&R','BWR'/
DATA X(1),TC(1),PC(1),DC(1),OMEG(1)/0.5,535.6,4.5E+6,3745.3,.329/
DATA X(2),TC(2),PC(2),DC(2),OMEG(2)/0.5,563.0,4.13E+6,3164.6,.263/
DATA Y1(1),Y2(1),Y3(1),Y4(1)/0.,0.,0.,0./
DATA Y1(2),Y2(2),Y3(2),Y4(2)/0.,0.,0.,0./
DATA BK(1,1),BK(2,2)/0.,0./
DATA T,P,IP/323.15,25.E+3,2/
WRITE(*,1000) T,P-1.E-6
1000 FORMAT(5X,'@ ',F6.2,' K  & ',F6.2,' MPA',/
     &5X,' EOS   PART.FUG.1  PART.FUG.2',
     &' FUG.MIX',/14X,'(-)',8X,'(-)',8X,'(-)')
C    RKS
     BK(1,2)=0.
     BK(2,1)=0.
     CALL RKS(T,P,IS,X,TC,PC,OMEG,Y1,Y2,Y3,Y4,BK,D,H,S,FUG)
     WRITE(*,1200) EOS(1),FUG(1),FUG(2)
```

```
C.......PR
        BK(1,2)=0.
        BK(2,1)=0.
        CALL PR (T,P,IP,IS,X,TC,PC,OMEG,Y1,Y2,Y3,Y4,BK,D,H,S,FUG)
        WRITE (*,1200) EOS(2),FUG(1),FUG(2)
C........BWR
        BK(1,2)=0.
        BK(2,1)=0.
        CALL BWR(T,P,IP,IS,X,TC,DC,OMEG,Y1,Y2,Y3,Y4,BK,D,H,S,FUG)
        WRITE (*,1200) EOS(3),FUG(1),FUG(2)
1200  FORMAT(5X,A6,F11.4,F12.4)
        END


   @ 323.15 K & 0.02 MPa
   EOS        PART.FUG.1      PART.FUG.2
                  (-)             (-)
   RKS          0.9911          0.9894
   PR           0.9909          0.9891
   BWR          0.9884          0.9855
```

❏

EXAMPLE 6.5 *Calculation of the Density, Fugacity Coefficients and Mixture Fugacity of (0.766 C_3H_8 + 0.234 CH_4) at (a) T = 260 K, P = 1.7 MPa, and (b) T = 420 K, P = 10 MPa. The Enthalpy and Entropy Differences Between the two States are also Calculated.*

The calling program and sample output for the first set of conditions are shown in Display 6.7. Data required are obtained from Appendix A. The binary interaction parameter for this mixture for the Soave [4] and Peng-Robinson [5] equations of state is set equal to zero in both cases, while $k_{12} = 0.022$ in the case of the Benedict-Webb-Rubin (Han-Starling) equation of state [16].

In Table 6.12 the results are shown together with results obtained with the Lee-Kesler scheme (as described in Example 5.2), and experimental enthalpy difference [19]. The superiority of the Lee-Kesler corresponding-states scheme, at least for the calculation of the enthalpy difference, is apparent. Since there are no experimental densities available with which to compare the results, we can only observe that the BWRS(HS) and Lee-Kesler methods gives results which are in close agreement with each other.

Table 6.12 Results for Example 6.5

	ρ kg·m^{-3}	h kJ·kg^{-1}	s kJ·kg^{-1}·K^{-1}	Φ_1	Φ_2	Φ_{mix}
			$T = 260$ K, $P = 1.7$ MPa			
RKS	457	-413.0	-1.55	6.176	0.183	0.416
PR	517	-408.8	-1.60	5.957	0.181	0.410
BWR	494	-414.8	-0.69	6.740	0.175	0.411
L-K	488	-405.8	-1.99			0.417
			$T = 420$ K, $P = 10$ MPa			
RKS	158	139.7	-0.32	1.245	0.584	0.697
PR	168	136.0	-0.35	1.199	0.550	0.660
BWR	163	141.3	-0.17	1.264	0.568	0.685
L-K	167	139.2	-0.45			0.689
	Δh kJ·kg^{-1}	Δs kJ·kg^{-1}·K^{-1}		Φ_1	Φ_2	Φ_{mix}
RKS	553	1.23				
PR	545	1.25				
BWR	556	0.52				
L-K	545	1.54				
Exp.[18]	542					

Component 1 is methane
$\Delta h = h(T = 420$ K, $P = 10$ MPa$) - h(T = 260$ K, $P = 1.7$ MPa$)$
$\Delta s = s(T = 420$ K, $P = 10$ MPa$) - s(T = 260$ K, $P = 1.7$ MPa$)$

```
                                          DISPLAY 6.7   Sample Input/Output
      PROGRAM EX0605
      PARAMETER (IS=2)
      DIMENSION X(IS),TC(IS),PC(IS),DC(IS),OMEG(IS)
      DIMENSION Y1(IS),Y2(IS),Y3(IS),Y4(IS),BK(IS,IS),FUG(IS)
      CHARACTER*4 EOS(3)
      DATA EOS/'RKS','P&R','BWR'/
      DATA X(1),TC(1),PC(1),DC(1),OMEG(1)/.224,190.55,4.599E+6,10080.6,0.011/
      DATA X(2),TC(2),PC(2),DC(2),OMEG(2)/.766,369.85,4.247E+6,4926.1,0.153/
      DATA Y1(1),Y2(1),Y3(1),Y4(1)/19.25,5.213E-2,1.197E-5,-1.132E-8/
      DATA Y1(2),Y2(2),Y3(2),Y4(2)/-4.224,3.063E-1,-1.586E-4,3.215E-8/
      DATA BK(1,1),BK(2,2)/0.,0./
      DATA T,P,IP/260.,1.7E+6,1/
      WMOL=X(1)*0.0160426+X(2)*0.0440962
      WRITE (*,1000) T,P*1.E-6
1000  FORMAT(5X,'@ ',F6.2,' K & ',F6.2,' MPA',/,
     &5X,' EOS   DENSITY ENTHALPY ENTROPY PART.FUG.1 PART.FUG.2',
     &' FUG.MIX.'/5X,'       (KG/M3) (KJ/KG) (KJ/KG/K)   (-)      (-)',
     &'      (-)')
C...  RKS
      BK(1,2)=0
```

```
      BK(2,1)=0
      CALL RKS(T,P,IP,IS,X,TC,PC,OMEG,Y1,Y2,Y3,Y4,BK,D,H,S,FUG)
      FUGMIX=EXP(X(1)*LOG(FUG(1))+X(2)*LOG(FUG(2)))
      WRITE (*,1200) EOS(1),D*WMOL,H/WMOL/1000,S/WMOL/1000,FUG(1),
     &FUG(2),FUGMIX
C     PR
      BK(1,2)=0
      BK(2,1)=0
      CALL PR (T,P,IP,IS,X,TC,PC,OMEG,Y1,Y2,Y3,Y4,BK,D,H,S,FUG)
      FUGMIX=EXP(X(1)*LOG(FUG(1))+X(2)*LOG(FUG(2)))
      WRITE (*,1200) EOS(2),D*WMOL,H/WMOL/1000,S/WMOL/1000,FUG(1),
     &FUG(2),FUGMIX
C     BWR
      BK(1,2)=0.022
      BK(2,1)=0.022
      CALL BWR(T,P,IP,IS,X,TC,DC,OMEG,Y1,Y2,Y3,Y4,BK,D,H,S,FUG)
      FUGMIX=EXP(X(1)*LOG(FUG(1))+X(2)*LOG(FUG(2)))
      WRITE (*,1200) EOS(3),D*WMOL,H/WMOL/1000,S/WMOL/1000,FUG(1),
     &FUG(2),FUGMIX
1200  FORMAT(5X,A6,F9.1,F9.1,F10.2,F11.4,F12.4,F11.4)
      END


@ 260.00 K & 1.70 MPa
EOS   DENSITY  ENTHALPY  ENTROPY  PART.FUG.1  PART.FUG.2  FUG.MIX
      (KG/M3)  (KJ/KG)   (KJ/KG/K)   (-)         (-)        (-)
RKS    457     -413.0    -1.55      6.176       0.183      0.416
PR     517     -408.8    -1.60      5.957       0.181      0.410
BWR    494     -414.8    -0.69      6.740       0.175      0.411
```

❑

EXAMPLE 6.6   *Calculation of the Density, Fugacity Coefficients and Mixture Fugacity of (0.324 n-C_8H_{18} + 0.676 C_6H_6) at (a) T = 470 K, P = 1.4 MPa, and (b) T = 590 K, P = 9.7 MPa. The Enthalpy and Entropy Differences Between the two States are also Calculated.*

The calling program for the first set of conditions is shown in Display 6.8. Data required were obtained, as in all other examples, from Table A1 in Appendix A. Since for this mixture, the binary interaction parameter was not known, it was set equal to zero in all cases.

In Table 6.13 the results are shown together with those obtained from the Lee-Kesler corresponding-states scheme (see Example 5.3), and the experimental enthalpy difference [20].

Table 6.13   Results for Example 6.6

	ρ kg·m^{-3}	h kJ·kg^{-1}	s kJ·kg^{-1}·K^{-1}	ϕ_1	ϕ_2	ϕ_{mix}
		$T = 470$ K, $P = 1.4$ MPa				
RKS	511	21.8	0.26	0.355	0.838	0.635
PR	579	23.3	0.24	0.348	0.823	0.623
BWR	601	17.3	-0.32	0.358	0.828	0.633
L-K	597	15.2	0.018			0.581
		$T = 590$ K, $P = 9.7$ MPa				
RKS	381	363.3	0.79	0.338	0.495	0.437
PR	418	363.0	0.77	0.312	0.469	0.411
BWR	433	360.6	-0.20	0.317	0.485	0.422
L-K	462	347.3	0.62			0.397
		Δh kJ·kg^{-1}	Δs kJ·kg^{-1}·K^{-1}	ϕ_1	ϕ_2	ϕ_{mix}
RKS		342	0.53			
PR		340	0.53			
BWR		343	0.12			
L-K		332	0.60			
Exp.[20]		315				

Component 1 is *n*-octane

$\Delta h = h(T = 590$ K, $P = 9.7$ MPa$) - h(T = 470$ K, $P = 1.4$ MPa$)$

$\Delta s = s(T = 590$ K, $P = 9.7$ MPa$) - s(T = 470$ K, $P = 1.4$ MPa$)$

In this case, the agreement is not as good as in the previous examples. Although there is good agreement between the Lee-Kesler density and that obtained from the BWR(HS) equation for the first set of conditions, this is not the case for the second set of conditions where a 6 per cent difference is noted. A larger difference is also noted between the density calculated from the RKS equation and that from the other equations of state - a difference that rises to 16 per cent relative to the BWR(HS) equation. Furthermore, it seems that calculated enthalpy differences obtained from the three equations of state rather overestimate the experimental result.

```
PROGRAM EX0606                                    DISPLAY 6.8  Sample Input/Output
PARAMETER (IS=2)
DIMENSION X(IS),TC(IS),PC(IS),DC(IS),OMEG(IS)
DIMENSION Y1(IS),Y2(IS),Y3(IS),Y4(IS),BK(IS,IS),FUG(IS)
CHARACTER*4 EOS(3)
DATA EOS/'RKS','P&R','BWR'/
DATA X(1),TC(1),PC(1),DC(1),OMEG(1)/0.324,568.95,2.49E+6,2032.5,0.398/
```

```
      DATA X(2),TC(2),PC(2),DC(2),OMEG(2)/0.676,562.06,4.895E+6,386.1,0,0.272/
      DATA Y1(1),Y2(1),Y3(1),Y4(1)/-6.096,0.7712,-4.195E-4,8.855E-8/
      DATA Y1(2),Y2(2),Y3(2),Y4(2)/-33.92,0.4739,-3.017E-4,7.13E-8/
      DATA BK(1,1),BK(2,2)/0.,0./
      DATA T,P,IP/470.,1.4E+6,1/
C     DATA T,P,IP/590.,9.7E+6,1/
      WMOL=X(1)*0.1142302+X(2)*0.0781134
      WRITE(*,1000) T,P*1.E-6
1000  FORMAT(5X,'@ ',F6.2,' K & ',F6.2,' MPA'/,
     &5X,' EOS  DENSITY ENTHALPY ENTROPY PART.FUG.1  PART.FUG.2',
     &' FUG.MIX.'/,5X,'      (KG/M3) (KJ/KG) (KJ/KG/K)  (-)'
     &,'       (-)     (-)')
C.....RKS
      BK(1,2)=0.
      BK(2,1)=0.
      CALL RKS(T,P,IP,IS,X,TC,PC,OMEG,Y1,Y2,Y3,Y4,BK,D,H,S,FUG)
      FUGMIX=EXP(X(1)*LOG(FUG(1))+X(2)*LOG(FUG(2)))
      WRITE (*,1200) EOS(1),D*WMOL,H/WMOL/1000.,S/WMOL/1000.,FUG(1)
     &,FUG(2),FUGMIX
C.....PR
      BK(1,2)=0.
      BK(2,1)=0.
      CALL PR (T,P,IP,IS,X,TC,PC,OMEG,Y1,Y2,Y3,Y4,BK,D,H,S,FUG)
      FUGMIX=EXP(X(1)*LOG(FUG(1))+X(2)*LOG(FUG(2)))
      WRITE (*,1200) EOS(2),D*WMOL,H/WMOL/1000.,S/WMOL/1000.,FUG(1),
     &,FUG(2),FUGMIX
C.....BWR
      BK(1,2)=0.
      BK(2,1)=0.
      CALL BWR(T,P,IP,IS,X,TC,DC,OMEG,Y1,Y2,Y3,Y4,BK,D,H,S,FUG)
      FUGMIX=EXP(X(1)*LOG(FUG(1))+X(2)*LOG(FUG(2)))
      WRITE (*,1200) EOS(3),D*WMOL,H/WMOL/1000.,S/WMOL/1000.,FUG(1),
     &,FUG(2),FUGMIX
1200  FORMAT(5X,A6,F9.1,F9.1,F10.2,F11.4,F12.4,F11.4)
      END
```

@ 470.00 K & 1.40 MPa

EOS	DENSITY (KG/M3)	ENTHALPY (KJ/KG)	ENTROPY (KJ/KG/K)	PART.FUG.1 (-)	PART.FUG.2 (-)	FUG.MIX (-)
RKS	511	21.8	0.26	0.355	0.838	0.635
PR	573	23.3	0.24	0.348	0.823	0.623
BWR	601	17.3	0.32	0.358	0.828	0.633

□

158   *Equilibrium Properties*

| EXAMPLE 6.7 |

Calculation of the Density of Water, (a) in the Liquid Phase at T = 303.15 K, P = 2.0 MPa, and (b) in the Vapour Phase at T = 423.15 K, P = 0.2 MPa.

This example is particularly interesting since water is a highly polar substance. The calling program for both sets of conditions is shown in Display 6.9. Data required are obtained from Appendix A.

The values of density recommended by the International Association for the Properties of Steam (IAPS84 FORTRAN package [21]) are 996.5 kg·m^{-3} for the liquid and 1.042 kg·m^{-3} for the vapour under the conditions specified. In the case of the liquid, the best prediction is produced by the BWR(HS) equation, while that from the RKS equation is too low by some 25 per cent (although volume translation could be used to force agreement with experiment). For the vapour phase, all three equations of state produce similar values in fair agreement with the IAPS recommended value.

DISPLAY 6.9 Sample Input/Output

```
      PROGRAM EX0607
      PARAMETER (IS=1)
      DIMENSION X(IS),TC(IS),PC(IS),DC(IS),OMEG(IS)
      DIMENSION Y1(IS),Y2(IS),Y3(IS),Y4(IS),BK(IS,IS),FUG(IS)
      CHARACTER*4 EOS(3)
      DATA EOS/'RKS','P&R','BWR'/
      DATA X(1),TC(1),PC(1),DC(1),OMEG(1)/1.,647.3,22.1E+6,17513.1,.344/
      DATA Y1(1),Y2(1),Y3(1),Y4(1),BK(1,1)/0.,0.,0.,0.,0./
      DATA T,P/303.15,2.E+6/
      DO 100 I=1,2
      IF (I.EQ.2) T=T+120.
      IF (I.EQ.2) P=P/10.
      IP=1
      WRITE (*,1000) T,P*1.E-6
 1000 FORMAT(1X,/,5X,'@ ',F6.2,' K & ',F6.2,' MPA',/,
     &5X,' EOS      DENSITY ',/,7X,'---------------------------')
C     RKS
      CALL RKS(T,P,IP,IS,X,TC,PC,OMEG,Y1,Y2,Y3,Y4,BK,D,H,S,FUG)
      WRITE (*,1200) EOS(1),D*0.0180152
C     PR
      CALL PR (T,P,IP,IS,X,TC,PC,OMEG,Y1,Y2,Y3,Y4,BK,D,H,S,FUG)
      WRITE (*,1200) EOS(2),D*0.0180152
C     BWR
      CALL BWR(T,P,IP,IS,X,TC,DC,OMEG,Y1,Y2,Y3,Y4,BK,D,H,S,FUG)
      WRITE (*,1200) EOS(3),D*0.0180152
  100 CONTINUE
 1200 FORMAT(5X,A6,F12.3,' KG/M3')
      END
```

```
@ 303.15 K  &  2.0 MPA
EOS              DENSITY
-------------------------------
RKS            754.287  KG/M3
P&R            847.182  KG/M3
BWR            960.685  KG/M3

@ 423.15 K  &  0.2 MPA
EOS              DENSITY
-------------------------------
RKS            1.036  KG/M3
P&R            1.037  KG/M3
BWR            1.037  KG/M3
```

❑

References

1. J.D. van der Waals, PhD thesis, Leyden (1873).
2. M. Benedict, G.B. Webb and L.C. Rubin, *J. Chem. Phys.* **8** (1940) 334.
3. O. Redlich and J.N.S. Kwong, *Chem. Review* **44** (1949) 233.
4. G. Soave, *Chem. Eng. Sci.* **27** (1972) 1197.
5. D.Y. Peng and D.B. Robinson, *Ind. Eng. Chem. Fundam.* **15** (1976) 59.
6. N.F. Carnahan and K.E. Starling, *AIChE J.* **18** (1972) 1184.
7. S.M. Walas, *Phase Equilibria in Chemical Engineering* (Butterworth Publishers, London, 1985).
8. O. Redlich, *Thermodynamics Fundamentals and Applications* (Elsevier, 1978).
9. M.S. Graboski and T.E. Daubert, *Ind. Eng. Chem. Process Des. Dev.* **17**:443 (1978).
10. T.E. Daubert, M.S. Graboski & R.P. Danner, *Documentation of the Basis for Selection of the Contents of Chapter 8 - Vapour-Liquid Equilibrium K-Values in Technical Data Book - Petroleum Refining* (API, No.8-78, 1978).
11. M. Benedict, G.B. Webb and L.C. Rubin, *J. Chem. Phys.* **8** (1940) 334.
12. M. Benedict, G.B. Webb and L.C. Rubin, *J. Chem. Phys.* **10** (1942) 747.
13. M. Benedict, G.B. Webb and L.C. Rubin, *Chem. Eng. Prog.* **47** (1951) 419.
14. M. Benedict, G.B. Webb and L.C. Rubin, *Chem. Eng. Prog.* **47** (1951) 449.
15. R. Holub and P. Vonka, *Chemical Equilibrium of Gaseous Systems* (Reidel, 1976).
16. K.E. Starling and M.S. Han, *Hydrocarb. Proc.* **51** (1972)192.
17. *Thermophysical Properties of Environmentally Acceptable Refrigerants* (Japanese Association of Refrigeration, Tokyo, 1991).
18. J.M. Smith and H.C. van Ness, *Introduction to Chemical Engineering Thermodynamics*, 4th Ed. (McGraw-Hill, 1987).
19. V.F. Yesavage, D.L. Katz and J.E. Powers, *J. Chem. Eng. Data* **14** (1969) 139.
20. J.M. Lenoir and K.E. Hayworth, *J. Chem. Eng. Data* **16** (1971) 280.
21. *IAPS84: Fortran Package for the Calculation of the Properties of Water/Steam* (Produced and Supplied by N.E.L., East Kilbride, Scotland, U.K.).

22. A. Péneloux, E. Rauzy and R. Freeze, *Fluid Phase Equilibria* **8** (1982) 7.
23. K. Magoulas and D. Tassios, *Fluid Phase Equilibria* **56** (1990) 119.
24. D.R. Douslin and R.H. Harrison, *J. Chem. Thermodyn.* **5** (1973) 491.
25. D.F. Friend, H. Ingham and J.F. Ely, *J. Phys. Chem. Ref. Data* **20** (1991) 275.
26. J. M. H. Levelt Sengers, *Experimental Thermodynamics, Vol II: Experimental Thermodynamics of Non-reacting Fluids*, Ed: B. Le Neindre and B. Vodar (Butterworth, London,1975). Ch. 14.
27. K.G. Wilson. *Phys. Rev.* **4** (1974) 3174, and *Phys. Rev.* **4** (1974) 3184.
28. J.C. Le Guillou and J. Zinn-Justin, *Phys. Rev B* **21** (1980) 3976.
29. A. Van Pelt, G. X Jin and J. V. Sengers, *Int. J. Thermophys.* **15** (1994) 687.
30. U. Setzmann and W. Wagner, *J. Phys. Chem. Ref. Data* **20** (1991) 1061.
31. D.G. Friend and J.F. Ely, *J. Phys. Chem. Ref. Data* **20** (1991) 275.

7
Activity Coefficient Models

An appropriate equation of state provides a thermodynamically consistent route to the fugacity of components in both vapour and liquid phases and thus offers a very convenient basis for phase-equilibrium calculations. Unfortunately, the simple cubic equations are only applicable to mixtures of molecules without strong specific interactions and they generally fail to give reliable results for the liquid phase of associating mixtures (*e.g.* acetone + methanol), although they may be used to obtain vapour-phase properties. In these cases, better results are obtained when the fugacity of components in the liquid phase is estimated from an activity-coefficient model. The combination of an equation of state for the vapour phase and an activity-coefficient model for the liquid phase thus offers a practical method for phase-equilibrium calculations in systems containing associating molecules. Activity coefficient models cannot be used to obtain any other properties of the liquid.

In this chapter, the use of activity-coefficient models as a tool for phase equilibrium calculations is introduced. All such models are empirical in nature and represent the activity coefficient of a component in a mixture (and hence its fugacity) in terms of an equation that contains a set of parameters. Two general approaches are employed.

(a) The parameters of the activity coefficient model are determined in a fit to experimental VLE data on binary mixtures, usually at a single temperature. Provided that parameters are determined for all possible binary pairs, the model may then be applied in the prediction of activity coefficients in a multi-component mixture over a range of temperature and pressure. In this sense the model is only a correlation for binary systems (although it may allow extrapolation with respect to temperature or pressure) but it is truly predictive for multi-component systems. Examples of this approach presented in this chapter are the Wilson, T-K-Wilson, NRTL, and UNIQUAC activity-coefficient models. This general approach is to be preferred from the point of view of reliability but the need for experimental data on binary sub-systems can be demanding.

(b) An alternative approach, which requires no experimental data, is one in which the parameters of the activity-coefficient model are estimated by a group-contribution method. Several such schemes have been developed with functional-group parameters determined by regression against a very large date base of experimental VLE results. In application, group-contribution models are predictive but generally offer inferior accuracy. Examples of this approach which will be mentioned in this chapter are the ASOG and UNIFAC models.

7.1. Activity and Activity Coefficients

The *activity* α_i of substance i in a mixture of n components is defined by the relation

$$\alpha_i(T, P, x_1, x_2 \cdots x_n) = f_i(T, P, x_1, x_2 \cdots x_n)/f_i^{\circ}(T) \tag{7.1}$$

in which $f_i(T, P, x_1, x_2 \cdots x_n)$ is the fugacity of component i under the specified conditions of temperature, pressure and composition, and $f_i^{\circ}(T)$ is the standard-state fugacity at the given temperature.

The activity is usually expressed in terms of the corresponding activity coefficient γ_i defined by

$$\gamma_i = \frac{\alpha_i}{x_i} = \frac{f_i}{x_i f_i^{\circ}} \tag{7.2}$$

in which, for brevity, the functional dependence of the various terms has been suppressed. This definition ensures that $RT \ln \gamma_i$ is identical with the excess chemical potential μ_i^E.

The standard state for a component in the liquid phase is defined as that of the pure saturated liquid at the given temperature. This definition is advantageous because the fugacity of the pure saturated liquid is identical with that of the pure saturated vapour which we may calculate from an equation of state. Thus

$$f_i^{\circ} \equiv f_i^{\sigma} = \phi_i^{\sigma} P_i^{\sigma}, \tag{7.3}$$

where P_i^{σ} is the pressure and ϕ_i^{σ} the fugacity coefficient of the pure saturated vapour at the given temperature. Difficulties can arise with this definition of the standard state when components are present in a liquid phase at a temperature above their critical temperature; in these cases alternative definitions may be adopted.

The objective with an activity coefficient model is to represent the dependence of γ_i on temperature, pressure and composition. By definition, $\gamma_i = 1$ for the pure saturated liquid. The effect of pressure may be accounted for by noting that

$$\left(\frac{\partial \ln f_i}{\partial P}\right)_T = \frac{V_i}{RT},$$ (7.4)

where V_i is the partial molar volume of component i. Thus the fugacity of component i in a liquid mixture may be written

$$f_i^{(L)} = x_i \gamma_i f_i^\circ \mathcal{F}_i,$$ (7.5)

where the activity coefficient refers to the specified temperature and composition but the saturated vapour pressure of the pure substance, and

$$\mathcal{F}_i = \exp\left[\int_{P_i^\sigma}^{P} \left(V_i^{(L)}/RT\right) dP\right]$$ (7.6)

is known as the Poynting Factor. Frequently, the partial molar volume is not known as a function of composition and pressure and is therefore approximated by the molar volume of the pure liquid at the specified temperature and, typically, at atmospheric pressure. In that approximation, one has

$$\mathcal{F}_i \approx \exp\left[\left(V_i^{*(L)}/RT\right)(P - P_i^\sigma)\right]$$ (7.7)

The standard-state fugacity $f_i^\circ(T)$ is obtained from Eq.(7.3), typically with the vapour pressure calculated from a correlation of experimental values and the fugacity coefficient calculated from an equation of state for the vapour.

The fugacity of component i in the vapour phase is given by

$$f_i^{(V)} = y_i \, \Phi_i^{(V)} \, P$$ (7.8)

where $\Phi_i^{(V)}$ is the fugacity coefficient of component i at the specified temperature, pressure and vapour-phase composition and y_i is the mole fraction in the vapour. Equality of fugacity at phase equilibrium then leads to the following condition connecting the composition of coexisting vapour and liquid phases:

$$y_i \, \Phi_i^{(V)} \, P \; = \; x_i \, \gamma_i \, \Phi_i^\sigma \, P_i^\sigma \, \mathscr{F}_i \qquad\qquad (7.9)$$

This relation forms the basic working equation for most vapour-liquid equilibrium problems. The rôle of the activity-coefficient models described in this chapter is to represent the activity coefficient γ_i as a function of temperature and composition at constant pressure.

7.2. Experimental Determination of Activity Coefficients

In order to apply many activity-coefficient models, it is necessary to have experimental values of activity coefficients in binary systems. Eq.(7.9) is easily rearranged into a working equation for the determination of activity coefficients from experimental measurements:

$$\gamma_i \; = \; \left(\frac{y_i}{x_i} \right) \left(\frac{P}{P_i^\sigma} \right) \left(\frac{\Phi_i^{(V)}}{\Phi_i^\sigma} \right) \left(\frac{1}{\mathscr{F}_i} \right) \qquad\qquad (7.10)$$

A typical experimental determination requires measurements of P, y_1 and x_1 for a binary system at vapour-liquid equilibrium at a given temperature. The saturated vapour pressure is either calculated from a correlation or also measured, while the ratio of fugacity coefficients is calculated from an equation of state for the vapour phase. Finally, the Poynting Factor may be estimated from Eq.(7.7) using values of the molar volume obtained either from experiment or from the literature. Eq.(7.10) then permits evaluation of the activity coefficients of each component in the liquid phase at the experimentally determined composition. By studying a range of binary mixtures with differing overall composition, results over a range of liquid-phase compositions may be obtained and used in the determination of parameters in a suitable activity-coefficient model as discussed in Section 7.3.4.

As an example, Figure 7.1 shows experimental P-x-y data at $T = 318.15$ K for the binary system (nitromethane + tetrachloromethane) while, in Figure 7.2, the corresponding activity coefficients of both components are plotted as a function of liquid composition. In this highly non-ideal system, both activity coefficients differ greatly from unity. However, since the pressures are low, the Poynting Factor is negligible and vapour-phase non-ideality is not very important. Under such circumstances it is reasonable to replace the ratio $\Phi_i^{(V)}/\Phi_i^\sigma$ by unity; indeed this approximation may be good even when the pressure is such that the fugacity

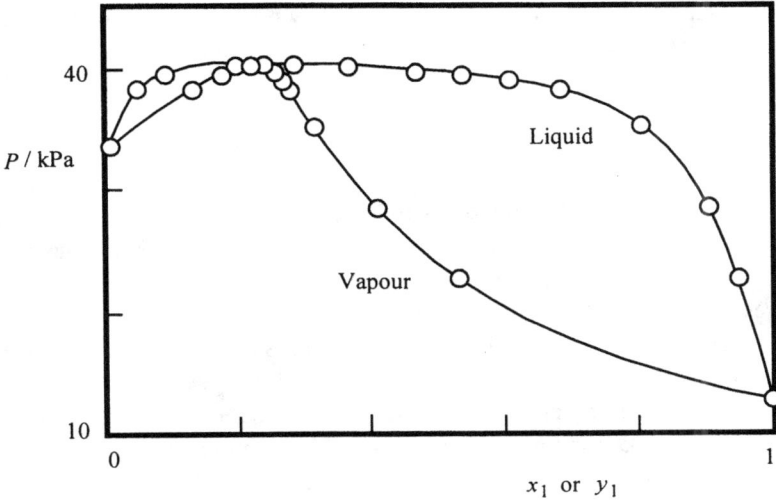

Figure 7.1. Vapour-liquid equilibrium in the system
{nitromethane(1) + tetrachloromethane(2)} at $T = 318.15$ K.
Symbols denote experimental values. Curves represent Wilson's equation with
$\Lambda_{12} = 0.1480$ and $\Lambda_{21} = 0.2745$.

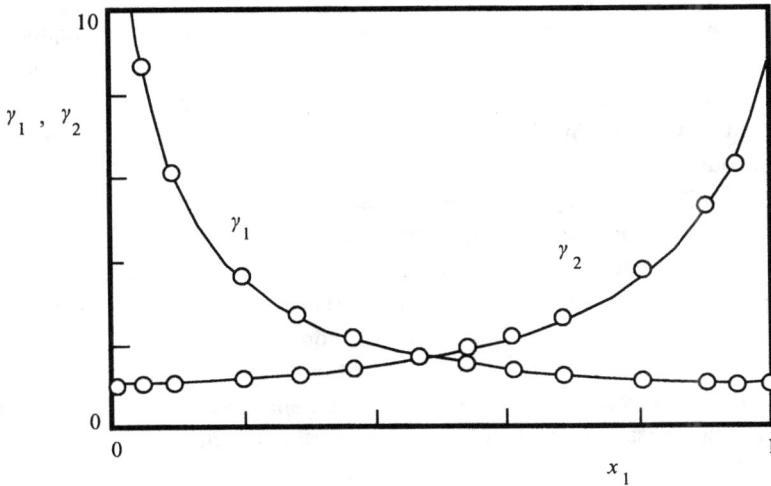

Figure 7.2. Activity coefficients for the system
{nitromethane(1) + tetrachloromethane(2)} at 318.15 K.
Symbols denote experimental values. Curves represent Wilson's equation with
$\Lambda_{12} = 0.1480$ and $\Lambda_{21} = 0.2745$.

coefficients themselves differ significantly from unity provided that the pressures do not differ much from the vapour pressures of the pure components.

7.3. Vapour Pressure Equations

Before discussing models for the activity coefficients, it would be as well to pass a few comments on the correlation and prediction of vapour pressures. The scientific and engineering literature contains a very large database of experimental vapour pressures covering a vast number of compounds. For practical purposes it is desirable to have a correlation of the data for each substance of interest and several well-known equations have been proposed for this purpose. We have already seen in Chapter 5 that, at least superficially, the dependence of vapour pressure on temperature is remarkably simple and we should therefore expect a correlation containing a small number of parameters to be successful. By far the most well-known empirical formula for the correlation of vapour pressure data is Antoine's equation:

$$\ln P^\sigma = A - \frac{B}{T+C}.$$ (7.11)

Values of the parameters A, B and C which appear in this equation are available for a large number of compounds and some of the results are given in Appendix A. The equation is usually capable of correlating experimental vapour pressures at temperatures from below the normal boiling temperature to the critical temperature with an accuracy of one per cent. Over a more limited range, the two parameter equation obtained by setting $C = 0$ may also be acceptable.

Antoine's equation may not be entirely satisfactory over the full range of vapour-liquid coexistence in cases where the reduced vapour pressure at the triple point is extremely small or where it is desired to represent high quality experimental data to within its uncertainty. More accurate formulae, such the six parameter equation due to Wagner [28], have been proposed for that purpose.

For use in connection with an activity-coefficient model in the estimation of liquid-phase fugacities, Eq.(7.11) is almost always satisfactory provided that the parameters may be determined for the substances in question. Vapour pressures for non-polar and weakly polar compounds may also be estimated from the critical constants and the normal boiling temperature by means of the correlation due to Lee and Kesler - Eq.(5.36). Other methods for estimating vapour pressures have been described by Reid, Prausnitz and Poling [26].

7.4. Activity-Coefficient Models

In order to represent activity coefficients as a function of temperature and composition at constant pressure, some kind of thermodynamic model is required. Typical activity-coefficient models contain parameters that are fitted to experimental data on binary mixtures. The most useful models contain semi-theoretical expressions which permit the prediction of activity coefficients for both binary and multi-component systems over a range of temperatures while requiring only experimental input from the binary sub-systems at one temperature.

The oldest of the models still in common use is that of Margules [1] published in 1895. Margules's proposal amounts to representing the logarithm of the activity coefficient by a power series in composition for each component. A model with some theoretical basis was developed by van Laar in the period 1910-1913 [2,3]. The van Laar equation was based on van der Waals' equation of state and, although it can represent experimental data rather well when the two constants are treated as empirical parameters, its predictive capabilities are limited. The modern development of activity-coefficient models began with the work of Wilson, published in 1964 [4], in which he introduced the "local composition" model. Later developments include the Non-Random Two-Liquid model (NRTL) of Renon [5,6], the 1975 modification of the Wilson equation, known as the T-K-Wilson model [7], and the UNIQUAC equations.

7.4.1. Wilson & T-K-Wilson Models

The model proposed by Wilson [4] in 1964 is based on the concept of *local composition* and leads to an expression for the Gibbs free energy from which the activity coefficients can be obtained. Wilson recognised that, in a mixture with specific interactions, the distribution of molecules is not purely random and that non-ideal mixing is associated with this fact. Consider a binary mixture of components 1 and 2 with bulk mole fractions x_1 and x_2. The composition in the immediate viscinity of a molecule of species 1 will not usually be the same as the mean bulk composition. Instead, Wilson suggested that the *local* compositions x_{11} and x_{21} of components 1 and 2 around a molecule of species 1 are given by Boltzmann-weighted averages of the bulk mole fractions. Thus

$$\frac{x_{11}}{x_{21}} = \frac{x_1 \exp(-\varepsilon_{11}/RT)}{x_2 \exp(-\varepsilon_{21}/RT)} = \frac{x_1}{x_2} \exp(\lambda_{12}/RT) \qquad (7.12)$$

where ε_{11} and ε_{21} are energies of interaction defined in a manner similar to the attractive part of the van der Waals potential (see Section 6.4) and $\lambda_{12} = \varepsilon_{21} - \varepsilon_{11}$. As one might expect, the compositional ordering of the fluid is determined not by the absolute magnitude of the molecular interactions but by the difference between the like and unlike interactions.

Eq.(7.12) is next used to evaluate the local volume fraction, z_1, of species 1 around itself, with the result

$$z_1 = \frac{x_{11} V_1}{x_{11} V_1 + x_{21} V_2} = \frac{x_1}{x_1 + \Lambda_{12} x_2} \tag{7.13}$$

where

$$\Lambda_{12} = V_{12} \exp\left(-\frac{\lambda_{12}}{RT}\right) \quad \text{and} \quad V_{12} = \frac{V_2}{V_1}. \tag{7.14}$$

Here, V_1 and V_2 are respectively the partial molar volumes of species 1 and 2 in the liquid mixture.

In a similar manner, the local volume fraction z_2 occupied by molecules of type 2 around a molecule of the same species is found to be

$$z_2 = \frac{x_{22} V_2}{x_{22} V_2 + x_{12} V_1} = \frac{x_2}{x_2 + \Lambda_{21} x_1} \tag{7.15}$$

where

$$\Lambda_{21} = V_{21} \exp\left(-\frac{\lambda_{21}}{RT}\right) \quad \text{and} \quad V_{21} = \frac{V_1}{V_2}. \tag{7.16}$$

It should be noted that Λ_{ij} cannot be negative, that $\Lambda_{11} = \Lambda_{22} = 1$, and that, in general, $\Lambda_{12} \neq \Lambda_{21}$.

In order to obtain an expression for the Gibbs free energy G_m of the mixture, Wilson employed the Flory-Huggins theory [8] according to which

$$G_m = \sum_{i=1}^{n} x_i \left(\mu_i^* + RT \ln z_i\right) \tag{7.17}$$

where μ_i^* is the chemical potential of pure i. Since the molar Gibbs free energy of an ideal mixture is given by (see Appendix B, Section B.3)

$$G_m^{id} = \sum_{i=1}^{n} x_i \left(\mu_i^* + RT \ln x_i \right) , \tag{7.18}$$

it follows that the excess molar Gibbs free energy of the mixture $G_m^E = G_m - G_m^{id}$, is given by

$$G_m^E/RT = \sum_{i=1}^{n} x_i \ln(z_i/x_i) . \tag{7.19}$$

Combining Eqs.(7.13) and (7.15) with Eq.(7.19) for a binary mixture, one obtains

$$G_m^E/RT = -x_1 \ln(x_1 + x_2 \Lambda_{12}) - x_2 \ln(x_1 \Lambda_{21} + x_2) \tag{7.20}$$

and, applying the thermodynamic identity

$$\mu_i^E = RT \ln \gamma_i = G_m^E + (1 - x_i) \left(\frac{\partial G_m^E}{\partial x_i} \right)_{T,P} , \tag{7.21}$$

the activity coefficients are found to be:

$$\left. \begin{array}{l} \ln \gamma_1 = 1 - \ln(x_1 + x_2 \Lambda_{12}) - \dfrac{x_1}{x_1 + x_2 \Lambda_{12}} - \dfrac{x_2 \Lambda_{21}}{x_1 \Lambda_{21} + x_2} \\[4mm] \ln \gamma_2 = 1 - \ln(x_1 \Lambda_{21} + x_2) - \dfrac{x_1 \Lambda_{12}}{x_1 + x_2 \Lambda_{12}} - \dfrac{x_2}{x_1 \Lambda_{21} + x_2} \end{array} \right\} . \tag{7.22}$$

Generalised expressions for multi-component mixtures are given in Table 7.1 and may be used to obtain the activity coefficient of each component i in a mixture of n components provided that all the binary parameters λ_{ij} and the partial molar volumes are known. No additional parameters are required. Typically, partial molar volumes are approximated by the molar volume of the pure liquids and, provided that this is done consistently in both parameter determination and application, the results are generally satisfactory.

Table 7.1   The Wilson Model, 1964 [4]

$$\ln\gamma_i = 1 - \ln\left(\sum_{j=1}^{n} x_j \Lambda_{ij}\right) - \sum_{k=1}^{n} \frac{x_k \Lambda_{ki}}{\sum_{j=1}^{n} x_j \Lambda_{kj}} \tag{7.23}$$

where

$$\Lambda_{ij} = V_{ij} \exp\left(-\frac{\lambda_{ij}}{RT}\right) \qquad V_{ij} = \frac{V_j}{V_i} \tag{7.24}$$

and

$$\Lambda_{ii} = \Lambda_{jj} = 1 \qquad \Lambda_{ij} \neq \Lambda_{ji} \tag{7.25}$$

In Appendix A, Table A4, values for the parameters λ_{12} and λ_{21} obtained from experimental measurements are given for about 70 binary pairs [9]. In most cases, these parameters are assumed to be independent of temperature although, for the best accuracy, they should not be used at temperatures too far from that at which they were determined. An extensive compilation of parameters can be found in the literature [10].

The binary parameters λ_{12} and λ_{21} are usually determined by fitting either experimentally determined excess Gibbs free energies to Eq.(7.20) or experimentally determined activity coefficients to Eqs.(7.22). In either case Eq.(7.24) is used to relate λ_{ij} and Λ_{ij}. Usually, results over a range of liquid compositions are employed and the optimum parameters found by a non-linear regression analysis. However, it is possible, though probably less reliable, to determine the parameters from a single pair of activity coefficients, for example those determined at an azeotrope, or from the values of γ_1 and γ_2 in the limits of infinite dilution.

The outstanding features of the Wilson equations include the generally superior representation of activity coefficients for both polar and non-polar mixtures, and the ability to treat multi-component systems with only binary parameters. One drawback of the model however, is its inability to handle either liquid-liquid equilibria (LLE) or vapour-liquid-liquid equilibria (VLLE). This is a fundamental restriction imposed by the form of the Wilson equations and many modifications have been suggested to overcome the problem. Perhaps the simplest useful modification is that of Tsubota & Katayama [7] which we refer to as the T-K-Wilson model.

| **EXAMPLE 7.1** | *Estimation of Activity Coefficients for the [Methanol(1)+ Ethyl Acetate(2)] System at T = 333.15 K and P = 0.1 MPa.* |

From Appendix A, Table A4, the Wilson parameters are found as: $\lambda_{12}/R = 518.39$ K and $\lambda_{21}/R = -87.28$ K at 333 K. We also estimate $V_{12} = 0.411$ from the BWR(HS) equation of state. Thus from Eq.(7.24), $\Lambda_{12} = 0.0867$ and $\Lambda_{21} = 1.8723$. Substituting these values in Eqs.(7.23), the activity coefficients can be evaluated as a function of the mole fraction x_1. The results of these calculations are compared with experimental data in Figure 7.3 and Table 7.2; the agreement is seen to be quite good.

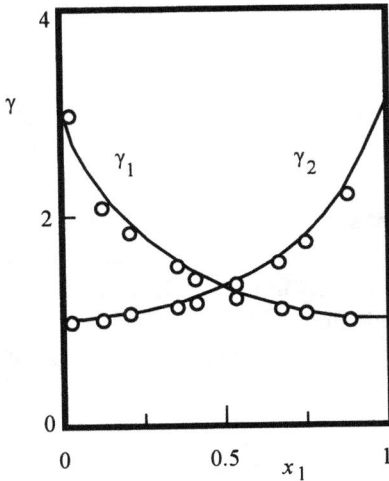

Figure 7.3

Table 7.2   Activity Coefficients Comparison

	Miller [11]		Calculated	
x_1	γ_1	γ_2	γ_1	γ_2
0.000			3.104	1.000
0.028	3.020	1.002	2.891	1.001
0.123	2.115	1.028	2.328	1.019
0.211	1.867	1.086	1.963	1.054
0.352	1.539	1.157	1.569	1.150
0.408	1.425	1.193	1.457	1.204
0.533	1.221	1.363	1.266	1.364
0.664	1.135	1.560	1.133	1.610
0.748	1.081	1.748	1.074	1.829
0.883	1.030	2.230	1.016	2.341
1.000			1.000	3.046

☐

The **T-K-Wilson equations** are based on the following modification of Eq.(7.20) for the excess Gibbs free energy:

$$\frac{G_m^E}{RT} = x_1 \ln\left(\frac{x_1 + x_2 V_{12}}{x_1 + x_2 \Lambda_{12}}\right) + x_2 \ln\left(\frac{x_1 V_{21} + x_2}{x_1 \Lambda_{21} + x_2}\right). \qquad (7.26)$$

Here, $V_{ij} = (V_j/V_i)$ and, when $V_{ij} = 1$, this expression reduces to Wilson's equation.

The resulting expressions for the activity coefficients in a binary system are

$$
\ln\gamma_1 = \ln\left(\frac{x_1 + x_2 V_{12}}{x_1 + x_2 \Lambda_{12}}\right) + x_1\left(\frac{1}{x_1 + x_2 V_{12}} - \frac{1}{x_1 + x_2 \Lambda_{12}}\right)
$$
$$
+ x_2\left(\frac{V_{21}}{x_1 V_{21} + x_2} - \frac{\Lambda_{21}}{x_1 \Lambda_{21} + x_2}\right)
$$
$$
\ln\gamma_2 = \ln\left(\frac{x_1 V_{21} + x_2}{x_1 \Lambda_{21} + x_2}\right) + x_1\left(\frac{V_{12}}{x_1 + x_2 V_{12}} - \frac{\Lambda_{12}}{x_1 + x_2 \Lambda_{12}}\right)
$$
$$
+ x_2\left(\frac{1}{x_1 V_{21} + x_2} - \frac{1}{x_1 \Lambda_{21} + x_2}\right)
$$

(7.27)

The corresponding formulae for a multi-component system are given in Table 7.3.

Although the T-K-Wilson expressions are able to correlate LLE data in binary systems rather well, it is essential to obtain the parameters from reliable experimental results if meaningful predictions are to be made for multi-component systems. Relatively few parameter sets have been reported in the literature but a selection of those available are given in Appendix A, Table A5.

Table 7.3 The T-K-Wilson Model, 1975 [7]

$$
\ln\gamma_i = \ln\left(\frac{\sum_{j=1}^{n} x_j V_{ij}}{\sum_{j=1}^{n} x_j \Lambda_{ij}}\right) + \sum_{k=1}^{n} x_k\left(\frac{V_{ki}}{\sum_{j=1}^{n} x_j V_{kj}} - \frac{\Lambda_{ki}}{\sum_{j=1}^{n} x_j \Lambda_{kj}}\right)
$$

(7.28)

where

$$
\Lambda_{ij} = V_{ij}\exp\left(-\frac{\lambda_{ij}}{RT}\right) \qquad V_{ij} = \frac{V_j}{V_i}
$$

(7.29)

and

$$
\Lambda_{ii} = \Lambda_{jj} = 1 \qquad \Lambda_{ij} \neq \Lambda_{ji}
$$

(7.30)

7.4.2. Non-Random Two-Liquid (NRTL) Model

The NRTL model was developed by Renon and Prausnitz [5,6] in an attempt to overcome the inadequacy of the Wilson equations in describing liquid-liquid equilibria. The NRTL model contains three parameters per binary interaction, compared with the two parameters of the Wilson and T-K-Wilson models, but is based on a similar local composition treatment. According to the NRTL model, the excess Gibbs free energy is correlated by

$$\frac{G^E}{RT} = x_1 x_2 \left[\frac{\tau_{21} G_{21}}{x_1 + x_2 G_{21}} + \frac{\tau_{12} G_{12}}{x_2 + x_1 G_{12}} \right], \tag{7.31}$$

where

$$G_{ij} = \exp\left(-\alpha \tau_{ij}\right) \quad \text{and} \quad \tau_{ij} = g_{ij}/RT . \tag{7.32}$$

Here g_{ij} and g_{ji} denote energy parameters, while α is best viewed as a purely empirical parameter; all three are supposed independent of temperature. The activity coefficients determined for a binary system from Eq.(7.31) are:

$$\left.\begin{aligned}
\ln\gamma_1 &= x_2^2 \left[\tau_{21}\left(\frac{G_{21}}{x_1 + x_2 G_{21}} \right)^2 + \tau_{12}\left(\frac{G_{12}}{(x_2 + x_1 G_{12})^2} \right) \right] \\[2ex]
\ln\gamma_2 &= x_1^2 \left[\tau_{12}\left(\frac{G_{12}}{x_2 + x_1 G_{12}} \right)^2 + \tau_{21}\left(\frac{G_{21}}{(x_1 + x_2 G_{21})^2} \right) \right]
\end{aligned}\right\} . \tag{7.33}$$

In Table 7.4 the corresponding expressions for multi-component systems are given.

In Appendix A, Table A6, values for the parameters g_{12}/R, g_{21}/R and α are given for a small selection of binary mixtures. A larger compilation of the NRTL parameters also exists in ref.[10]. As with the other models discussed, it is essential to determine the parameters from reliable experimental data on binary systems if meaningful predictions are to be made.

Table 7.4 The NRTL (Renon) Model, 1968 [5]

$$\ln\gamma_i = \frac{\sum\limits_{k=1}^{n}\tau_{ki}\, x_k\, G_{ki}}{\sum\limits_{k=1}^{n} x_k\, G_{ki}} + \sum\limits_{j=1}^{n}\frac{x_j G_{ij}}{\sum\limits_{k=1}^{n} x_k\, G_{kj}}\left(\tau_{ij} - \frac{\sum\limits_{k=1}^{n} x_k\, \tau_{kj}\, G_{kj}}{\sum\limits_{k=1}^{n} x_k\, G_{kj}}\right) \qquad (7.34)$$

where
$$G_{ij} = \exp\!\left(-\alpha_{ij}\tau_{ij}\right) \qquad (7.35)$$

$$\tau_{ii} = \tau_{jj} = 0, \qquad \text{and} \qquad G_{ii} = G_{jj} = 1 \qquad (7.36)$$

and
$$\tau_{ij} = g_{ij}/RT \qquad (7.37)$$

7.4.3. The UNIQUAC Equations

The UNIQUAC (Universal Quasi-Chemical) equations, were developed by Abrams and Prausnitz [17] and are based on a semi-theoretical approach to the mixture problem that includes a local composition model. It was also recognised that the non-ideality of liquid mixtures has contributions not only from specific interactions but also from the differences in the size and shape of the molecules. Consequently, in the UNIQUAC model, the Gibbs free energy of the mixture is correlated by the sum of two separate terms,

$$G_m^E/RT = (\Delta G_m^c + \Delta G_m^r)/RT \qquad (7.38)$$

which comprise:
- a contribution ΔG_m^c, known as the *configurational term*, due to differences in sizes and shapes; and
- a contribution ΔG_m^r, known as the *residual term*, due to energetic interactions between the molecules.

Consequently, the logarithm of the activity coefficient for each component also has configurational and residual contributions:

$$\ln\gamma_i = \ln\gamma_i^c + \ln\gamma_i^r . \qquad (7.39)$$

The original formulation of the free energy terms was modified by Andersen *et al.* [18] who arrived at the following expressions for a binary mixture:

$$\Delta G_m^c / RT = x_1 \left[\ln\left(\frac{\varphi_1}{x_1}\right) + \frac{1}{2} q_1 z \ln\left(\frac{\theta_1}{\varphi_1}\right) \right] + x_2 \left[\ln\left(\frac{\varphi_2}{x_2}\right) + \frac{1}{2} q_2 z \ln\left(\frac{\theta_2}{\varphi_2}\right) \right] \Bigg\}$$

$$\Delta G_m^r / RT = -x_1 q_1' \ln(\theta_1' + \theta_2' \tau_{21}) - x_2 q_2' \ln(\theta_1' \tau_{12} + \theta_2')$$

(7.40)

The quantities φ_i, θ_i and θ_i' which appear here are pure-substance parameters given by

$$\varphi_i = \frac{x_i r_i}{x_1 r_1 + x_2 r_2}, \quad \theta_i = \frac{x_i q_i}{x_1 q_1 + x_2 q_2}, \quad \text{and} \quad \theta_i' = \frac{x_i q_i'}{x_1 q_1' + x_2 q_2'}, \quad (7.41)$$

where r_i is determined by the volume and q_i by the surface area of the molecule both of which may be deduced in principle from crystallographic measurements on the pure solid. These area and volume parameters are however usually obtained by a group contribution method and tables exist [19, 20] which permit values to be determined for a very large selection of molecules. Usually, $q_i' = q_i$ so that $\theta_i' = \theta_i$ but for water and alcohols modifications to q_i' have been made to improve agreement with experimental results. z is a co-ordination number to which the value 10 is usually, but not necessarily, assigned. Finally, the model contains two adjustable binary energy parameter u_{12} and u_{21} that enter the residual part of the excess Gibbs free energy through the quantities τ_{12} and τ_{21} which are defined by

$$\tau_{ij} = \exp(-u_{ij} / RT) . \qquad (7.42)$$

The contributions to the activity coefficient of component i which result from this formulation are:

$$\ln\gamma_i^c = \ln\left(\frac{\varphi_i}{x_i}\right) + \frac{z}{2} \ln\left(\frac{\theta_i}{\varphi_i}\right) + \varphi_j (l_i - r_i l_j / r_j)$$

$$\ln\gamma_i^r = -q_i' \ln(\theta_i' + \theta_j' \tau_{ji}) + \theta_j' q_i' \left(\frac{\tau_{ji}}{\theta_i' + \theta_j' \tau_{ji}} - \frac{\tau_{ij}}{\theta_i' \tau_{ij} + \theta_j'} \right)$$

(7.43)

where

$$l_i = \frac{z}{2}(r_i - q_i) - (r_i - 1) . \qquad (7.44)$$

The corresponding results for a multi-component system are given in Table 7.5. Table A8 of Appendix A gives values of the parameters r_i, q_i and q'_i for a selection of compounds, while Table A7 gives some values of the binary parameter.

The UNIQUAC model is one of the most successful two-parameter models in use today and extensive tabulations of the pure-component and binary parameters are available [10]. Only pure-component and binary parameters are required for application to multi-component systems and the model is useful in both VLE and LLE problems.

In Section 7.8.4, a computer program will be described for the calculation of activity coefficients for multi-component systems from the UNIQUAC model. Example 7.8 illustrates the application of the model.

Table 7.5   UNIQUAC Equations [18]

$$\ln\gamma_i = \ln\gamma_i^c + \ln\gamma_i^r \qquad (7.45)$$

where

$$\ln\gamma_i^c = \ln\left(\frac{\varphi_i}{x_i}\right) + \frac{z}{2}\ln\left(\frac{\theta_i}{\varphi_i}\right) + l_i - \frac{\varphi_i}{x_i}\sum_{j=1}^{n}x_j l_j \qquad (7.46)$$

$$\ln\gamma_i^r = q'_i\left[1 - \ln\left(\sum_{k=1}^{n}\theta'_k\tau_{ki}\right) - \sum_{j=1}^{n}\frac{\theta'_j\tau_{ij}}{\sum_{k=1}^{n}\theta'_k\tau_{kj}}\right] \qquad (7.47)$$

and $\tau_{ij} = \exp\left(-\frac{u_{ij}}{RT}\right)$, $\tau_{ii} = \tau_{jj} = 1$, $l_i = \frac{z}{2}(r_i - q_i) - (r_i - 1)$,   and $z = 10$ (7.48)

$$\varphi_i = \frac{x_i r_i}{\sum_{j=1}^{n}x_j r_j} \qquad \theta_i = \frac{x_i q_i}{\sum_{j=1}^{n}x_j q_j} \qquad \theta'_i = \frac{x_i q'_i}{\sum_{j=1}^{n}x_j q'_j} \qquad (7.49)$$

Note: $q'_i = q_i$ except in the cases of water and alcohols.

7.4.4. Determination of Model Parameters from Experimental Activity Coefficients

The most reliable procedure for the determination of parameters in an activity-coefficient model involves a fit to experimental data over a range of liquid compositions. The experimental quantities required to determine activity coefficients were outlined in Section 7.2. The solution of the model for the parameters which best represent the data is a standard problem in non-linear regression analysis which may be solved rather easily with modern computer software. When the experimental data cover only a limited composition range, problems with multiple roots, similar to those discussed below in connection with the use of infinite-dilution activity coefficients, may arise and care should be taken in these cases. If only a single composition is used, the problem of finding the parameters reduces to the solution of a pair of non-linear simultaneous equations. Computer subroutines for that purpose are given in Section 7.8.2. Again, the possibility of multiple roots should be considered.

7.4.5. Determination of Model Parameters from Infinite-Dilution Activity Coefficients

When there is insufficient experimental data for the evaluation of model parameters, and new experimental studies are not feasible, one method of estimating the parameters is by analysis of infinite-dilution activity coefficients. These may have been measured for the binary systems of interest and, if they have not, group-contribution estimation schemes are available as a last resort.

The activity coefficient of component i in the limit of infinite dilution ($x_i \to 0$) is denoted γ_i^∞ and is a function of temperature only at constant pressure. The expressions for the activity coefficients in terms of model parameters become rather simple in the limit of infinite dilution, allowing solutions for the parameters to be obtained easily. Those for the models considered previously are given in Table 7.6.

Both parameters of a two-parameter model may be evaluated from the two infinite-dilution activity coefficients. In the case of a three-parameter model, one of the parameters must be constrained to a 'typical' value unless activity coefficients are available over a range of compositions. For example, in the NRTL model one might constrain α to be 0.2.

In solving the equations of Table 7.6, the situation is not as simple as it looks. Although one has two equations with two unknowns, the non-linearity of the

Table 7.6 Binary Activity Coefficient Correlations at Infinite Dilution

---

Wilson model

$$1 - \ln\gamma_1^\infty = \ln\Lambda_{12} + \Lambda_{21}$$
$$1 - \ln\gamma_2^\infty = \Lambda_{12} + \ln\Lambda_{21} \quad\Bigg\}$$

(7.50)

T-K-Wilson model

$$\ln\left(\frac{V_2}{V_1\gamma_1^\infty}\right) + \frac{V_1}{V_2} = \Lambda_{21} + \ln\Lambda_{12}$$

$$\ln\left(\frac{V_1}{V_2\gamma_2^\infty}\right) + \frac{V_2}{V_1} = \Lambda_{12} + \ln\Lambda_{21} \quad\Bigg\}$$

(7.51)

NRTL model

$$\ln\gamma_1^\infty = \tau_{21} + \tau_{12}\exp(-a_{12}\tau_{12})$$
$$\ln\gamma_2^\infty = \tau_{12} + \tau_{21}\exp(-a_{12}\tau_{21}) \quad\Bigg\}$$

(7.52)

UNIQUAC model

$$\ln\gamma_1^\infty = \ln\left(\frac{r_1}{r_2}\right) + \frac{z}{2}\ln\left(\frac{q_1 r_2}{q_2 r_1}\right) + l_1 - \frac{r_1}{r_2}l_2 - q_1'\ln\tau_{21} + q_1'(1 - \tau_{12})$$

$$\ln\gamma_2^\infty = \ln\left(\frac{r_2}{r_1}\right) + \frac{z}{2}\ln\left(\frac{q_2 r_1}{q_1 r_2}\right) + l_2 - \frac{r_2}{r_1}l_1 - q_2'\ln\tau_{12} + q_2'(1 - \tau_{21}) \quad\Bigg\}$$

(7.53)

---

equations is such that multiple roots can exist. This may be demonstrated for the Wilson model if Eqs.(7.50) are rewritten for one of the parameters, for example

$$\Lambda_{12} = \frac{1}{\gamma_1^\infty}\exp\left[1 - \frac{1}{\gamma_2^\infty}\exp(1 - \Lambda_{12})\right]$$

(7.54)

and plotted as a function of the other parameter for various values of, say, γ_2^∞. Such a plot is shown in Figure 7.4 with emphasis in the region where multiple roots exist. Clearly when γ_2^∞ exceeds unity, only one set of parameters is obtained, but as many as three sets may be obtained when either or both γ_i^∞ are less than unity. In order to

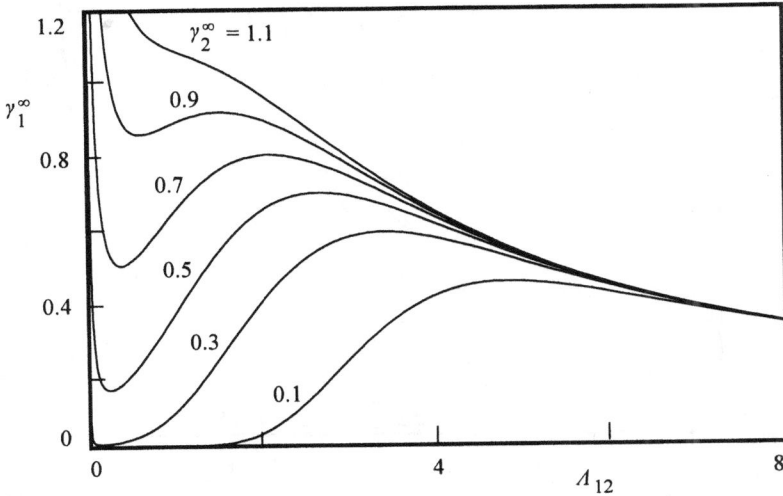

Figure 7.4 Plot of the infinite-dilution Wilson equations

be sure that one has the 'correct' solution, one should have results over a range of finite mole fractions with which to compare the predictions of the equation for each set of root. When only the infinite-dilution activity coefficients are available one can only note that the Λ_{ij}'s must be positive, that activity coefficients are usually single-valued functions of mole fraction and that, according to Silverman and Tassios [12], the correct set of parameters is usually the one for which the sum of $|\lambda_{12}|+|\lambda_{21}|$ is a minimum.

An equivalent plot for the NRTL equations with $\alpha = 0.2$ is shown in Figure 7.5. For positive values of α, three roots exist when an infinite-dilution activity coefficient is less than unity, and only one root otherwise. When α is negative, the reverse situation prevails. Where multiple root exist, Tassios [13] suggests that the physically correct solution is the one for which $|\tau_{12}|+|\tau_{21}|$ is a minimum. This proposition will also be tested in Example 7.3.

When experimental values of the infinite dilution activity coefficients are unavailable, estimation schemes may be used with due caution. Various methods have been proposed including that of Pierotti *et al.* [14] for polar mixtures, based on the structure of the two molecules, and the scheme of Helpinstill and van Winkle [15] which is an extension of the Scatchard and Hildebrand equations applied to polar systems. More recently, Thomas and Eckert [16] proposed the MOSCED

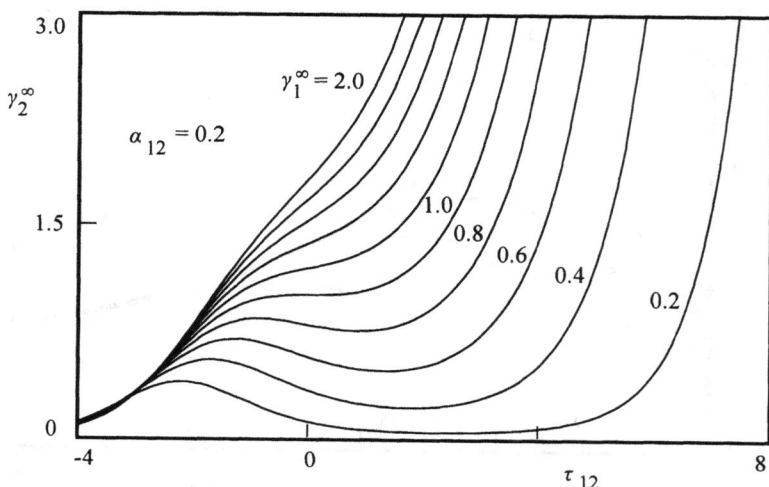

Figure 7.5 Plot of the infinite-dilution NRTL equations (with $\alpha = 0.2$)

(modified separation of cohesive energy density) model for predicting infinite-dilution activity coefficients from pure component parameters only. Here we discuss only the simple correlation devised by Pierotti *et al.* [14] which has been found to be reasonably accurate.

In Table 7.7, the correlation of the infinite-dilution activity coefficients due to Pierotti *et al.* [13] is given; the method is stated by the authors to be accurate to within ±8 per cent and, while this is probably better than nothing, one must proceed with some caution when attempting to predict phase-equilibrium conditions using model parameters based on such values. The correlation is valid for molecules containing a polar group bonded to two or three alkyl chains and expresses $\log_{10}\gamma_1^{\infty}$ as the sum of group contributions.

The following definitions are used:
1. n_1 (n_2) is the number of carbon atoms in molecule 1 (molecule 2).
2. n', n'', and n''' are the numbers of carbon atoms in the respective branches counted from the polar group on molcule 1. These are illustrated in Figure 7.6 for two examples. When three branches are present, as in the case of *tert*-alcohols, all three primed parameters are included and Eq.(7.56) is used. When only two branches are present, as in the case of ketones, only the first two primed parameters are included and Eq. (7.55) applies. Use of the scheme is illustrated in Example 7.2.

Table 7.7 Correlations for Infinite-Dilution Activity Coefficients, Pierotti *et al* [14]

$$\log_{10}\gamma_1^\infty = A + Bn_1 + C\frac{1}{n_1} + D\frac{1}{n_2} + E\left(\frac{1}{n'} + \frac{1}{n''}\right) \qquad (7.55)$$

	T/K	A	B	C	D	E
n-alcohol(1) + water(2)	298.15	-0.995	0.622	0.558	-	-
	333.15	-0.755	0.583	0.460	-	-
	373.15	-0.420	0.517	0.230	-	-
water(1) + *n*-alcohol(2)	298.15	0.760	-	-	-0.630	-
	333.15	0.680	-	-	-0.440	-
	373.15	0.617	-	-	-0.280	-
sec-alcohol(1) + water(2)	298.15	-1.220	0.622	-	-	0.170
	333.15	-1.023	0.583	-	-	0.252
	373.15	-0.870	0.517	-	-	0.400
water(1) + *sec*-alcohol(2)	353.15	1.208	-	-	-	-0.690
ketone(1) + water(2)	298.15	-1.475	0.622	-	-	0.500
	333.15	-1.040	0.583	-	-	0.330
	373.15	-0.621	0.517	-	-	0.200
water(1) + ketone(2)	289.15	1.857	-	-	-	-1.019
	333.15	1.493	-	-	-	-0.730
	373.15	1.231	-	-	-	-0.557

$$\log_{10}\gamma_1^\infty = A + Bn_1 + C\frac{n_1}{n_2} + D(n_1 - n_2)^2 + E\left(\frac{1}{n'} + \frac{1}{n''}\right) + F\left(\frac{1}{n'''} - 3\right) \qquad (7.56)$$

	T/K	A	B	C	$10^5 D$	E	F
alcohol(1) + alkane(2)	298.15	1.960	-	-	-49	0.475	0.475
	333.15	1.460	-	-	-57	0.390	0.390
	373.15	1.070	-	-	-61	0.340	0.340
alkane(1) + ethanol(2)	298.15	0.570	0.088	-	-49	-	-
	323.15	0.580	0.073	-	-55	-	-
	363.15	0.610	0.059	-	-61	-	-
ketone(1) + alkane(2)	298.15	0.088	-	-	-49	0.757	-
	333.15	0.016	-	-	-57	0.680	-
	373.15	-0.067	-	-	-61	0.605	-
alkane(1) + ketone(2)	298.15	-	-	0.1821	-49	0.402	-
	333.15	-	-	0.1145	-57	0.402	-
	363.15	-	-	0.0746	-61	0.402	-

Note: n', n'' and n''' refer to the alcohol or the ketone molecule 1 accordingly.

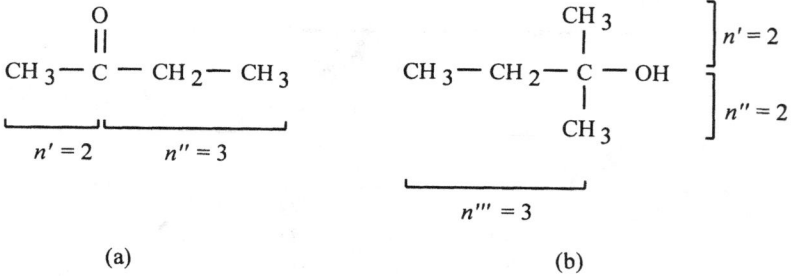

Figure 7.6   Examples of evaluation of n', n'', and n'''

EXAMPLE 7.2	*Calculation of the Infinite-Dilution Activity Coefficients of:*

(1) 1-propanol+water at T = 333.15 K,
(2) ethanol+n-hexane at T = 298.15 K, and
(3) methyl ethyl ketone+n-hexane at T = 333.15 K.

For all the calculations the equations of Table 7.7 will be employed.

1. a) 1-propanol(1) + water(2) at $T = 333.15$ K.
       Parameters: $n_1=3$ $\gamma_1^\infty = 14.04$
    b) water(1) + 1-propanol(2) at $T = 333.15$ K.
       Parameters: $n_2=3$ $\gamma_1^\infty = 3.41$

2. a) ethanol(1) + n-hexane(2) at $T = 298.15$ K.
       Parameters: $n_1=2$, $n_2=6$, $n'=2$, $n''=n'''=1$ $\gamma_1^\infty = 51.84$
    b) n-hexane(1) + ethanol(2) at $T = 298.15$ K.
       Parameters: $n_1=6$, $n_2=2$ $\gamma_1^\infty = 12.31$

3. a) methyl ethyl ketone(1) + n-hexane(2) at $T = 333.15$ K.
       Parameters: $n_1=4$, $n_2=6$, $n'=2$, $n''=3$ $\gamma_1^\infty = 3.81$
    b) n-hexane(1) + methyl ethyl ketone(2) at $T = 333.15$ K.
       Parameters: $n_1=6$, $n_2=2$, $n'=2$, $n''=2$, $n'''=3$ $\gamma_1^\infty = 4.67$

❑

| **EXAMPLE 7.3** | *Calculation of the Activity Coefficients of the Mixture (0.8 n-Hexane + 0.2 Diethyl Ketone) at T = 298.15 K from the NRTL model with (a) α = -1 and (b) α = 0.2.* |

For the mixture n-hexane(1) + diethyl ketone(2), NRTL parameters are not tabulated and are estimated here from infinite-dilution activity coefficients predicted by the method of Pierotti *et al.*

(a) Infinite-dilution activity coefficients
From the expressions in Table 7.7: for n-hexane $n = 6$, while for diethyl ketone $n = 5$ and $n' = n'' = 3$. Thus, at $T = 298.15$ K, we obtain $\gamma_1^\infty = 2.54$ and $\gamma_2^\infty = 3.91$.

(b) NRTL parameters τ_{12}, τ_{21} with α = -1.
For the case α = -1, Eqs.(7.52) can be solved but, as α is negative and the infinite-dilution activity coefficients are greater than unity, multiple roots are to be expected. Indeed in the range $-5 \leq \tau_{12} \leq +5$, the following three roots are found:

Sol.	τ_{12}	τ_{21}	$|\tau_{12}| + |\tau_{21}|$
(1)	-2.330	1.159	3.489
(2)	0.253	0.606	0.859
(3)	1.403	-4.776	6.179

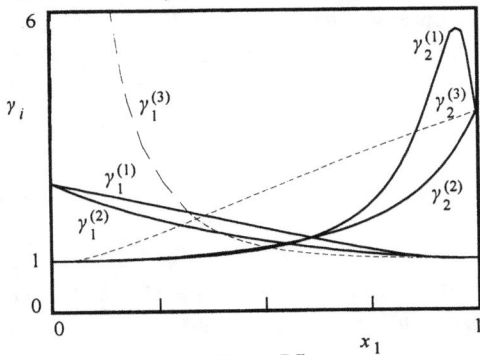

Figure 7.7

According to the Tassios criterion, the physically correct solution is the one for which $|\tau_{12}| + |\tau_{21}|$ is minimum: solution (2). To further illustrate the point, activity coefficients calculated[1] with each set of parameters are plotted as a function of composition in Figure 7.7. It can be seen that, while solution (2) gives reasonable results, solutions (1) and (3) give unexpected behaviour. Thus adopting solution (2), we find at $x_1 = 0.8$, that $\gamma_1 = 1.071$ and $\gamma_2 = 2.013$.

(c) NRTL parameters τ_{12}, τ_{21} when α = 0.2.
In this case only one solution is found with $\tau_{12} = 1.6070$ and $\tau_{21} = -0.2331$. The activity coefficients calculated with these parameters at $x_1 = 0.8$, are $\gamma_1 = 1.069$ and $\gamma_2 = 2.069$.

---

[1] See Example 7.7 for the computer program used to perform this calculation.

Remarkably, the two predictions are not far apart, although both are subject to uncertainties arising from the estimation of the infinite-dilution activity coefficients. To examine such an influence, the value of γ_1^∞ in case (a) was arbitrary increased by 10 per cent. Again solution (2) was selected and the resulting values of the activity coefficients at $x_1 = 0.8$ were, $\gamma_1 = 1.068$ and $\gamma_2 = 2.072$; there are still quite close to those obtained above.

❑

7.5. Group-Contribution Methods

The activity-coefficient models presented in the previous sections are widely used but in some cases it is difficult or impossible to obtain the necessary binary parameters. In such cases, group contribution methods may be used as the method of last resort. Although such methods often provide reasonable results, and are being continuously improved, they are not based on any sound theory and in some cases they predict entirely incorrect phase behaviour. Amongst group-contribution methods, the UNIFAC scheme is the most popular and this will be described along with the earlier ASOG model.

7.5.1. ASOG Model

The ASOG (Analytical Solution of Groups) model was proposed by Wilson and Deal [21] and Wilson [22] and has been further developed into a working tool by Kojima and Toshigi [23].

The ASOG model is based on a treatment of the excess Gibbs free energy which is in many ways similar to that of the UNIQUAC model. The configurational part is obtained by a fairly-simple group contribution method while the residual part is represented by a Wilson-type equation. Binary parameters involved in the latter are represented within the ASOG model by group-contribution rules which were developed in a fit to a large database of experimental activity coefficients for many different binary systems. The necessary constants have been tabulated by Kojima and Toshigi [23]. The ASOG model is restricted to temperatures near ambient and has been largely superseded by the UNIFAC method which will be described next.

7.5.2. UNIFAC Method

The UNIFAC (UNIQUAC Functional Group Activity Coefficients) method was developed largely by Fredenslund *et al.* [24,254] and has been updated almost continuously since [19]. Its prime use is as a prediction tool for the activity coefficients in multi-component systems where there is insufficient data to apply a better-founded method. Application of the model is restricted to the temperature range 273 to 373 K.

The model is based on the UNIQUAC equations with all pure-substance and binary parameters determined by empirical group-contribution methods. Since the equations employed are relatively complicated and the tables of parameters very large, its application is usually restricted to large commercial phase-equilibrium and flow-sheeting packages.

A full description of the UNIFAC model is beyond the scope of this book. For an introduction to the UNIFAC equations tables of parameters, the reader is referred to Walas [8], while a full description is given by Fredenslund [25]. Revised versions of the method appear frequently in the chemical engineering literature.

7.6. Supercritical Components

Activity coefficient models were originally developed for applications in which the temperature is well below the critical temperature of every component. Nevertheless, problems frequently arise in which it is desirable to model a mixed liquid phase by one of the methods described in this chapter but where the temperature is above the critical temperature of one or more of the components. This raises several difficulties including the definition of the standard state and the question of how best to calculate Poynting factors. Two approaches may be adopted in these cases. The first is to make some alternative definition of the standard state for some or all of the components, while the second involves separate treatment of the supercritical components in accordance with Henry's law. The latter appears to be the more logical approach for systems containing components of widely differing volatility although one then requires methods for correlating and predicting Henry's constant.

7.6.1. Alternative Standard States

The standard state adopted elsewhere in this chapter for liquid-phase components is that of the pure liquid at a standard pressure $P°$ equal to the saturated vapour pressure at the given temperature. Clearly, this cannot be applied to any component with a critical temperature less than the temperature of interest. Several methods may be adopted to circumvent this problem the simplest and most brutal of which is to define the standard pressure $P°$ by an extrapolation of the vapour pressure curve to the temperature of interest. This might be achieved, for example, by extrapolating the Antoine equation to supercritical temperatures and, provided that some precise definition is adopted for each component, it is at least possible to proceed with the calculations.

The fugacity of the pure supercritical fluid at the standard pressure may be evaluated in the normal way from experimental PVT data or from an equation of state valid at all pressures up to $P°$. It should however be noted that, at temperatures far in excess of the critical, the corresponding standard pressure will be very large and the standard-state fugacity may then be the subject of a large uncertainty. Furthermore, the calculation of Poynting factors presents serious difficulties when the pressure of interest differs greatly from the standard pressure.

This approach really has only one merit; namely, that it permits calculations to proceed by means of a 'normal' activity-coefficient model despite the presence of supercritical components. It should not be regarded as providing better than a guess at the true fugacity for supercritical components although there may, of course, be situations in which this is deemed to be acceptable.

7.6.2. Application of Henry's Law

When faced with the presence of supercritical components dissolved in an otherwise normal mixed liquid phase, a more logical approach is to treat those components separately by means of a generalisation of Henry's law. The fugacity of a component treated in this way is written

$$f_i = x_i \, \mathcal{H}_i \, \mathcal{F}_i \qquad (7.57)$$

where $\mathcal{H}_i$ is Henry's constant for the component at the standard pressure $P°$ and $\mathcal{F}_i$ is a Poynting factor calculated from either Eq.(7.6) or, more usually, from Eq.(7.7) with P_i^σ replaced by $P°$. The standard pressure for supercritical components is usually chosen as either zero or as 0.1 MPa (either is acceptable as long as it is used consistently). Henry's constant may be determined from experimental VLE data in much the same way as activity coefficients but, unfortunately, there have been relatively few determinations except at temperature close to ambient. A number of models have been proposed for the correlation and prediction of Henry's constant in pure and mixed solvents and the interested reader is referred to the literature [29].

7.7. Incorporation of Activity-Coefficient Models in Equations of State

In Chapter 6, we saw that an equation of state may be used to calculate the residual properties, including partial fugacity coefficients, for both gaseous and liquid phases. One limitation of equations of state with 'conventional' mixing rules is that they cannot model accurately the properties of mixtures that contain strongly associating substances and/or highly polar compounds. This limitation can be largely overcome by the adoption of more suitable mixing rules for the parameters in the equation of state.

A number of approaches to the problem have been proposed in which attempts are made to incorporate an activity coefficient model in the mixing rules used with a cubic equation of state [30]. The objective of these methods is to determine one or both of the parameters a and b in the cubic equation of state for the mixture by equating at a particular pressure the predicted excess Gibbs free energy G_m^E with that given by an activity-coefficient model. The equation of state would then give component fugacities that approach ideal-gas behaviour in the dilute vapour phase but agree with the activity coefficient model in the liquid phase at the chosen pressure. If that chosen pressure is suitable then it should be possible to exploit the existing database of activity-coefficient-model parameters. As this subject is still an active field of research, we shall restrict our discussion to an outline of the main principles involved; examples illustrating the success of the methods may be found in the literature.

7.7.1. Excess Functions From a Cubic Equation of State

The basis of all current methods of implanting an activity-coefficient model in an equation of state is an expression for the excess Helmholtz free energy A_m^E of the mixture at temperature T and pressure P. This quantity may be properly defined as the difference between the molar Helmholtz free energy of the mixture and that of an ideal mixture having the same temperature, pressure and composition. Since the independent variables in an equation of state are usually $(T, V_m, x_1, x_2 \cdots x_n)$, rather then $(T, P, x_1, x_2 \cdots x_n)$, we write the excess Helmholtz free energy in the form

$$A_m^E = A_m(T, V_m, x_1, x_2 \cdots x_n) - \sum_i x_i \left[A_i^*(T, V_i^*) + RT \ln x_i \right], \qquad (7.58)$$

where A_m is the molar Helmholtz free energy of the mixture at $(T, V_m, x_1, x_2 \cdots x_n)$ and A_i^* is the molar Helmholtz free energy of a pure component i at the same temperature but at molar volume V_i^*. In order that this expression be consistent with the definition of the excess Helmholtz free energy of mixing at *constant pressure*, the molar volumes V_m and V_i^* must be those obtained by solving the equation of state (for the mixture and for each pure component) at pressure P. The excess Gibbs free energy of mixing is then obtained by adding to A_m^E the product of the pressure and the excess molar volume of mixing V_m^E at $(T, P, x_1, x_2 \ldots x_n)$:

$$G_m^E = A_m^E + PV_m^E = A_m^E + P\left(V_m - \sum_i x_i V_i^* \right). \qquad (7.59)$$

Starting from Eq.(6.34), the generic cubic equation of state, it is easy to show that the Helmholtz free energy at temperature T and molar volume V_m is given by

$$A_m/RT = -\ln\left(\frac{V_m}{V_m - b} \right) - \frac{(a/bRT)}{c_1 - c_2} \ln\left(\frac{V_m + c_1 b}{V_m + c_2 b} \right) + A_m^{pg}/RT \qquad (7.60)$$

and hence that G_m^E is given by

$$G_m^E/RT = \sum_i x_i \ln\left(\frac{V_i^* - b_i}{V_m - b} \right) - \left(\frac{a/bRT}{c_1 - c_2} \right) \ln\left(\frac{V_m + c_1 b}{V_m + c_2 b} \right)$$

$$+ \sum_i x_i \left(\frac{a_i/b_i RT}{c_1 - c_2} \right) \ln\left(\frac{V_i^* + c_1 b_i}{V_i^* + c_2 b_i} \right) + \left(\frac{PV_m^E}{RT} \right). \qquad (7.61)$$

Here, the equation-of-state parameters a and b refer to the mixture, while a_i and b_i refer to the pure component i.

7.7.2. *The Huron-Vidal Method*

To proceed further, it is necessary to solve the equation of state for the molar volumes that appear in Eqs.(7.61). To simplify the problem, Huron and Vidal [31] chose the special case $P \to \infty$ which corresponds to setting $V_m = b$ and $V_i^* = b_i$ and leads to

$$G_m^E / RT = \left(-\alpha + \sum_i x_i \alpha_i \right) \left(\frac{1}{c_1 - c_2} \right) \ln \left(\frac{1 + c_1}{1 + c_2} \right) + \left(\frac{P V_m^E}{RT} \right), \qquad (7.62)$$

where $\alpha = (a/RTb)$ and $\alpha_i = (a_i/RTb_i)$. It was further assumed that $b = \Sigma_i x_i b_i$ so that $V_m^E \to 0$ and, it was argued, $P V_m^E \to 0$ as $P \to \infty$. Then, inserting a model expression for G_m^E and setting $P V_m^E = 0$, Eq.(7.62) may be solved to obtain the mixture parameter α, and hence a, as a function of composition.

The Huron and Vidal method is successful in combination with a model such as the NRTL equation but only when the parameters of the latter are refitted to binary VLE data. Wong and Sandler [32] have followed a similar approach but with an alternative mixing rule for b chosen such that the correct composition dependence of the mixture second virial coefficient is obtained. Meanwhile, several schemes have been developed based on the adoption of a low-pressure state at which to match the equation of state with an activity coefficient model.

7.7.3. *Zero-Pressure Methods*

It is possible to adopt any suitable reference pressure, insert a model expression for G_m^E which is valid at that pressure and solve Eq.(7.61) for either a or b with V_m^E evaluated from the liquid roots of the equation of state at the given temperature and pressure. Typically, one assumes the linear mixing rule $b = \Sigma_i x_i b_i$ and solves for a. Difficulties can arise if the reference pressure is below the bubble pressure of the mixture and/or below the saturation pressure of any component because thermodynamically stable liquid roots are absent. Otherwise, this approach satisfies the objectives of matching the equation of state to a model for G_m^E in the region of the phase diagram where the latter was optimised.

Several investigators have adopted a reference pressure of zero so that $P V_m^E$ again vanishes and one can show that the procedure reduces to the solution of the following non-linear equation [33]:

$$q(\alpha) = (G_m^E/RT) + \sum_i x_i \left[q(\alpha_i) - \ln(b_i/b) \right]. \qquad (7.63)$$

Here, the function $q(\alpha)$ is defined by

$$q(\alpha) = -1 - \ln\left(\frac{1-u}{u}\right) - \left(\frac{\alpha}{c_1 - c_2}\right) \ln\left(\frac{1 + c_1 u}{1 + c_1 u}\right) \qquad (7.64)$$

where $u = b/V_m^{(L)}$ and $V_m^{(L)}$ is the liquid root of the equation of state at zero pressure. The same formula is applied to determine $q(\alpha_i)$ from the values of α_i and u_i for each pure component. Once a model for G_m^E is adopted, and a combining rule for b is assumed, all of the terms on the right-hand side of Eq.(7.63) are known and it only remains to solve Eq.(7.64) for a. This may be achieved by noting that, since solution of the equation $P(V_m) = 0$ gives

$$\alpha = \frac{(1 + c_1 u)(1 + c_2 u)}{u(1 - u)}, \qquad (7.65)$$

Eq.(7.64) may be solved instead for u. α may then be recovered from Eq.(7.65).

Low-pressure VLE predictions based on the resulting equation-of-state models have been found to agree almost exactly with those of the corresponding activity-coefficient model [33,34]. However, the method is limited to the temperature range within which the activity-coefficient model is valid. A more fundamental difficulty arises from the fact that the liquid root of the equation of state at zero pressure is at best a metastable one and, below a certain value of α (i.e. above a certain temperature), the equation $P(V_m) = 0$ has no real roots at all.

Several procedures have been proposed both to simplify the application of the method and extend its range to temperatures above those at which the zero-pressure equation can be solved. We mention here only the MHV2 method of Dahl and Michelsen [34] in which Eq.(7.64) is replaced by the polynomial

$$q(\alpha) = q_0 + q_1\alpha + q_2\alpha^2 . \qquad (7.66)$$

The coefficients in this polynomial have been determined by fitting the results obtained from Eq.(7.64) in the range of its validity. The constant term q_0 cancels out when Eqs.(7.65) and (7.63) are combined; but for the other coefficients, Dahl and Michelsen suggest $q_1 = -0.478$ and $q_2 = -0.0047$ for RKS-type equations ($c_1 = 1$, $c_2 = 0$), while Huang and Sandler [35] propose $q_1 = -0.4347$ and $q_2 = -0.003654$ for PR-type equations ($c_1 = c_2 = 1 \pm \sqrt{2}$).

7.7.4. *Discussion*

Although numerical examples of the application of these methods are beyond the scope of this chapter, both the infinite-pressure method, based on Eq.(7.62), and the zero-pressure method, based on Eq.(7.63), may be applied with any of the activity-coefficient models described in this chapter. Expressions for the partial fugacity coefficients of components in the mixture may be derived [33]; however, they are more complicated than those given in Chapter 6 because of the modified composition dependence of a implied by the usual models for G_m^E. The zero-pressure methods do not generally require refitting of the parameters in the activity-coefficient/G_m^E model and may even be used in combination with UNIFAC. However, the strict implementation of the method involves some computational effort and it has a restricted range of applicability. On the other hand, the MHV2 approximation is almost as accurate, may be applied more easily and has solutions at any temperature.

7.8. Summary

Activity-coefficient models are applied in the calculation of liquid-phase fugacities as a part of phase-equilibrium calculations for systems that cannot be adequately treated using an equation of state for both phases. In this chapter, four simple activity-coefficients models (Wilson, T-K-Wilson, NRTL and UNIQUAC) were presented in detail. The ASOG and UNIFAC group-contribution models were also outlined as methods of last resort. All of the model discussed correlate low-pressure activity coefficients as a function of temperature and composition. If activity coefficients or fugacities are required at elevated pressures then inclusion of the Poynting factor is essential. All methods are restricted to a fairly narrow temperature range because the dependence of the excess Gibbs free energy is not accurately modelled in any case

In comparing the methods, one should distinguish between correlation and prediction of activity coefficients. Most models offer satisfactory correlations of activity coefficients for binary systems and many permit extrapolations with respect to composition and temperature. The most useful models will permit reasonable predictions to be made for multi-component systems. All of the models described in this chapter purport to fall into that category.

Examining the Wilson, T-K-Wilson, NRTL and UNIQUAC models, the following points can be made:

- Although parameters for the Wilson model are easily found, its inability to describe liquid-liquid equilibrium restricts its use.
- Both T-K-Wilson and the NRTL models are simple to use and can be employed for both vapour-liquid and liquid-liquid equilibrium calculations. The simplicity of these models makes them most suitable for repetitive calculations such as those involved in simulating a distillation column. More parameters are available for the NRTL model but, where parameters must be estimated, the greater simplicity of the two-parameter T-K-Wilson model is advantageous.
- The UNIQUAC model is a little more complicated to apply but gives good results for both VLE and LLE when the binary parameters are optimised using accurate experimental results on binary systems.

In respect of the group contribution methods, the following points can be made

- UNIFAC is the most widely used and it can be applied in principle to almost any system without reference to experimental results.
- All such methods are highly empirical and, while they work in many cases, poor results will be obtained in some cases.

7.9. Computational Implementation

In this section, computer routines for the calculation of the activity coefficients of a multi-component system from the Wilson, T-K-Wilson, NRTL and UNIQUAC models are described. Routines are also provided for the determination of model parameters from activity-coefficient data.

7.9.1. Activity-Coefficient Models

Subroutine TKWILSON, shown in Display 7.1, calculates the activity coefficient of each component in a liquid mixture according to either the Wilson or T-K-Wilson equations. Parameter IOPT selects the model to be used: for the Wilson one IOPT=1, while for the T-K-Wilson model IOPT=2. Other inputs required are:
- IS, the number of components
- X(IS) a one-dimensional array of length IS containing the mole fractions of each component
- the temperature, T in Kelvin,
- array SLR(IS,IS) containing the model parameters λ_{ij}/R
- array V(IS,IS) containing the liquid molar volume ratios V_{ij}.

```
                                        DISPLAY 7.1  TKWILSON List
      SUBROUTINE TKWILSON(IS,X,T,SLR,V,G,IOPT)
C
C     The subroutine calculates activity coefficients G(I) for the
C     Wilson (IOPT=1) and the T-K-Wilson (IOPT=2) models, for a mixture
C     at temperature T [K], of IS components each of mole fraction X(I)
C
C     Other input required is
C     - SLR(I,J)  lamda(I,J)/R and
C     - V (I,J)   molar volume of J / molar volume of I
C
      DIMENSION X(IS),SLR(IS,IS),V(IS,IS),G(IS)
      DIMENSION AL(5,5),VV(5,5)
      DO 100 I=1,IS
      DO 100 J=1,IS
      AL(I,J)=V(J,I)*EXP(-SLR(I,J)/T)
      IF (IOPT.EQ.1) VV(J,I)=1.
100   IF (IOPT.EQ.2) VV(J,I)=V(J,I)
      DO 500 I=1,IS
      SXV=0.
      SXL=0.
      DO 200 J=1,IS
      SXV=SXV+X(J)*VV(I,J)
200   SXL=SXL+X(J)*AL(I,J)
      G11=LOG(SXV/SXL)
      SXK=0.
      DO 400 K=1,IS
      SXV=0.
      SXL=0.
      DO 300 J=1,IS
      SXV=SXV+X(J)*VV(K,J)
300   SXL=SXL+X(J)*AL(K,J)
400   SXK=SXK+X(K)*((VV(K,I)/SXV)-(AL(K,I)/SXL))
500   G(I)=EXP(G11+SXK)
      RETURN
      END
```

Display 7.2 shows subroutine NRTL which performs the same tasks as the previous routine but uses the NRTL model (Table 7.4). Inputs required by this subroutine are:
- IS, the number of components
- X(IS) a one-dimensional array of length IS containing the mole fractions of each component
- the temperature, T in Kelvin,
- array GR(IS,IS) containing the model parameters g_{ij}/R
- array ALJ(IS,IS) containing the model parameters α_{ij}.

```
                                                    DISPLAY 7.2  NRTL List
        SUBROUTINE NRTL(IS,X,T,GR,AIJ,G)
C
C       The subroutine calculates activity coefficients G(I) for the NRTL model,
C       for a mixture at temperature T [K], of IS components each of
C       mole fraction X(I).
C       Other input required is
C       - GR(I,J) : g(I,J)/R  and
C       - AIJ(I,J) : NRTL parameter a12 for binary mixture of I and J
C
        DIMENSION X(IS),GR(IS,IS),AIJ(IS,IS),G(IS)
        DIMENSION GC(5,5),TA(5,5)
        DO 100 I=1,IS
        DO 100 J=1,IS
        TA(I,J)=GR(I,J)/T
100     GC(I,J)=EXP(-AIJ(I,J)*TA(I,J))
        DO 500 I=1,IS
        STXG=0.
        SXG=0.
        DO 200 K=1,IS
        STXG=STXG+TA(K,I)*X(K)*GC(K,I)
200     SXG=SXG+X(K)*GC(K,I)
        G11=STXG/SXG
        S2PART=0.
        DO 400 J=1,IS
        SXG=0.
        SXTG=0
        DO 300 K=1,IS
        SXG=SXG+X(K)*GC(K,J)
300     SXTG=SXTG+X(K)*TA(K,J)*GC(K,J)
400     S2PART=S2PART+(X(J)*GC(I,J)/SXG)*(TA(I,J)-(SXTG/SXG))
500     G(I)=EXP(G11+S2PART)
        RETURN
        END
```

EXAMPLE 7.4   *Calculation of the Activity Coefficients of the Mixture (0.736 Methyl Acetate + 0.264 Methanol) at T = 400 K and P = 0.79 MPa.*

To demonstrate the use of subroutines TKWILSON and NRTL, the activity coefficients will be calculated from all three models. In the present case we use model parameters determined in Example 7.6 in a fit to the experimental data at the given temperature and pressure. Consequently, all three models give identical results in this example, although they would not do so if applied at another temperature. Since the Poynting factor is not included in the routines used, the effect of pressure is neglected here but should be included in general.

The parameters used in this example are as follows (1 = methyl acetate, 2 = methanol):

	λ_{12}/R	λ_{21}/R	g_{12}/R	g_{21}/R	α_{12}
(1) Wilson	760.70	-363.51			
(2) T-K-Wilson	723.27	-425.21			
(3) NRTL			-299.33	684.72	0.200

The liquid volume ratio, V_{12}, was estimated to be 0.5107 from the BWR(HS) equation of state. In Display 7.3, the calling program and its output are shown for all three cases; as expected, the results are identical.

DISPLAY 7.3   TKWILSON Sample Input/Output

```
PROGRAM EX0704
CHARACTER*10 FNAM(3)
PARAMETER (IS=2)
DIMENSION SLR(IS,IS),GR(IS,IS),V(IS,IS),AIJ(IS,IS),X(IS),G(IS)
DATA FNAM/'WILSON    ','T-K-WILSON','NRTL    '/
DATA SLR,GR/8*0./
DATA V(1,1),V(1,2),V(2,1),V(2,2)/1.,0.5107,1.9581,1./
DATA AIJ/4*0.2/
DATA X(1),X(2),T/0.736,0.264,400./
C........Wilson model (IOPT=1)
    SLR(1,2)= 760.70
    SLR(2,1)=-363.51
    CALL TKWILSON(IS,X,T,SLR,V,G,1)
    WRITE (*,1000) FNAM(1),G(1),G(2)
1000 FORMAT(2X,A10,5X,'G1 =',F8.4,5X,'G2 =',F8.4)
C........T-K-Wilson model (IOPT=2)
    SLR(1,2)= 723.27
    SLR(2,1)=-425.21
    CALL TKWILSON(IS,X,T,SLR,V,G,2)
    WRITE (*,1000) FNAM(2),G(1),G(2)
C........NRTL model
    GR(1,2)=-299.33
    GR(2,1)= 684.72
    CALL NRTL(IS,X,T,GR,AIJ,G)
    WRITE(*,1000) FNAM(3),G(1),G(2)
    END

WILSON        G1 = 1.0224    G2 = 1.3983
T-K-WILSON    G1 = 1.0224    G2 = 1.3983
NRTL          G1 = 1.0224    G2 = 1.3983
```

□

| EXAMPLE 7.5 | *Calculation of the the Activity Coefficients of the Ternary Mixture (0.30 Ethyl Acetate + 0.10 Ethanol + 0.60 Water) at T = 343.15 K and P = 0.1 MPa.* |

For this mixture, ethyl acetate(1) + ethanol(2) + water(3), NRTL parameters obtained by Renon and Prausnitz [6] are tabulated in Appendix A, Table A6. The calling program and its output are shown in Display 7.4. This example demonstrates simplicity of application; unfortunately, there are no experimental data with which to compare the results.

```
                                      DISPLAY 7.4   NRTL Sample Input/Output
      PROGRAM EX0705
      PARAMETER (IS=3)
      DIMENSION GR(IS,IS),AIJ(IS,IS),X(IS),G(IS)
      DATA GR(1,1),GR(1,2),GR(1,3)/     0.     162.04,   671.80/
      DATA GR(2,1),GR(2,2),GR(2,3)/   151.47,    0.      44.28/
      DATA GR(3,1),GR(3,2),GR(3,3)/  1263.07,  491.14,    0.  /
      DATA X,T/0.30,0.10,0.60,343.15/
      DATA AIJ/9*0.3/
      AIJ(1,3)=0.4
      AIJ(3,1)=0.4
      CALL NRTL(IS,X,T,GR,AIJ,G)
      WRITE(*,1000) G(1),G(2),G(3)
1000  FORMAT(5X,'G1 =',F6.3,/,5X,'G2 =',F6.3,/,5X,'G3 =',F6.3)
      END

  G1 = 2.530
  G2 = 1.411
  G3 = 1.626
```

□

7.9.2. *Parameters of Activity-Coefficient Models*

In principle, two parameters in an activity-coefficient model may be obtained from a pair of activity coefficients measured in a binary system at a single composition. Provided that due attention is paid to the possibility of multiple roots, the resulting parameters may permit useful predictions to be made over a range of compositions.

Here, two non-linear simultaneous equations must be solved for the two model parameters (in the case of the NRTL model, a value for parameter α is assumed) and the Newton-Raphson method is employed. For this method the partial derivatives of the two equations with respect to the unknown parameters are required.

FORTRAN subroutine WILSPAR, shown in Display 7.5, determines the parameters λ_{12}/R and λ_{21}/R parameters in the Wilson (IOPT = 1) or T-K-Wilson model (IOPT = 2). Additional inputs required by the subroutine are:

- temperature (T) in Kelvin and the mole fraction of component 1 (X1).
- activity coefficients at this composition for the 2 components (GAM1, GAM2)
- liquid molar volume ratio (V12).
- initial guesses for the parameters Λ_{12} and Λ_{21} (SL12, SL21).

```
                                             DISPLAY 7.5   WILSPAR List
        SUBROUTINE WILSPAR(T,X1,GAM1,GAM2,V12,SL12,SL21,SL12R,SL21R,IOPT)
C
C       The subroutine calculates Wilson (IOPT=1) and T-K-Wilson (IOPT=2) parameters
C         SLij=Vji*exp[-SLIJR/T]    for given activity coefficients GAM1, GAM2
C       at mole fraction X1 and temperature T [K].
C       - Guess starting values of parameters SL12, SL21 are required
C         (usually both at 0.5)
C
        X2=1.-X1
        IF (IOPT.EQ.1) V12E=1.
        IF (IOPT.EQ.2) V12E=V12
        V21E=1./V12E
C........Wilson & T-K-Wilson equations
    70  F1=LOG(GAM1)-LOG((X1+X2*V12E)/(X1+X2*SL12))
        F1=F1-X1*((1./(X1+X2*V12E))-(1./(X1+X2*SL12)))
        F1=F1-X2*((V21E/(X1+V21E+X2))-(SL21/(X1*SL21+X2)))
        F2=LOG(GAM2)-LOG((X1*V21E+X2)/(X1*SL21+X2))
        F2=F2-X1*((V12E/(X1+X2*V12E))-(SL12/(X1+X2*SL12)))
        F2=F2-X2*((1./(X1*V21E+X2))-(1./(X1*SL21+X2)))
C..........Derivatives
        F1L12=SL12*(X2/(X1+X2*SL12))**2
        F1L21=(X2/(X2+X1*SL21))**2
        F2L12=(X1/(X1+SL12*X2))**2
        F2L21=SL21*(X1/(X2+SL21*X1))**2
C..........Solution
        D=F2L12*F1L21-F1L12*F2L21
        H=(F1*F2L21-F2*F1L21)/D
        G=(F1L12*F2-F2L12*F1)/D
        SL12=SL12+H
        SL21=SL21+G
        CHK=ABS(H/SL12)+ABS(G/SL21)
        IF (CHK.GE.0.0001) GOTO 70
C..........Small greek lamda(I,J)/R
        SL12R=-T*LOG(SL12*V12)
        SL21R=-T*LOG(SL21/V12)
        RETURN
        END
```

In the case of the NRTL model, FORTRAN subroutine NRTLPAR, shown in Display 7.6, calculates the parameters g_{12}/R and g_{21}/R for a given value of α (A12). Additional inputs required by this subroutine are:

- temperature (T) in Kelvin and the mole fraction of component 1 (X1).
- activity coefficients at this composition for the 2 components (GAM1, GAM2)
- initial guesses for the parameters τ_{12} and τ_{21} (T12, T21).

```
                                              DISPLAY 7.6   NRTLPAR List
      SUBROUTINE NRTLPAR(T,X1,GAM1,GAM2,A12,T12,T21,G12R,G21R)
C
C     The subroutine calculates NRTL parameters G12R=g12/R and G21R=g21/R
C       Gij=exp(-aij*Tij) where Tij=(gij/R)/T     for given activity coefficients
C     GAM1, GAM2 at mole fraction X1, temperature T (K) and NRTL parameter A12.
C     - Guess starting values of parameters T12, T21 are required
C       (usually both at 0.5 for A12=0.2)
C
      X2=1.-X1
C......NRTL equations
60    G12=EXP(-A12*T12)
      G21=EXP(-A12*T21)
      GX1=X1+X2*G21
      GX2=X2+X1*G12
      F1=LOG(GAM1)-X2*X2*T21*(G21/GX1)**2-X2*X2*T12*G12/(GX2**2)
      F2=LOG(GAM2)-X1*X1*T12*(G12/GX2)**2-X1*X1*T21*G21/(GX1**2)
C.....Derivatives
      F1L12=(1.-A12*T12)+2.*G12*A12*(X1*X2+X1*X1)/GX2**2)
      F1L12=-X2*X2*G12*F1L12/(GX2**2)
      F1L21=(G21/2.*A12*T12)+2.*G21*G21*A12*(X1*X2+X2*X2)/(GX1**2)
      F1L21=-X2*X2*G21*F1L21/(GX1**2)
      F2L12=(1.-A12*T21)+2.*G21*A12*(X2*X1+X2*X2)/(GX1**2)
      F2L12=-X1*X1*G21*F2L12/(GX1**2)
      F2L21=(G12/2.*A12*T21)+2.*G12*G12*A12*(X2*X1+X1*X1)/(GX2**2)
      F2L21=-X1*X1*G12*F2L21/(GX2**2)
C     Solution
      D=F2L12*F1L21-F1L12*F2L21
      H=(F1*F2L21-F2*F1L21)/D
      G=(F1L12*F2-F2L12*F1)/D
      T12=T12+H
      T21=T21+G
      CHK=ABS(H/T12)+ABS(G/T21)
      IF (CHK.GE.0.0001) GOTO 60
      G12R=T12*T
      G21R=T21*T
      RETURN
      END
```

> **EXAMPLE 7.6**

Calculation of the Parameters of the Wilson, T-K-Wilson and NRTL Models for the Mixture (0.736 Methyl Acetate + 0.264 Methanol) at T = 400 K and P = 0.78 MPa. The Experimental Activity Coefficients (p.199, ref.[8]) are known to be 1.0224 and 1.3983 respectively.

For both Wilson and T-K-Wilson models, the ratio of the molar volumes of the two compounds, methyl acetate(1) and methanol(2), is required. This is readily estimated with the BWR(HS) equation of state with the result $V_{12} = 0.5107$. In the case of the NRTL model, the parameter α is assumed to be 0.2.

In Display 7.7 the calling program and its output are shown.

DISPLAY 7.7   WILSPAR Sample Input/Output

```
        PROGRAM EX0706
        CHARACTER*10 FNAM(3)
        DATA X1,T,GAM1,GAM2,V12,A12/0.736,400.,1.0224,1.3983,0.5107,0.2/
        DATA GUES12,GUES21/0.5,0.5/
        DATA FNAM/'WILSON    ','T-K-WILSON','NRTL      '/
        WRITE(*,1000) X1,V12,T,GAM1,GAM2
1000    FORMAT(3X,'@ X1    = ',F9.4,5X,' V12    = ',F9.4,5X,' T [K] =',
      &F7.2,/3X,' G1(EXPT) =',F9.4,/3X,' G2(EXPT) =',F9.4,/)
C       Wilson model (IOPT=1)
        CALL WILSPAR(T,X1,GAM1,GAM2,V12,GUES12,GUES21,SL12R,SL21R,1)
        WRITE(*,1100) FNAM(1),SL12R,SL21R
1100    FORMAT(5X,A10,' PARAMETERS:',3X,'L12/R =',F9.2,4X,'L21/R =',F9.2)
C       T-K-Wilson model (IOPT=2)
        CALL WILSPAR(T,X1,GAM1,GAM2,V12,GUES12,GUES21,SL12R,SL21R,2)
        WRITE(*,1100) FNAM(2),SL12R,SL21R
C       NRTL model
        CALL NRTLPAR(T,X1,GAM1,GAM2,A12,GUES12,GUES21,G12R,G21R)
        WRITE(*,1200) FNAM(3),G12R,G21R,A12
1200    FORMAT(5X,A10,' PARAMETERS:',3X,'T12/R =',F9.2,4X,'T21/R =',F9.2,
      &1X,'(A12 =',F7.4,')')
        END


@ X1      =     0.7360     V12 =    0.5107     T [K]  =   400.00
  G1 (EXPT) =     1.0224
  G2 (EXPT) =     1.3983


WILSON     PARAMETERS:   L12/R =    760.70    L21/R =   -363.51
T-K-WILSON PARAMETERS:   L12/R =    723.27    L21/R =   -425.21
NRTL       PARAMETERS:   T12/R =   -299.33    T21/R =    684.72   (A12=0.2000)
```

❑

7.9.3. Parameters from Infinite-Dilution Activity Coefficients

In Example 7.3, a procedure for the calculation of NRTL parameters τ_{12} and τ_{21} was described based on infinite-dilution activity coefficients. The latter were estimated from the correlation of Pierotti *et al.* and no experimental values were employed. Program EX0707 is the code used in that calculation and Example 7.7 shows its application for the case $\alpha = -1$.

The listing and its output is shown in Display 7.8. Inputs to the program are:
- the two infinite-dilution coefficients (G1INF, G2INF)
- the parameter α (A12) and the range of τ_{12} to search (T12FIRST, T12LAST).

In order to obtain all the possible pairs of the parameters, initial values of τ_{12} in the range -5 to +5 are employed and the desired solution is obtained by the criterion of Tassios [13].

EXAMPLE 7.7   *Calculation of the NRTL Parameters for (0.80 n-Hexane + 0.20 Diethyl Ketone) at T = 298.15 K when $\alpha_{12} = -1$.*

Data required are obtained from Example 7.3 where the results were also discussed; only the program and its output are shown here.

```
                                         DISPLAY 7.8  NRTLROOT List
      PROGRAM EX0707
      DIMENSION T12F(5),T21F(5),G1S(5),G2S(5)
      DATA G1INF,G2INF,A12,T12FIRST,T12LAST/2.54,3.91,-1,-5,5./
      SUMABS=9999.
      IJ=0
      T12=T12FIRST
      WRITE (*,1000)
1000  FORMAT(2X,'SOLUTION NO   T12    T21   |T12|+|T21|  A12')
C     Scan range for solutions
50    T12=T12+.0001
      T21=LOG(G1INF)-T12*EXP(-A12*T12)
      T12N=LOG(G2INF)-T21*EXP(-A12*T21)
      IF ((ABS((T12-T12N)/T12N)*100.).GT.0.05) GOTO 70
      IJ=IJ+1
      SUM=ABS(T12)+ABS(T21)
      WRITE (*,1100) IJ,T12,T21,SUM
1100  FORMAT(6X,I2,5X,2F10.4,F11.3)
      T12F(IJ)=T12
      T21F(IJ)=T21
      IF (SUM.GT.SUMABS) GOTO 60
      SUMABS=SUM
```

```
        ISOL=IJ
   60   T12=T12+0.1
   70   IF (T12.GT.T12LAST) GOTO 100
        GOTO 50
C.........Optimum solution (Tassios criterion)
   100  WRITE(*,1200) T12F(ISOL),T21F(ISOL),A12
  1200  FORMAT(2X,'OPTIMUM SET >>',2F10.4,16X,F7.4,/)
C..........Tables of all activity coefficients
        WRITE(*,1400) T12F(1),T21F(1),T12F(2),T21F(2),T12F(3),T21F(3)
  1400  FORMAT(1X/,10X,3(5X,'T12',6X,'T21   '),/,7X,3(5X,F7.4,2X,F7.4),/,
       &1X,70('-'),/,2X,'MOLEFR.',3(7X,'G1',7X,'G2   '))
        DO 140 J=1,21
        X1=FLOAT(J-1)*0.05
        DO 120 I=1,IJ
        X2=1.-X1
        G12=EXP(-A12*T12F(I))
        G21=EXP(-A12*T21F(I))
        G1=(X2*X2*T21F(I)*G21*G21)/((X1+X2*G21)**2)
        G1=G1+(X2*X2*T12F(I)*G12)/((X2+X1*G12)**2)
        G1S(I)=EXP(G1)
        G2=(X1*X1*T12F(I)*G12*G12)/((X2+X1*G12)**2)
        G2=G2+(X1*X1*T21F(I)*G21)/((X1+X2*G21)**2)
   120  G2S(I)=EXP(G2)
   140  WRITE(*,1500) X1,G1S(1),G2S(1),G1S(2),G2S(2),G1S(3),G2S(3)
  1500  FORMAT(3X,F4.2,3(5X,F7.4,2X,F7.4))
        STOP
        END
```

SOLUTION NO	T12	T21	|T12|+|T21|	A12
1	-2.3301	1.1589	3.489	
2	0.2533	0.6058	0.859	
3	1.4032	-4.7759	6.179	
OPTIMUM SET >>	0.2533	0.6058		-1.0000

	T12	T21	T12	T21	T12	T21
	-2.3301	1.1589	0.2533	0.6058	1.4032	-4.7759
MOLEFR.	G1	G2	G1	G2	G1	G2
0.00	2.5400	1.0000	2.5400	1.0000	2.5400	1.0000
0.05	2.4525	1.0009	2.3600	1.0019	43.8744	1.0138
0.10	2.3636	1.0039	2.1996	1.0076	14.6386	1.1065
0.15	2.2735	1.0096	2.0559	1.0174	6.8501	1.2317
0.20	2.1822	1.0184	1.9267	1.0315	4.0470	1.3758
0.25	2.0898	1.0313	1.8102	1.0504	2.7883	1.5319
0.30	1.9966	1.0493	1.7047	1.0746	2.1311	1.6954
0.35	1.9028	1.0740	1.6091	1.1049	1.7500	1.8634
0.40	1.8087	1.1073	1.5222	1.1423	1.5121	2.0334
0.45	1.7146	1.1520	1.4433	1.1882	1.3555	2.2040
0.50	1.6210	1.2121	1.3715	1.2444	1.2484	2.3737

0.55	1.5285	1.2937	1.3064	1.3132	1.1733	2.5418
0.60	1.4378	1.4057	1.2475	1.3978	1.1197	2.7075
0.65	1.3497	1.5622	1.1946	1.5027	1.0810	2.8702
0.70	1.2657	1.7862	1.1475	1.6338	1.0532	3.0297
0.75	1.1872	2.1158	1.1062	1.7999	1.0332	3.1856
0.80	1.1167	2.6147	1.0708	2.0134	1.0193	3.3377
0.85	1.0575	3.3823	1.0418	2.2930	1.0099	3.4860
0.90	1.0149	4.5094	1.0196	2.6673	1.0040	3.6304
0.95	0.9960	5.6225	1.0052	3.1815	1.0009	3.7709
1.00	1.0000	3.9055	1.0000	3.9102	1.0000	3.9075

❑

7.9.4. UNIQUAC Equations

Subroutine UNIQUAC, shown in Display 7.9, solves the UNIQUAC equations for the activity coefficients of each component in a multi-component system. Parameters required for a selection of compounds are tabulated in Tables A7 and A8 in Appendix A.

Additional inputs required by the subroutine UNIQUAC are

- the number of components (IS) and the mole fraction of each (array X)
- the temperature (T) in Kelvin
- volume parameter r_i, for component i, as R(IS)
- area parameter q_i, Q(IS), and
- secondary area parameter q'_i, as QPR(IS).

Note, that for compounds other than water or alcohols $q'_i = q_i$. An application of this subroutine is presented in the next example.

```
                                                        DISPLAY 7.9  UNIQUAC List
        SUBROUTINE UNIQUAC(IS,T,X,R,Q,QPR,U,G)
C
C
C       The subroutine calculates activity coefficients G(I) from the
C       UNIQUAC equations for a mixture at temperature T [K], of IS
C       components each of mole fraction X(I).
C
C
C       Other input required is
C       - R(I)    : r(I), volume parameter of component i
C       - Q(I)    : q(i), area parameter of component i
C       - QPR(I)  : q'(i), area parameter for water and alcohols
C                          for all other compounds equal to q(i), and
C       - U(I,J)  : u(I,J)/R
C
        DIMENSION U(IS,IS),X(IS),R(IS),Q(IS),QPR(IS),G(IS)
        DIMENSION PHI(5),THI(5),THIPR(5),WL(5),TA(5,5)
        SPHI=0
```

```
          STHI=0.
          STHIPR=0.
          DO 100 J=1,IS
          SPHI=SPHI+X(J)*R(J)
          STHI=STHI+X(J)*Q(J)
  100   STHIPR=STHIPR+X(J)*QPR(J)
          DO 120 I=1,IS
          PHI(I)=X(I)*R(I)/SPHI
          THI(I)=X(I)*Q(I)/STHI
          THIPR(I)=X(I)*QPR(I)/STHIPR
  120   WL(I)=5.*(R(I)-Q(I))-(R(I)-1.)
          DO 140 I=1,IS
          DO 140 J=1,IS
  140   TA(I,J)=EXP(-U(I,J)/T)
C........Combinatorial part
          DO 220 I=1,IS
          SXL=0.
          STHTJI=0.
          DO 160 J=1,IS
          SXL=SXL+X(J)*WL(J)
  160   STHTJI=STHTJI+TA(J,I)*THIPR(J)
          GC=LOG(PHI(I)/X(I))+5.*Q(I)*LOG(THI(I)/PHI(I))
          GC=GC+WL(I)-PHI(I)*SXL/X(I)
C........Residual part
          FS=0.
          DO 200 J=1,IS
          STHT=0.
          DO 180 K=1,IS
  180   STHT=STHT+THIPR(K)*TA(K,J)
  200   FS=FS+THIPR(J)*TA(I,J)/STHT
          GR=QPR(I)*(1.-LOG(STHTJI)-FS)
C........Activity coefficient
  220   G(I)=EXP(GC+GR)
          RETURN
          END
```

EXAMPLE 7.8   *Calculation of the Activity Coefficients of the Mixture (0.30 Ethyl Acetate + 0.10 Ethanol + 0.60 Water} at T = 343.15 K, P = 0.1 MPa.*

All required pure-component parameters are obtained from Table A8 in Appendix A, while binary interaction parameters are obtained from Table A7 in the same Appendix. The calling program and its output are shown in Display 7.10.

The three activity coefficients obtained are shown in the table together with the those obtained from the NRTL model in Example 7.5. As can be seen, the agreement is rather poor but as no experimental results are available, no comments can be made as to which method is better.

Mixture Component	Mole fraction	Activity Coefficient	
		Example 7.8 UNIQUAC	Example 7.5 NRTL
ethyl acetate (1)	0.30	1.92	2.53
ethanol (2)	0.10	1.65	1.41
water (3)	0.60	1.32	1.63

DISPLAY 7.10   UNIQUAC Sample Input/Output

```
      PROGRAM EX0708
      PARAMETER (IS=3)
      DIMENSION U(IS,IS),X(IS),R(IS),Q(IS),QPR(IS),G(IS)
      DATA U(1,1),U(1,2),U(1,3)/ 0. ,571.73,1081.40/
      DATA U(2,1),U(2,2),U(2,3)/-167.61, 0. ,-81.94/
      DATA U(3,1),U(3,2),U(3,3)/-222.02,437.92, 0. /
      DATA R/ 3.48,2.11,0.92/
      DATA Q/ 3.12,1.97,1.40/
      DATA QPR/3.12,0.92,1.00/
      DATA X,T/0.3000,0.1000,0.6000,343.15/
      CALL UNIQUAC(IS,T,X,R,Q,QPR,U,G)
      WRITE(*,1000) G(1),G(2),G(3)
1000  FORMAT(3X,'G1 =',F6.2,/,3X,'G2 =',F6.2,/,3X,'G3 =',F6.2)
      END

   G1 = 1.92
   G2 = 1.65
   G3 = 1.32
```

❑

EXAMPLE 7.9   *Calculation of the Activity Coefficients of the Mixture (0.047 Acetone + 0.953 n-Pentane) at T = 307 K, P = 0.1 MPa.*

All parameters required are obtained from Tables A7 and A8 in Appendix A. The calling program and its output are shown in Display 7.11.

For this mixture Reid *et al.* [26] have calculated the activity coefficients by the UNIFAC method and quote them in comparison with the experimental values obtained by Lo *et al.* [27]. A comparison of all these set of values is shown in the table.

Mixture	Mole	Activity Coefficient		
Component	fraction	UNIQUAC	UNIFAC [26]	Expt. Lo [27]
acetone (1)	0.047	4.97	5.07	4.41
n-pentane (2)	0.953	1.01	1.01	1.11

Although the agreement with experiment is not that good, the example is rather typical of the accuracy to be expected with this kind of prediction.

DISPLAY 7.11   UNIQUAC Sample Input/Output

```
      PROGRAM EX0709
      PARAMETER (IS=2)
      DIMENSION U(IS,IS),X(IS),R(IS),Q(IS),QPR(IS),G(IS)
      DATA U(1,1),U(1,2)/  0. ,-22.83/
      DATA U(2,1),U(2,2)/266.31,  0. /
      DATA R/ 2.57,3.82/
      DATA Q/ 2.34,3.31/
      DATA QPR/2.34,3.31/
      DATA X,T/0.047,0.953,307./
      CALL UNIQUAC(IS,T,X,R,Q,QPR,U,G)
      WRITE(*,1000) G(1),G(2)
1000  FORMAT(3X,'G1 =',F6.3,/,3X,'G2 =',F6.3)
      STOP
      END

  G1 = 4.972
  G2 = 1.005
```

❑

References

1. S. Margules, *Math. Naturw. Akad. Wiss.* (Vienna) **104** (1895) 1243.
2. J.J. van Laar, *Z. Physik Chem.* **72** (1910) 723.
3. J.J. van Laar, *Z. Physik Chem.* **83** (1913) 599.
4. G.M. Wilson, *J. Am. Chem. Soc.* **86** (1964) 127.
5. H. Renon and J.M. Prausnitz, *AIChE J.* **14** (1968) 135.
6. H. Renon and J.M. Prausnitz, *Ind. Eng. Chem. Process Des. Dev.* **8** (1969) 413.
7. T. Tsuboka and T. Katayama, *J. Chem. Eng. Japan* **8** (1975) 181.
8. S.M. Walas, *Phase Equilibria in Chemical Engineering* (Butterworth Publishers, Boston, 1985).
9. M.J. Holmes and M. van Winkle, *Ind. Eng. Chem.* **62** (1970) 21.
10. DECHEMA *Vapour-Liquid & Liquid-Liquid Equilibrium Data Collection* .
11. A. Miller and T. Bliss, *Ind. Eng. Chem.* **32** (1940) 123.

12. N. Silverman and D. Tassios, *Ind. Eng. Chem. Process Des. Dev.* **16** (1977) 13.
13. D. Tassios, *Ind. Eng. Chem. Process Des. Dev.* **18** (1979) 182.
14. G.J. Pierotti, C.H. Deal and E.L. Derr, *Ind. Eng. Chem.* **51** (1959) 95.
15. J.G. Helpinstill and M. van Winkle, *Ind. Eng. Chem. Process Des. Dev.* **7** (1968) 213.
16. E.R. Thomas and C.A. Eckert, *Ind. Eng. Chem. Process Des. Dev.* **23** (1984) 194.
17. D.S. Abrams and J.M. Prausnitz, *AIChE J.* **21** (1975) 116.
18. T.F. Anderson, D.S. Abrams and E.A. Grens, *AIChE J.* **24** (1978) 20.
19. J.M. Prausnitz, T. Anderson, E. Grens, C. Eckert, R. Hsieh and J. O'Connell, *Computer Calculations for Multicomponent Vapor-Liquid and Liquid-Liquid Equilibria* (Prentice Hall, London, 1980).
20. J. Gmehling, P. Rasmussen and F. Fredenslund, *Ind. Eng. Chem. Process Des. Dev.* **21** (1982) 118.
21. G.M. Wilson and C.H. Deal, *Ind. Eng. Chem. Fundam.* **1** (1962) 20.
22. G.M. Wilson, *J. Am. Chem. Soc.* **86** (1964) 127.
23. K. Kojima and K. Toshigi, *Prediction of Vapor-Liquid Equilibria by the ASOG Method, Physical Sciences Data 3* (Elsevier Publishing Company, Tokyo, 1979).
24. A. Fredenslund, R.L. Jones and J.M. Prausnitz, *AIChE J.* **21** (1975) 1086.
25. A. Fredenslund, J. Gmehling and P. Rasmussen, *Vapor-Liquid Equilibria Using UNIFAC* (Elsevier, 1977).
26. R.C. Reid, J.M. Prausnitz and B.E. Poling, *The Properties of Gases and Liquids* (McGraw-Hill, Singapore, 1988).
27. T.C. Lo, H.H. Bierber and A.E. Karr, *J. Chem. Eng. Data* **7** (1962) 327.
28. W. Wagner, *Cryogenics* **13** (1973) 470.
29. J.M. Prausnitz, R.N. Lichtenthaler and E Gomes de Azevedo, *Molecular Thermodynamics of Fluid Phase Equilibria,* 2nd edition (Prentice-Hall, Englewood Cliffs, N.J., 1986) Ch. 8 and references therein.
30. R.A. Heidemann, *7th Int. Conf. on Fluid Properties and Phase Equilibria for Chemical Process Design* (Snowmass, Colorado 1995) p. 295.
31. M.J. Huron and J. Vidal, *Fluid Phase Equilibria* **3** (1979) 255.
32. D.S.H. Wong and S.I. Sandler, *AIChE J* **38** (1992) 671.
33. R.A. Heidemann and S.L. Kokal, *Fluid Phase Equilibria* **56** (1990) 17.
34. S. Dahl and M.L. Michelsen, *AIChE J* **36** (1990) 1829.
35. H. Huang and S.I. Sandler, *Ind. Eng. Chem. Res.* **32** (1993) 1498.

8

Phase-Equilibrium Calculations

The quantitative prediction of phase behaviour is the central problem in chemical-engineering thermodynamics and a very important consideration in the design of chemical process plant. The almost infinite number of possible mixtures and wide ranges of temperature and pressure encountered in process engineering is such that no single thermodynamic model is ever likely to be applicable in all cases. Consequently, knowledge and judgement are required to select the most appropriate methods by which to estimate the conditions under which two or more phases will be in equilibrium.

In this chapter, some basic algorithms for the determination of vapour-liquid equilibrium conditions in mixtures are discussed together with their application to certain kinds of flash calculation. A short discussion of liquid-liquid equilibrium is also included. For every topic considered, a computer routine will be given and used as the basis of an example. As elsewhere in this book, the computational methods used are not the most efficient possible (instead, they are designed to be easy to understand and modify) and this may be a consideration if the routines are to be called many times as a part of a larger calculation.

8.1. Elementary Phase Behaviour

Before embarking on the quantitative aspects of phase equilibrium, it would be as well to have a qualitative understanding of the most common types of phase behaviour. The subject of phase behaviour in mixtures is a complicated one and we shall consider here only the most elementary kinds of phase diagram for vapour-liquid and vapour-liquid-liquid systems.

The simplest kind of vapour-liquid equilibrium (VLE) in a binary system is illustrated in the phase diagrams of Figure 8.1. Here, we show either the pressures at which vapour and liquid coexist under conditions of constant temperature (the P-x-y

phase diagram) or the coexistence temperatures at constant pressure (the *T-x-y* phase diagram). The boundary between vapour and (vapour + liquid) is known as the dew curve, while that between liquid and (liquid + vapour) is known as the bubble curve. Figures 8.1 depict conditions under which the variable held constant (*T* or *P*) is below the critical value of that quantity (*T*^{c} or *P*^{c}) for both pure components. Phase equilibrium is then possible over the entire composition range and, in the system represented in diagrams (a) and (b), complete separation of the components by distillation is possible in principle. Often, as illustrated by the *T-x-y* diagrams (c) and (d), the dew and bubble curves touch at some intermediate composition where the coexisting phases have the same composition. Such a point is known as an azeotrope and, although the two phases coexist there at the same temperature, pressure and composition, it is distinguished from a critical point by a difference in the density of the phases. Complete separation by distillation is not possible under

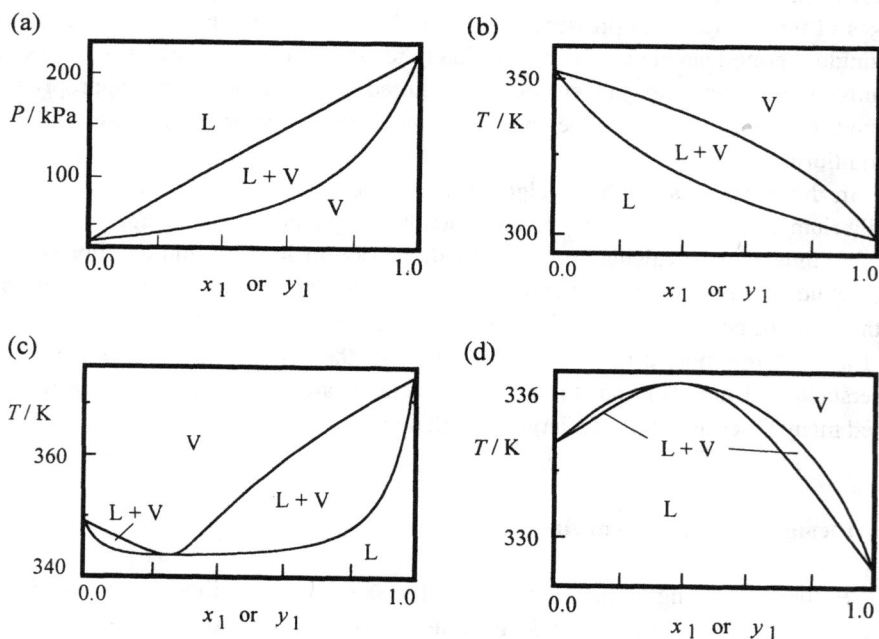

Figure 8.1. Vapour-liquid equilibrium in binary systems: (a) {2-methylbutane(1) + cyclo-hexane(2)} at *T* = 325 K; (b) {2-methylbutane(1) + cyclohexane(2)} at *P* = 0.1 MPa; (c) {nitromethane(1) + tetrachloromethane(2)} at *P* = 0.1 MPa; (d) {acetone(1) + trichlorometha-ne(2)} at *P* = 0.1 MPa. All results are simulated from thermodynamic models: (a) and (b) from the Peng-Robinson equation of state; (c) from the Wilson equations with ideal vapour; (d) from the UNIQUAC equations with ideal vapour.

conditions where azeotropy exists. An example of the *P-x-y* diagram for a system with a minimum-boiling azeotrope was given in the previous chapter (Figure 7.1).

Figure 8.2 illustrates the high-pressure phase behaviour of the non-azeotropic system {methane(1) + propane(2)}. Figure 8.2(a) is a *P-x-y* diagram for this system showing the dew and bubble curves for two isotherms that lie between the critical temperatures of the pure components. Phase equilibrium is now only possible over part of the composition range and each isotherm contains a critical point at which the dew and bubble curves join *and* the two phases become identical. This point always corresponds to a maximum in the *P-x-y* diagram. As shown in Figure 8.2(b), where the mixture critical temperature and pressure are plotted against composition, the critical pressure may exceed that of the pure components but in this system the critical temperature is a monatomic function of composition. If one takes a constant-composition section through the phase diagram and plots the corresponding P-ρ_n isotherm then, if phase separation occurs, results similar to those illustrated in Figure

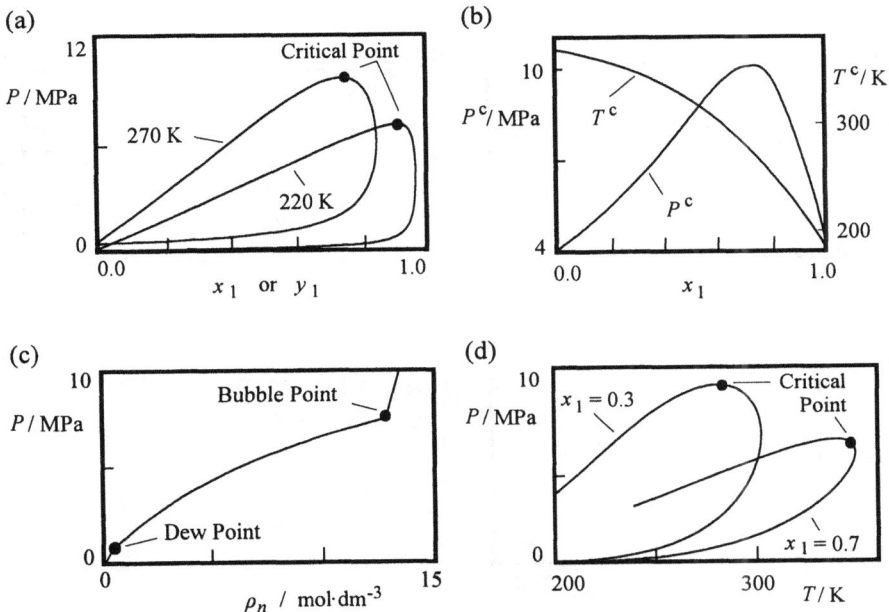

Figure 8.2. Vapour-liquid equilibrium in the system {methane(1) + propane(2)}: (a) *P-x-y* isotherms at $T = 220$ K and $T = 270$ K; (b) loci of mixture critical constants; (c) P-ρ_n isotherm at $T = 270$ K showing dew and bubble points; (d) *P-T* phase envelopes for selected compositions. Results simulated from the Peng-Robinson equation of state.

8.2(c) are usually obtained. As the vapour is compressed, the pressure rises until the dew point is reached; there liquid first starts to condense. As the system is further compressed, more liquid forms until, at the bubble point, all the vapour is gone. Thereafter the isotherm rises rapidly because the compressibility of the liquid is low. P-ρ_n isotherms for pure fluids and for binary and multi-component mixtures show rather similar behaviour. In the case of a pure fluid, the dew and bubble pressures are exactly the same and equal to the saturated vapour pressure at the given temperature, while in mixtures, the bubble-point pressure exceeds the dew-point pressure at the same temperature. Since the overall composition is fixed, the dew and bubble points refer to the same composition. Along some pathways, *retrograde* behaviour may be observed in which either the dew curve or the bubble curve is intersected twice. For example, a constant-composition isotherm may cross the dew curve twice when the mole fraction of the more volatile component exceeds its critical value. A liquid phase then starts to form at the lower-pressure dew point as the fluid is compressed but complete condensation is never achieved and, with

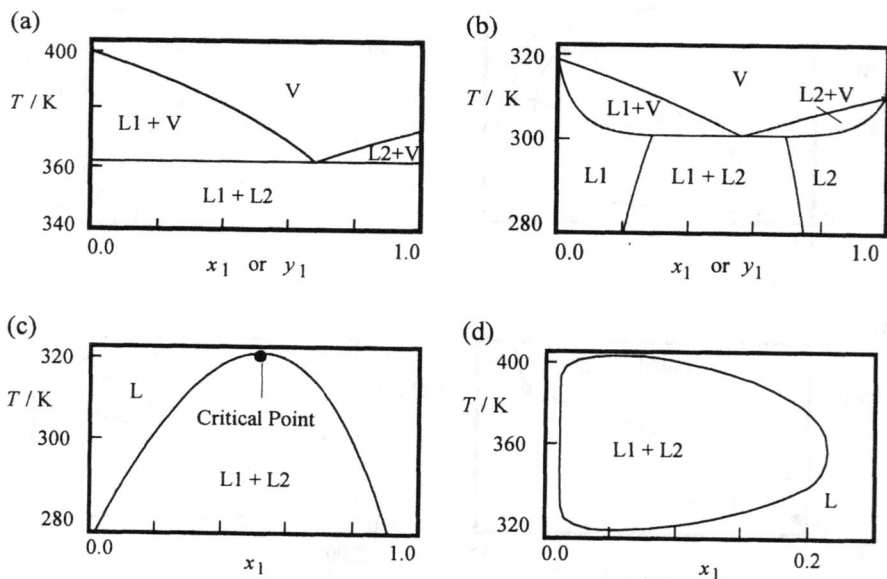

Figure 8.3. Vapour-liquid-liquid and liquid-liquid equilibrium in binary systems: (a) {water(1) + n-octane(2)} at $P = 0.1$ MPa; (b) {methanol(1) + cyclohexane(2)} at $P = 0.03$ MPa; (c) {methanol(1) + cyclohexane(2)} at $P = 0.1$ MPa. (d) {C$_6$H$_6$(1) + CH$_2$(OH)CH$_2$OC$_4$H$_9$(2)} at $P = 0.1$ MPa.

further compression, the liquid phase shrinks and disappears altogether at the upper pressure dew point.   A simple phase diagram, useful for both binary and multi-component mixtures, is obtained by plotting constant-composition dew and bubble curves on a *P-T* diagram such as that shown in Figure 8.2(d). A family of curves is obtained, one for each composition. A more detailed depiction of phase behaviour in ternary systems may be afforded by triangular diagrams showing, for example, dew or bubble curves under isothermal or isobaric conditions.

The phase behaviour of liquid mixtures in equilibrium with their vapour may be characterised by complete miscibility in the liquid phase, total immiscibility or by partial miscibility.   Complete miscibility in the liquid phase corresponds to the illustrations already given while examples of the isobaric *T-x-y* phase diagrams for vapour-liquid-liquid equilibrium (VLLE) in binary systems are given in Figures 8.3 (a) and (b).   Diagram (a) refers to the case of total immiscibility in the liquid phase and L1 and L2 are therefore pure liquid phases.   In diagram (b), there is partial miscibility with a 'miscibility gap' in which two liquid phases coexist covering only a part of the composition range; L1 and L2 are now mixed liquid phases.   When either the pressure is high enough or the temperature is low enough, only liquid-liquid equilibrium (LLE) need be considered.   Usually, the mutual solubility of liquids increases with temperature so that the miscibility gap narrows with increasing temperature and, provided that the pressure is high enough to prevent vaporisation, it may vanish as illustrated in Figure 8.3(c) at an upper critical solution temperature (UCST) above which there is complete miscibility in the liquid phase.   It is also possible for the miscibility gap to narrow at low temperatures and, if freezing does not occur first, it may vanish altogether at a lower critical solution temperature (LCST).   Some systems exhibit a closed-loop miscibility gap with both a UCST and a LCST; an example of that behaviour is shown in Figure 8.3(d).

It is important to note that, although the phase diagrams illustrated here are very common, they are by no means representative of VLE, VLLE and LLE in all systems.   For a more comprehensive review of phase behaviour, the reader is referred to the literature [1-3].

8.2. Bubble-Point and Dew-Point Calculations

The basic 'engine' of most phase equilibrium calculations is an algorithm to calculate the dew or bubble point for a mixture of specified composition. The kind of calculation to be made (dew or bubble) may be specified by giving the *vapour fraction* β which is defined as the amount of substance in the vapour phase divided by the total amount of substance. This quantity is unity at a dew point and zero at a bubble point. The phase rule requires specification of either the temperature or the pressure in addition to the composition of the bulk phase. It is then our task to calculate the remaining variables; these are either P (for specified T) or T (for specified P) and the composition of the coexisting phase at the dew or bubble point. This problem should have either one solution, when two phases are possible under the specified conditions, or no solution when they are not. Whether of not this is the case with a particular thermodynamic model remains to be seen.

We begin by recalling the basic conditions for equilibrium between vapour and liquid phases in a system of n components which require, in addition to equality of temperature and pressure, that:

$$f_i^{(V)} = f_i^{(L)} . \qquad (i = 1,2,\cdots n) \qquad (8.1)$$

This is a set of n (usually non-linear) simultaneous equations which must be solved to obtain the solution of the phase equilibrium problem. In terms of fugacity coefficients, these equations become

$$y_i \, \phi_i^{(V)} = x_i \, \phi_i^{(L)} \qquad (i = 1,2,\cdots n) \qquad (8.2)$$

and these are more useful than Eqs.(8.1) because the fugacity coefficients are usually slowly varying functions of temperature, pressure and composition. When we have an equation of state from which we may calculate the fugacity coefficients of all components in both phases, Eqs (8.2) may be solved more or less directly.

In cases where it is preferable to obtain the fugactiy of components in the liquid phase from an activity coefficient model, we write Eqs (8.1) as

$$y_i \, \phi_i^{(V)} P = x_i \, \gamma_i \, \phi_i^\sigma \, P_i^\sigma \, \mathscr{F}_i , \qquad (i = 1,2,\cdots n) \qquad (8.3)$$

where γ_i is the activity coefficient of component i at the specified temperature and liquid composition and at pressure P_i^σ, while $\mathscr{F}_i$ is the Poynting Factor:

$$\mathscr{F}_i = \exp\left[\int_{P_i^\sigma}^{P} \left(V_i^{(L)}/RT \right) dP \right] . \qquad (8.4)$$

The accuracy of the computed dew- or bubble-point depends entirely upon the accuracy of the thermodynamic model used to calculate the fugacity coefficients and/or activity coefficients and Poynting Factor. Such models have been discussed in detail in the previous chapters.

The state of the vapour and liquid phases in contact at a given temperature and pressure may be conveniently specified by the vapour fraction β and the *vaporisation equilibrium ratio* K_i, defined by $K_i = y_i/x_i$. When the two phases are in thermodynamic equilibrium, K_i is given by

$$K_i = \phi_i^{(L)}/\phi_i^{(V)} \qquad (8.5)$$

or, in terms of an activity-coefficient model instead of $\phi_i^{(L)}$, by

$$K_i = \frac{\gamma_i \, \phi_i^{\sigma} \, P_i^{\sigma} \, \mathcal{F}_i}{\phi_i^{(V)} P} . \qquad (8.6)$$

The concept of ideality in vapour and liquid mixtures is often useful as a means of obtaining an initial approximation to the solution of a VLE problem. For an ideal vapour mixture, all fugacity coefficients are unity while, for an ideal liquid mixture, all activity coefficients and Poynting factors are unity. Eqs.(8.3) then reduce to Raoult's law:

$$y_i P = x_i P_i^{\sigma} . \qquad (8.7)$$

Consequently, the total pressure in an isothermal ideal vapour-liquid system is a linear function of the mole fractions in the liquid phase; alternatively, the inverse of the total pressure is a linear function of the mole fractions in the vapour phase. Figure 8.1 (a) is an example of only slight deviations from Raoult's law.

8.2.1. Bubble-Point Calculations

In this section we describe two variants of a basic algorithm for determining the bubble-point of a fluid mixture, the one based on Eqs.(8.2) with an equation of state for both phases, and the other on Eqs.(8.3) with an activity-coefficient model for the liquid phase and an equation of state for the vapour. There are many ways in which one might set about solving the phase-equilibrium problem but the strategy outlined below in Figures 8.4 and 8.5 is a simple and reliable approach to the problem.

The first algorithm shown in Figure 8.4 is the simplest as it treats both phases using the same thermodynamic model. It involves the following steps:

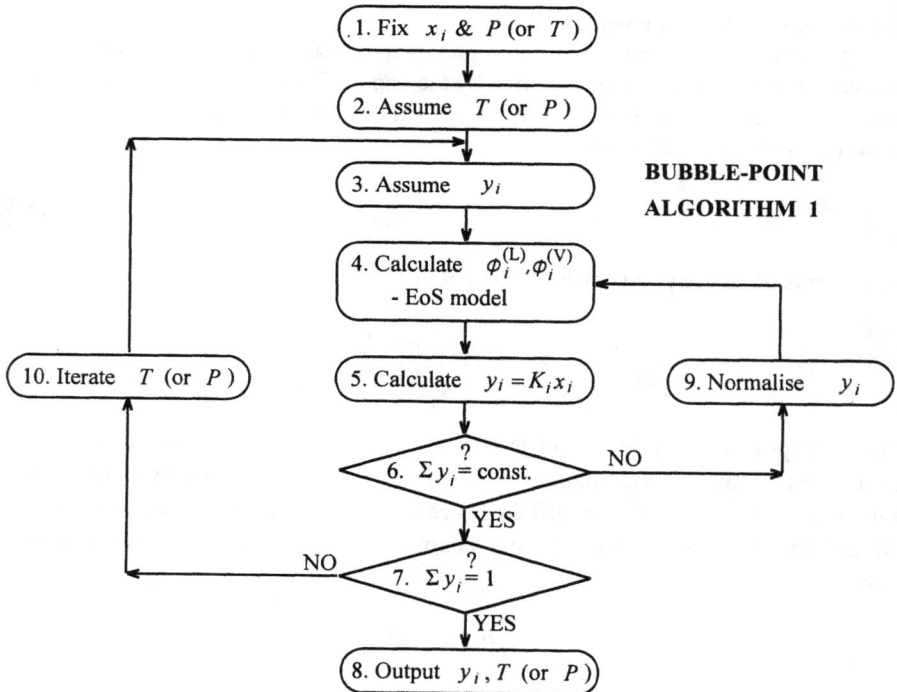

Figure 8.4. Bubble-point algorithm using an equation of state for both phases.

1. The liquid composition x_i $(i = 1,2,\cdots n)$ and either the pressure P or the temperature T must be specified.

2. An initial value is assumed for the unknown bubble-point temperature or pressure; often, Raoult's law is employed for this purpose.

3. Initial values for vapour composition y_i $(i = 1,2,\cdots n)$ are assumed. Unless the system is known to exhibit nearly ideal behaviour (in which case Eqs.(8.7) may be applied), one might as well set $y_i = x_i$. The sum $s = \Sigma y_i$ should be initialised at this stage.

4. Next, the fugacity coefficients $\phi_i^{(V)}$ and $\phi_i^{(L)}$ of each component in the vapour and liquid phases are calculated subject to the currently assumed temperature, pressure and phase compositions. This involves first solving the equation of state for the molar volume of each phase and then calculating the required values of $\phi_i^{(V)}$ and $\phi_i^{(L)}$.

5. New approximations to the vapour mole fractions are now calculated: $y_i = x_i K_i$ with K_i given in terms of the fugacity coefficients by Eq.(8.5).

6. The new sum $s = \Sigma y_i$ is calculated. If this is equal to that for the previous iteration then proceed to step 7; otherwise, go to step 9.

7. Once a constant value of s is obtained subject to the presently assumed estimate of the unknown bubble-point temperature or pressure, we test to see if $s = 1$. If this condition is satisfied then proceed to step 8; otherwise to step 10.

8   A solution has been found which satisfies the thermodynamic requirements for thermal, hydrostatic and phase equilibrium.

9. Normalised values of the vapour-phase mole fractions are calculated, $y_i' = y_i/s$, and used in another iteration starting at step 4.

10. A new estimate of the unknown bubble-point temperature or pressure must be made. If $s > 1$ then the assumed temperature (pressure) is too high (low) while, if $s < 1$ then the reverse applies. The simplest method for updating the unknown T or P is by means of a bisection algorithm; this requires that upper and lower limits of the unknown be established at the start of the procedure.

This or a similar procedure is used in conjunction with a suitable equation of state for the majority of VLE problems. For the best results with multi-component systems, the interaction parameters (k_{ij}'s) in the equation of state should be optimised to in a fit to dew- or bubble-point pressures for the binary sub-systems. Program BUBLPR, given in Section 8.5 and applied in Examples 8.1 and 8.2, implements this procedure with the Peng-Robinson equation of state.

An activity coefficient method may be preferred for the liquid phase in cases involving associating or, possibly, highly-polar substances, although the accuracy of these methods is not good at high pressures. Essentially the same algorithm as described above may be used to solve the phase-equilibrium conditions but a small modification permits faster convergence. This modified algorithm, shown in Figure 8.5, differs from Figure 8.4 in the following steps:

3. Vapour-phase mole fractions are not required in the calculation of the liquid-phase fugacity. Advantage of this fact is taken by assuming initially $\phi_i = 1$ for each component. This is usually a much better approximation than assuming $y_i = x_i$. Setting $s = 1$ will force step 9 on the first iteration. At this stage, the vapour pressure and fugacity coefficient at saturation may be evaluated for each component by means of a vapour-pressure equation and the equation of state for the vapour.

4. The liquid-phase is treated according to an activity-coefficient model.

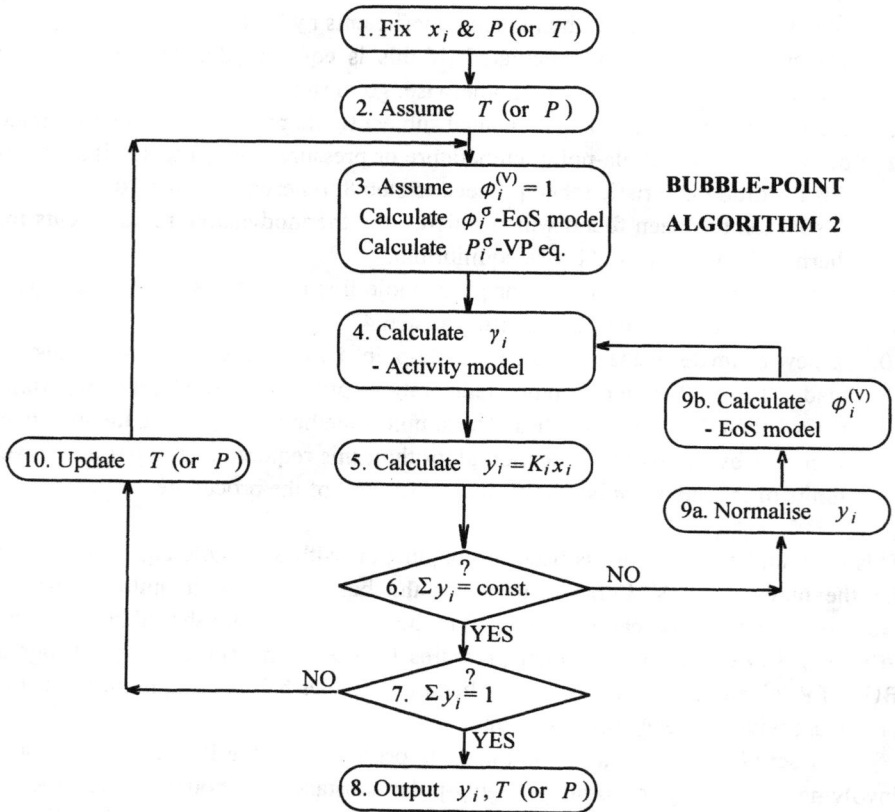

Figure 8.5. Bubble-point algorithm using an activity-coefficient model for the liquid phase.

5. K_i is given by Eq.(8.6).
9b. The vapour-phase fugacity coefficients are determined from the equation of state in this step.

Program BUBLVW, given in Section 8.5 and applied in Examples 8.3 to 8.5, implements this procedure with the T-K-Wilson activity-coefficient model for the liquid and the virial equation of state for the vapour.

8.2.2. Dew-Point Calculations

The determination of the dew-point temperature or pressure and the composition of the coexisting liquid is almost identical to the bubble-point problem. In this case, the vapour composition is specified and iterations are performed over the liquid mole fractions and over the unknown temperature or pressure. The algorithms shown in Figures 8.4 and 8.5 may be used after obvious changes. At the expense of a little complication, a composite algorithm can be devised for the calculation of either the dew or the bubble point at a specified temperature or pressure. This is left as an exercise for the reader. It might be interesting to note that, since a bubble-point routine returns the composition of the coexisting vapour, it may be used as it stands to generate points on the dew-point surface (although not at pre-determined vapour compositions).

8.3. Flash Calculations

The modelling of flash processes is probably the most important application of chemical engineering thermodynamics. A flash process is one in which a fluid stream of known overall composition and flow rate passes through a throttle, turbine or compressor and into a vessel (flash drum) where liquid and vapour phase are separated before passing through the appropriate outlet. Such a process may be operated under many different sets of conditions including the following:

1.   constant temperature and pressure (isothermal flash).
2.   constant enthalpy and pressure (isenthalpic flash).
3.   constant entropy and pressure (isentropic flash).

The thermodynamic modelling of these processes requires in each case determination of the vapour fraction and the vaporisation equilibrium ratio. It is also important in general to determine the thermal power (heat duty) absorbed or liberated in the flash process, although this is zero by definition in an isenthalpic or isentropic flash.

In the following sections we shall consider algorithms for the thermodynamic modelling of each of the three idealised flash processes defined above. As before, we may choose to employ an equation of state for both phases or, where necessary, an activity-coefficient model for the liquid and an equation of state for the vapour. For purposes of example, we shall adopt the latter approach with the isothermal flash and the former for the other two cases.

8.3.1. Isothermal Flash (constant T & P)

This process, illustrated schematically in Figure 8.6, is one of the most common encountered in chemical engineering. The feed, at temperature T_F and pressure P_F, passes through a throttle and enters the flash vessel, where liquid and vapour phases may separate. The operating pressure P of the unit is controlled in some way and heat is supplied or removed at rate Q though a heat exchanger so as to maintain isothermal conditions at temperature T. The molar flow rate F of the feed to the unit is specified, together with the overall composition (mole fractions z_i) and the temperature and pressure at which the unit operates. The objectives of the calculation are to determine the compositions (y_i and x_i) and the molar flow rates (F_V and F_L) of the vapour and liquid streams leaving the unit.

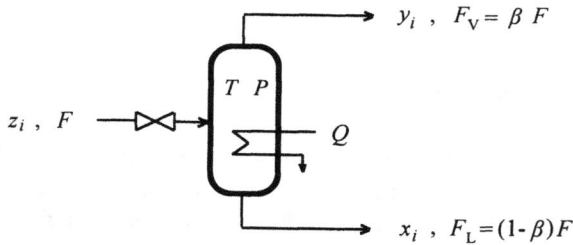

Figure 8.6. Isothermal flash unit.

Since only the overall composition of the mixture is known, a material balance for each of the n components is required:

$$F z_i = F_L x_i + F_V y_i \tag{8.8}$$

with

$$y_i = K_i x_i . \tag{8.9}$$

Combining these two equations and eliminating the flow rates in terms of the vapour fraction $\beta = F_V/F$, the flash condition may be written

$$f(\beta) = \sum_{i=1}^{n} x_i - 1 = \sum_{i=1}^{n} \frac{z_i}{1+\beta(K_i-1)} - 1 = 0 . \tag{8.10}$$

This may be solved easily for β by the Newton-Raphson algorithm which in this case gives

$$\beta_{k+1} = \beta_k + \left[\left(\sum_{i=1}^{n}\left(\frac{z_i}{1+\beta_k(K_i-1)}\right)\right)-1\right]\left[\sum_{i=1}^{n}\left(\frac{(K_i-1)z_i}{[1+\beta_k(K_i-1)]^2}\right)\right]^{-1}. \quad (8.11)$$

Successive approximations starting with $\beta_1=1$ generally converge rapidly. The phase compositions are given by

$$x_i = \frac{z_i}{1+\beta(K_i-1)} \quad \text{and} \quad y_i = K_i x_i. \quad (8.12)$$

Of course both β and the K_i's are unknown and the latter are therefore evaluated during each cycle of Eq.(8.11). An algorithm for solving this flash problem is shown in Figure 8.7 for the case in which an activity coefficient model is applied for the liquid phase.

The isothermal flash algorithm involves the following steps:
1. The temperature, pressure and overall mixture composition are specified.
2. Initial values of unity are assumed for the vapour-phase fugacity coefficient and liquid-phase activity coefficient of each component. β is initialised with the value unity.
3. A first approximation to the K_i is calculated for each component from Eq.(8.6) with P_i^σ determined from a suitable vapour-pressure equation and ϕ_i^σ calculated from the equation of state model.
4. A new value of β is determined from a single iteration of Eq.(8.11).
5. The compositions of each phases are determined from Eqs.(8.12).
6. The mole fractions are normalised so that $\Sigma x_i = \Sigma y_i = 1$.
7. New vaporisation equilibrium ratio's are calculated from Eq.(8.6) with γ_i determined from the activity coefficient model and ϕ_i determined from the equation of state.
8. We now test to see if the new vapour composition differs from that of the previous iteration. If it does, begin a new iteration at step 4, otherwise go to step 9.
9. A solution to the problem has been found.

Program FLASHTP, given in Section 8.5 and applied in Example 8.6, implements this procedure with the T-K-Wilson activity-coefficient model for the liquid and the virial equation of state for the vapour.

One rather obvious point that should not be forgotten is that two phases will only form when the specified pressure lies between the dew point and the bubble

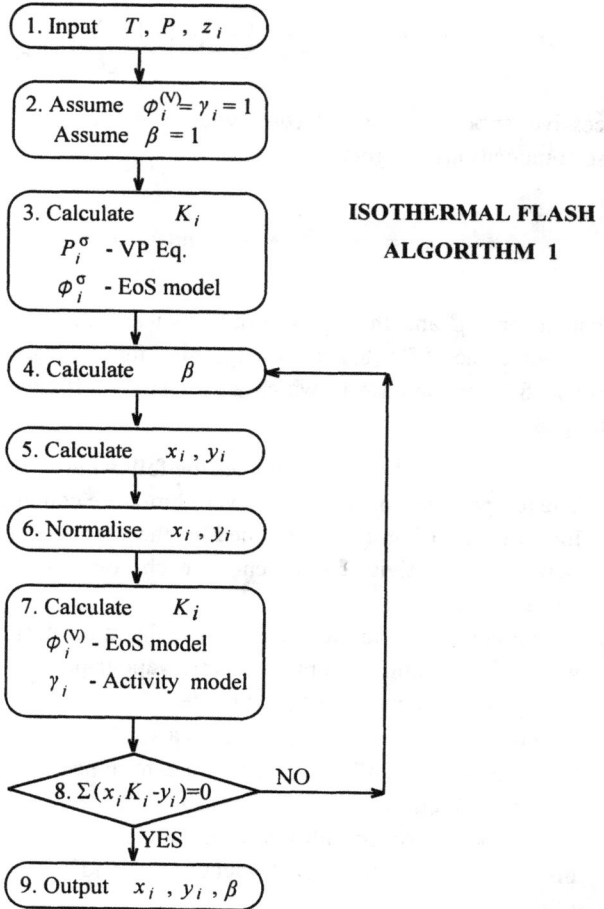

Figure 8.7. Isothermal flash algorithm using an activity-coefficient model for the liquid phase.

point for the given temperature and feed composition. Usually the heat duty Q on the flash unit is also required. Q (which is positive for heat supplied to the unit) may be determined from the molar flow rates and the molar enthalpy of the feed and product streams as follows:

$$Q = \left(F_V H_m^{(V)} + F_L H_m^{(L)} \right) - F H_m^{(F)} . \tag{8.13}$$

This is a more involved calculation. The enthalpy of the vapour stream may be evaluated easily from the equation of state and the perfect-gas heat capacity while that of the liquid stream should be determined by application of the thermodynamic relation:

$$-\left(\frac{H_i^{res}}{RT^2}\right) = \left(\frac{\partial \ln f_i}{\partial T}\right)_{P,x_1 \cdots x_n} . \tag{8.14}$$

In order to calculate the enthalpy of the feed, one must know the temperature T_F, pressure P_F and vapour fraction β_F at the inlet to the throttle. Assuming that the

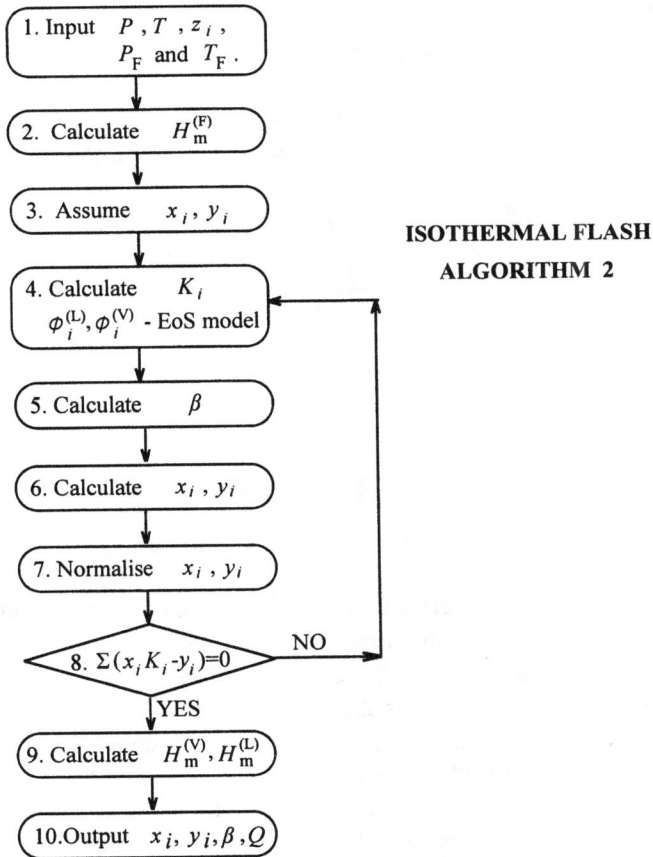

Figure 8.8. Isothermal flash algorithm using an equation-of-state model for both phases.

vapour and liquid in the inlet stream are in equilibrium, the enthalpy may be evaluated by the same method as used for the outlet streams. Often the feed will be a single phase and, if this is liquid, one must assume it to be at its bubble-point pressure, rather than the actual pressure, in order to determine the enthalpy from the activity-coefficient model and the vapour-pressure equations alone. Alternatively, an equation-of-state algorithm could be used for both phases in this calculation.

The method is a good deal simpler if an equation of state model is applied consistently to both phases during the entire calculation. Indeed it is straightforward to modify the algorithm in Figure 8.7 so that an equation-of-state model is used throughout and the heat duty is calculated; the procedure is given in Figure 8.8.

8.3.2. Isenthalpic Flash (constant H & P)

This type of flash, illustrated in Figure 8.9, involves operating of the unit under adiabatic conditions ($Q = 0$) and, as no work is done on the fluid, the process is isenthalpic.

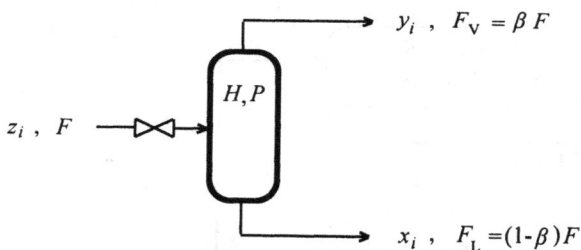

Figure 8.9. Isenthalpic flash unit.

The objective of the flash calculation is to find the temperature, vapour fraction and product compositions for the case in which the operating pressure and the temperature, pressure and composition of the feed are specified. A convenient method for solving this problem is shown in Figure 8.10.

The unknown temperature T of the products is determined by trial such that the combined enthalpy of the product streams is equal to that of the feed; that is, so that $Q = 0$. A bisection algorithm which adjusts T downwards when $Q > 0$, and vice versa, will suffice. Program FLASHHP, given in Section 8.5 and applied in Examples 8.7, implements this procedure with the Peng-Robinson equation of state.

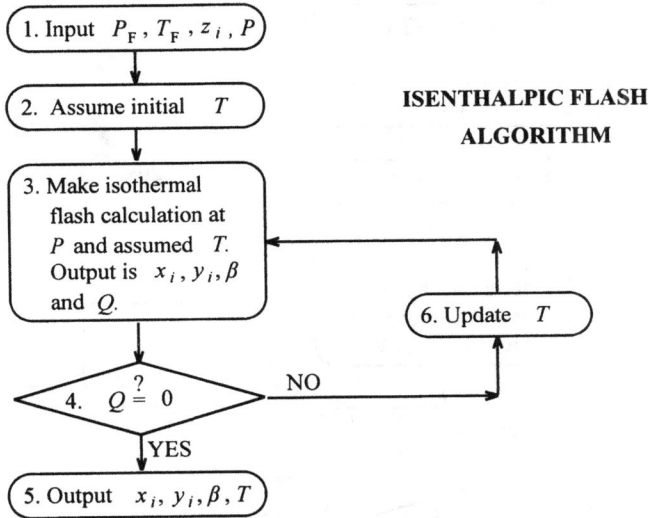

Figure 8.10. Isenthalpic flash algorithm.

8.3.3. Isentropic Flash (constant S and P)

If, instead of expanding though a throttle, the feed is compressed or expanded adiabatically and reversibly before entering the adiabatic flash vessel then the process is isentropic. An isentropic flash unit is illustrated schematically in Figure 8.11.

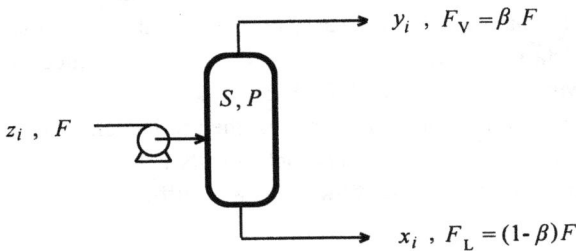

Figure 8.11. Isentropic flash.

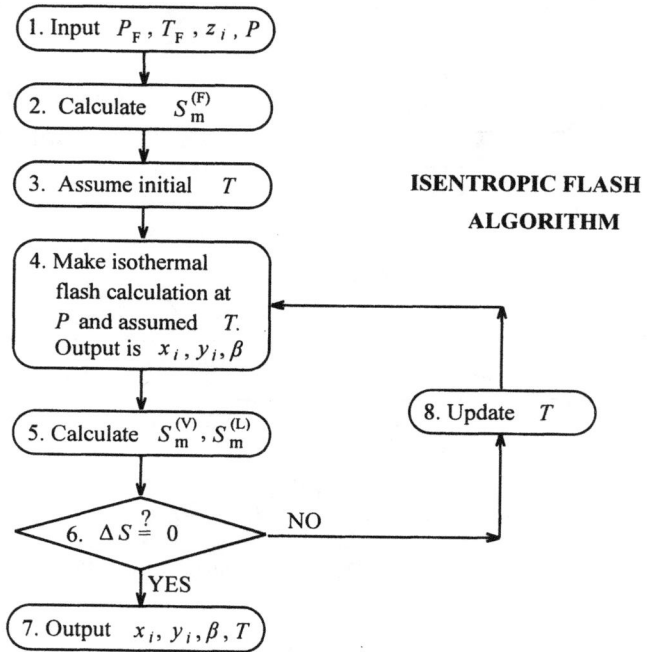

Figure 8.12. Isenthalpic flash algorithm.

The objective of the flash calculation is to find the temperature, vapour fraction and product compositions for the case in which the operating pressure and the temperature, pressure and composition of the feed are specified. This problem too can be solved by an algorithm the kernal of which is an isothermal flash routine; Figure 8.12 is an example. In this case, the molar entropy of the feed should be evaluated at the start and the molar entropy of the two product streams should be evaluated following each isothermal flash calculation. The temperature T is then adjusted by trial until the combined entropy of the product streams is equal to that of the feed. Again, a bisection algorithm (which adjusts T downwards when the entropy of the product streams exceed that of the feed) will suffice.

8.4. Liquid-Liquid Equilibrium

The general problem of LLE and VLLE is a complicated one, especially when more than two liquid phases are involved. The discussion here is limited mainly to the case of two liquid phases coexisting at pressures above the bubble curve; there is then no vapour phase present.

The thermodynamic problem is to determine the composition and the fraction of the material in each of the two coexisting liquid phases formed at a specified temperature and pressure when the overall composition is also specified. This is a liquid-liquid isothermal flash problem. The equality of fugacity between the two liquid phases is written

$$x_{1,i} \, \gamma_{1,i} = x_{2,i} \, \gamma_{2,i} , \qquad (8.15)$$

where subscripts 1 and 2 denote the phase. Introducing a distribution ratio $K_{d,i}$, defined by

$$K_{d,i} = x_{2,i}/x_{1,i} , \qquad (8.16)$$

the material balance is

$$z_i = \beta x_{1,i} + (1-\beta)x_{2,i} = \left[\beta + K_{d,i}(1-\beta)\right]x_{1,i} \qquad (8.17)$$

where β is the fraction of the total amount of substance in phase 1. Solving this for $x_{1,i}$ and summing over all components, the equilibrium condition becomes

$$f(\beta) = \sum_{i=1}^{n} x_{1,i} - 1 = \sum_{i=1}^{n} \frac{z_i}{\beta + K_{d,i}(1-\beta)} - 1 = 0 , \qquad (8.18)$$

and the Newton-Raphson algorithm to solve the problem is

$$\beta_{k+1} = \beta_k + \left[\sum_{i=1}^{n}\left(\frac{z_i}{\beta_k + K_{d,i}(1-\beta_k)}\right) - 1\right]\left[\sum_{i=1}^{n}\left(\frac{(1-K_{d,i})z_i}{[\beta_k + K_{d,i}(1-\beta_k)]^2}\right)\right]^{-1} . \qquad (8.19)$$

Thus if the distribution ratios are known, β can be obtained by successive approximations. In practice, the distribution ratios must be estimated for each trial value of β. The phase compositions can then be obtained from the expressions:

$$x_{1,i} = \frac{z_i}{\beta + K_{d,i}(1-\beta)} \qquad \text{while} \qquad x_{2,i} = K_{d,i} \, x_{1,i} . \qquad (8.20)$$

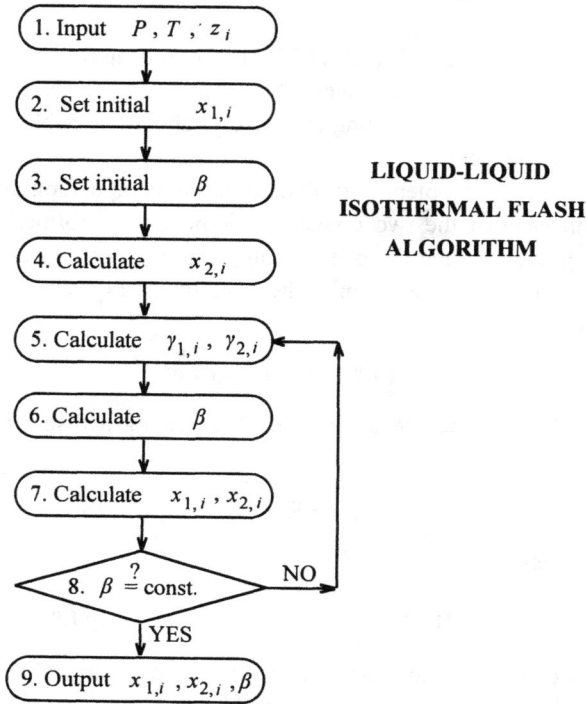

Figure 8.13. Liquid-liquid isothermal flash algorithm.

An algorithm for the solving the problem is given in Figure 8.13. The algorithm involves the following steps:

1. The temperature, pressure and overall composition are specified inputs.
2. Initial values for the mole fractions in liquid phase 1 are assumed.
3. An initial value for β must also be assumed.
4. The initial mole fractions for liquid phase 2 are calculated through Eq.(8.17).
5. The activity coefficients, $\gamma_{1,i}$ and $\gamma_{2,i}$, for each component in both phases are now calculated from an appropriate thermodynamic model (usually an activity-coefficient model but an equation of state could be used). The distribution coefficients $K_{d,i}$ are also evaluated in this step.
6. The fraction of the material in phase 1 is now obtained by successive iterations of Eq. (8.18), subject to the present values of the distribution ratios.
7. Once step 6 has converged on a solution for β, new mole fractions can be

calculated through Eqs.(8.20).

8.   If the value of β changed during step 6 then start a new iteration at step 5; otherwise go to step 9.

9.   A solution of the phase-equilibrium conditions has been found.

Notice that the strategy applied here involves separate iteration loops for determining β and the distribution ratios. This contrasts with the combined iteration loop used in the corresponding VLE flash problem. Of course, the existence of two liquid phases must be known in advance and it is certainly helpful to have a rough idea of the compositions of the phases so that a solution can be found in relatively few iterations. Program FLASH2L, given in Section 8.5 and applied in Example 8.8, implements this procedure with the UNIQUAC activity-coefficient model.

In the case where the pressure is below the bubble pressure, a vapour phase will be present. If the pressure is low, as it typically is in common processes involving VLLE, and the vapour volume is small then the presence of the additional phase will have little effect. The bubble pressure and composition of the vapour phase may be determined by considering equilibrium between the vapour and one of the liquid phases as in section 8.2. Since liquids L1 and L2 are in equilibrium, equilibrium between L1 and vapour V implies also equilibrium between L2 and V. The general three-phase vapour-liquid-liquid flash problem may be solved by an extension of the methods described above but we shall not explore the matter further here.

8.5. Computational Implementation

In this section, computer routines implementing the algorithms described in this chapter are presented. We begin with bubble-point routines which use either an equation of state or an activity-coefficient model to evaluate component fugacities. Next, programs implementing the same models are given for isothermal and isenthalpic vapour-liquid flash calculations. Finally, a routine that exploits an activity-coefficient model for liquid-liquid flash calculations is presented. Use of the routines is illustrated through several examples.

8.5.1. VLE Calculations

Display 8.1 lists a simple computer routine which implements the algorithm shown in Figure 8.4 in a calculation of the bubble point for a fluid mixture of specified composition at a specified temperature or pressure. The Peng-Robinson equation of state (subroutine PR, Display 6.2) is used for the evaluation of component fugacities.

For simplicity, a bisection algorithm is used to update the unknown bubble-point temperature or pressure after each iteration (step 10 in Figure 8.4). Upper and lower limits for this quantity are required inputs and the initial estimate is taken as the mean of these values.

Display 8.1   BUBLPR List

```
      SUBROUTINE BUBLPR
     $(T,P,ITP,WLO,WHI,IS,X,Y,TC,PC,BK,OMEG,FUGL,FUGV)
      DIMENSION X(IS),Y(IS),TC(IS),PC(IS),OMEG(IS),FUGV(IS),FUGL(IS)
      DIMENSION BK(IS,IS)
C
C
C     The subroutine calculates vapour mole fractions Y(I) of a mixture
C     of IS components, and
C     a) ITP=1, finds bubble-point temperature T in [K],
C               given the pressure P in [Pa], or
C     b) ITP=2, finds bubble-point pressure P in [MPa]
C               given the temperature T in [K].
C     Method adopted is based on the calculation of the equilibrium
C     constant from the liquid fugacity coefficients, FUGL(I), and the
C     vapour fugacity coefficient FUGV(I) from the Peng & Robinson EoS
C     - Additional input to the subroutine are liquid mole fractions
C     X(I), critical constants TC(I) in [K], and PC(I) in [Pa], the
C     acentric factor OMEG(I), and the PR binary interaction parameters
C     BK(I,J).
C     - Also first and last value, WLO and WHI, for the temperature or
C     pressure iteration must be provided.
C     - Output includes the liquid and vapour fugacity coefficients.
C
C.......Initial Guesses for Y(I)
      S1=0.
      DO 100 I=1,IS
      Y(I)=X(I)
 100  S1=S1+Y(I)
 20   PROP=(WLO+WHI)/2.
      IF (ITP.EQ.1) T=PROP
      IF (ITP.EQ.2) P=PROP
      WRITE (*,*) PROP
 30   CALL PR(T,P,2,IS,Y,TC,PC,OMEG,Y1,Y2,Y3,Y4,BK,D,H,S,FUGV)
      CALL PR(T,P,1,IS,X,TC,PC,OMEG,Y1,Y2,Y3,Y4,BK,D,H,S,FUGL)
```

```
      CHKT=ABS(FUGV(1)-FUGL(1))*100./FUGV(1)
      IF (CHKT.GT.0.001) GOTO 40
      WRITE (*,*) ' NO CONVERGENCE - TRY CLOSER TEMPERATURE LIMITS'
      STOP
  40  SY=0.
      DO 120 I=1,IS
      Y(I)=X(I)*FUGL(I)/FUGV(I)
 120  SY=SY+Y(I)
      DO 140 I=1,IS
 140  Y(I)=Y(I)/SY
      IF (ABS(SY-S1).LE.0.000001) GOTO 50
      S1=SY
      GOTO 30
C     Temperature (ITP=1) or Pressure (ITP=2) iteration
  50  IF (SY.GE.0.99995.AND.SY.LE.1.00005) GOTO 60
      IF (ITP.EQ.1.AND.SY.GT.1.00005) WHI=T
      IF (ITP.EQ.1.AND.SY.LT.0.99995) WLO=T
      IF (ITP.EQ.2.AND.SY.GT.1.00005) WLO=P
      IF (ITP.EQ.2.AND.SY.LT.0.99995) WHI=P
      GOTO 20
  60  RETURN
      END
```

Inputs to the subroutine are:
-   the specified temperature, T, in Kelvin or the specified pressure, P, in Pascal.
-   flag ITP set to 1 for specified pressure or 2 for specified temperature.
-   upper and lower limits for the unknown bubble-point temperature or pressure specified as WLO and WHI in units of Kelvin or Pascal.
-   the number of components of the mixture, IS
-   liquid-phase mole fractions as array X(IS),
-   critical temperature in Kelvin, critical pressure in Pascal and acentric factor for each component, in arrays TC(IS), PC(IS) and OMEG(IS).
-   The Peng-Robinson binary interaction parameters, BK(IS,IS) should also be given.

The output is as follows:
-   the vapour mole fractions, as array Y(IS).
-   the bubble-point temperature T or pressure P (according to flag ITP), and
-   the liquid and vapour fugacity coefficients for every component, as arrays FUGL(IS) and FUGV(IS).

Examples 8.1 and 8.2 demonstrates the application of this subroutine.

| EXAMPLE 8.1 | *Calculation of the Bubble-Point Temperature and Vapour Composition for the Liquid Mixture {0.055 Ethane(1) + 0.945 n-Heptane(2)} at P = 1.0 MPa.* |

Subroutine BUBLPR is employed with flag I T P = 1. The critical constants and the acentric factors are obtained from Table A1, Appendix A. The Peng - Robinson binary interaction parameter is 0.01 [7].

The calling program and its output are shown in Display 8.2 (only the last few iterations over the temperature are shown). Note that by a small change, the bubble-point pressure can be obtained. The bubble-point temperature produced by the corresponding-states package SUPERTRAPP [8] for this system (see also Chapter 10), is 427 K with equivalent vapour mole fractions 0.548 for ethane and 0.451 for n-heptane. The agreement is good.

DISPLAY 8.2 BUBLPR Sample Input/Output

```
      PROGRAM EX0801
      PARAMETER (IS=2)
      DIMENSION X(IS),Y(IS),TC(IS),PC(IS),OMEG(IS),FUGV(IS),FUGL(IS)
      DIMENSION BK(IS,IS)
      DATA (TC (I),I=1,IS)/305.33,540.15/
      DATA (PC (I),I=1,IS)/4.871E+6,2.735E+6/
      DATA (OMEG(I),I=1,IS)/0.099,0.349/
      DATA BK(1,1),BK(1,2),BK(2,1),BK(2,2)/0.,0.01,0.01,0./
      DATA ITP,P,WLO,WHI,X/1,1.0E+6,410,,450,,0.055,0.945/
C     DATA ITP,T,WLO,WHI,X/2,430.13,0.3E+6,1.5E+6,0.055,0.945/
      CALL BUBLPR(T,P,ITP,WLO,WHI,IS,X,Y,TC,PC,BK,OMEG,FUGL,FUGV)
   40 WRITE (*,1100) T,P*1.E-6
 1100 FORMAT(3X,/,4X,'>>>  T=',F7.2,'K  &  P=',F6.3,'MPA',/,4X,70('-'),/,
     &11X,'X(I)',4X,'Y(I)',10X,'FUG.LIQ.',6X,'FUG.VAP.',11X,'K')
      DO 60 J=1,IS
   60 WRITE (*,1200) J,X(J),Y(J),FUGL(J),FUGV(J),FUGL(J)/FUGV(J)
 1200 FORMAT(4X,'(',I1,')',F8.4,F9.4,2X,2(4X,F10.5),6X,F10.5)
      END
```

```
428.5938
428.6719
428.7109


   >>>  T= 428.71K  &  P=1.000MPA
---------------------------------------------------------------------
            X(I)         Y(I)          FUG.LIQ.      FUG.VAP.         K
   (1)    0.0550       0.5316         9.75669       1.00934      9.66645
   (2)    0.9450       0.4684         0.38259       0.77192      0.49563
```

□

EXAMPLE 8.2   *Calculation of the Binary Interaction Parameter k_{12} in the Peng-Robinson Equation of State for the System $(x_1\ CO_2 + x_2\ C_2H_6)$. Experimental VLE Data are Available at 263.15 K.*

Binary interaction parameters in equations of state are usually determined in a fit to binary VLE data at one or more temperatures. In the present case, experimental dew- and bubble-point data are available across the full composition range at 263.15 K [9] as shown in Figure 8.14. The system is highly non-ideal with an azeotrope at $x_1 = 0.69$. In order to determine k_{12} from these data, subroutine BUBLPR is employed at 263.15 K for various trial values of k_{12}. For each specified liquid composition, the routine returns the pressure and the vapour composition thus allowing both the dew and bubble curves to be constructed. With $k_{12} = 0$, nearly ideal mixture behaviour is predicted as shown by the broken line in Figure 8.14.

Table 8.1   Azeotropic compositions for the system $\{CO_2(1) + C_2H_6(2)\}$ predicted by the Peng-Robinson equation of state with various values of k_{12}.

k_{12}	0.05	0.10	0.15	0.124
x_1	0.77	0.72	0.67	0.69

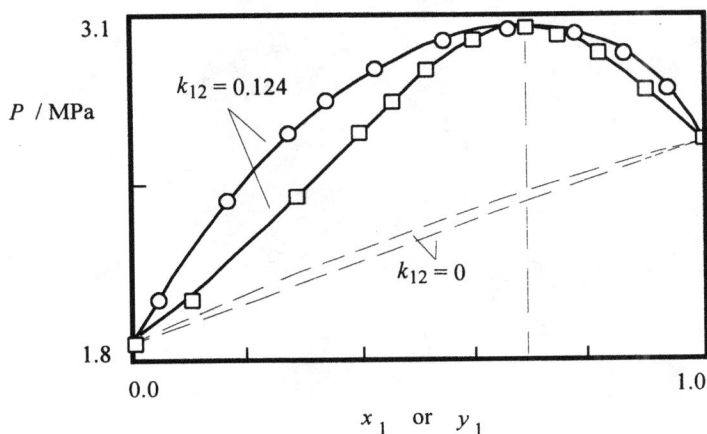

Figure 8.14. Vapour-liquid equilibrium in the system $\{CO_2(1) + C_2H_6(2)\}$. Calculated from the Peng-Robinson equation with $k_{12} = 0.124$ and $k_{12} = 0$. Symbols are experimental results [9].

Positive values of k_{12} lead to the prediction of an azeotrope with the compositions given in Table 8.1; the optimum value appears to be $k_{12} = 0.124$. The full dew and bubble curves computed with $k_{12} = 0.124$ are shown in Figure 8.14 and agree closely with the experimental data. VLE predictions made using this value of k_{12} at other temperatures also agree well with experiment while predictions based on the 'default' value $k_{12} = 0$ are clearly of little value.

The calling program employed for the optimisation of the binary interaction parameters and its output is shown in Display 8.3 for the final case of $k_{12} = 0.124$. Note that in this example the temperature is given as 263.15 K and the iteration is done to obtain the pressure.

```
                                      DISPLAY 8.3  BUBLPR Sample Input/Output
        PROGRAM EX0802
        PARAMETER (IS=2)
        DIMENSION X(IS),Y(IS),TC(IS),PC(IS),OMEG(IS),FUGV(IS),FUGL(IS)
        DIMENSION BK(IS,IS)
        DATA (TC  (I),I=1,IS)/304.1,305.33/
        DATA (PC  (I),I=1,IS)/7.380e+6,4.871E+6/
        DATA (OMEG(I),I=1,IS)/0.239,0.099/
        DATA BK(1,1),BK(1,2),BK(2,1),BK(2,2)/0.,0.124,0.124,0./
        DATA ITP,T,WLO,WHI,X/2,263.15,2.9E+6,3.3E+6,0.69,0.31/
        CALL BUBLPR(T,P,ITP,WLO,WHI,IS,X,Y,TC,PC,BK,OMEG,FUGL,FUGV)
   40   WRITE (*,1100) T,P*1.E-6
 1100   FORMAT(3X,/,4X,'>>>  T=',F7.2,'K  &  P=',F6.3,'MPA',/,4X,70('-'),/,
       &11X,'X(I)    Y(I)',10X,'FUG.LIQ',6X,'FUG.VAP.',11X,'K')
        DO 60 J=1,IS
   60   WRITE (*,1200) J,X(J),Y(J),FUGL(J),FUGV(J),FUGL(J)/FUGV(J)
 1200   FORMAT(4X,'(',I1,')',F8.4,F9.4,2X,2(4X,F10.5),6X,F10.5)
        END


 3100000.
 3000000.
 3050000.
 3025000.


  >>>  T= 263.15K  &  P=3.025MPA
  ------------------------------------------------------------
           X(I)         Y(I)          FUG.LIQ.      FUG.VAP.         K
    (1)    0.6900       0.6900        0.77723       0.77729       0.99993
    (2)    0.3100       0.3100        0.70573       0.70570       1.00005
```

Multiple calls on the routine BUBLPR can be used to construct easily both *P-x-y* and *T-x-y* phase diagrams for binary systems or to plot *P-T* phase envelopes for multi-component systems.

❑

For those cases in which an activity-coefficient model is preferred for the liquid phase, the routine given in Display 8.4 may be used. This program implements the algorithm shown in Figure 8.5 in a calculation of the bubble point for a fluid mixture of specified composition at a specified temperature or pressure. The T-K-Wilson activity-coefficient model is used for the liquid phase and the virial equation of state truncated after the second virial coefficient for the vapour phase. Routine TKWILSON (Display 7.1) is used to evaluate the activity coefficients and, for simplicity, the Poynting Factor is set equal to unity. Pure component and interaction second virial coefficients are estimated from the corresponding-states correlation of Tsonopoulos with Lorentz-Berthelot mixing rules for the critical constants (Table 4.3).

```
                                              DISPLAY 8.4  BUBLVW List
       SUBROUTINE BUBLVW
      &(IS,P,T,ITP,WLO,WHI,X,Y,TC,PC,OMEG,SLR,V,PSAT,EK,FPUR,F,G,VP)
C
C      The subroutine calculates vapour mole fractions Y(I) of a mixture
C      of IS components, and
C      a) ITP=1, finds bubble-point temperature T in [K],
C               given the pressure P in [Pa], or
C      b) ITP=2, finds bubble-point pressure P in [MPa],
C               given the temperature T in [K].
C
C      Method adopted is based on:
C      1. Activity coeficients, G(I),(supplied by another subroutine e.g. TKWILSON), and
C      2. Virial equation truncated to second coefficient (Tsonopoulos
C               expressions with Lorentz-Berthelot mixing rules)
C      - Additional input to the subroutine are liquid mole fractions
C      X(I), critical constants TC(I,I) in [K], and PC(I,I) in [Pa]
C      acentric factor OMEG(I,I), and Antoine equation coeffs. VP(IS,3).
C      Also first and last value, WLO and WHI, for the temperature or
C      pressure iteration must be provided.
C
C      - Output includes vapour pressure PSAT(I), equilibrium constant
C      EK(I),pure fugacity coef.FPUR(I) and component fugacity coef.F(I).
C
       DIMENSION X(IS),Y(IS),PSAT(IS),EK(IS),FPUR(IS),F(IS),G(IS)
       DIMENSION TC(IS,IS),PC(IS,IS),OMEG(IS,IS),VP(IS,3)
       DIMENSION SLR(IS,IS),V(IS,IS)
       DIMENSION B(10,10),VC(10,10),ZC(10,10)
       R=8.3145
C......Mixture Critical TC(I,J), PC(I,J) (Lorentz-Berthelot Mixing Rules)
       DO 100 I=1,IS
 100   VC(I,I)=(0.291-0.080*OMEG(I,I))*R*TC(I,I)/PC(I,I)
       DO 120 I=1,IS
       DO 120 J=1,IS
```

```
      OMEG(I,J)=(OMEG(I,I)+OMEG(J,J))/2.
      ZC(I,J)=0.291-0.080*OMEG(I,J)
      VC(I,J)=(1./8.)*((VC(I,I)**(1./3.))+(VC(J,J)**(1./3.)))**3
      TC(I,J)=SQRT(TC(I,I)*TC(J,J))
120   PC(I,J)=ZC(I,J)*R*TC(I,J)/VC(I,J)
C.....   Activity Coef. G(I), Virial Coef. B(I,J) (Tsonopoulos Equations)
10    PROP=(WLO+WHI)/2.
      IF (ITP.EQ.1) T=PROP
      IF (ITP.EQ.2) P=PROP
      WRITE(*,*) PROP
      CALL TKWILSON(IS,X,T,SLR,V,G,1)
      DO 140 I=1,IS
      DO 140 J=1,IS
      TRI=TC(I,J)/T
140   B(I,J)=R*TC(I,J)/PC(I,J)*(0.1445-0.33*TRI-0.1385*(TRI**2)-0.0121*
     &(TRI**3)-0.000607*(TRI**8)+OMEG(I,J)*(0.0637+0.331*(TRI**2)-0.423*
     &(TRI**3)-0.008*(TRI**8)))
C.....   First Y(I)'s
      S1=0.
      DO 160 J=1,IS
      FPUR(J)=EXP(B(J,J)*P/R/T)
      PSAT(J)=1000.*(10.**(VP(J,1)-VP(J,2)/(T+VP(J,3))))
      Y(J)=G(J)*PSAT(J)*FPUR(J)/P
160   S1=S1+Y(J)
      DO 180 J=1,IS
180   Y(J)=Y(J)/S1
C........  Pure Fugacity Coef. FPUR(I), Comp Fugacity Coef. F(I),
C          Sat.Pressure PSAT(I), Equil.Constant EK(I), & Y(I)s
20    BMIX=0.
      DO 200 I=1,IS
      DO 200 J=1,IS
200   BMIX=BMIX+Y(I)*Y(J)*B(I,J)
      SY=0
      DO 240 I=1,IS
      FTERM=0.
      DO 220 K=1,IS
220   FTERM=FTERM+Y(K)*B(K,I)
      F(I)=EXP(P/R/T*(2.*FTERM-BMIX))
      FPUR(I)=EXP(B(I,I)*P/R/T)
      PSAT(I)=1000.*(10.**(VP(I,1)-VP(I,2)/(T+VP(I,3))))
      EK(I)=G(I)*FPUR(I)*PSAT(I)/(P*F(I))
      Y(I)=EK(I)*X(I)
240   SY=SY+Y(I)
      DO 260 I=1,IS
260   Y(I)=Y(I)/SY
      IF (ABS(SY-S1).LE.0.000001) GOTO 30
      S1=SY
      GOTO 20
```

```
C        Temperature (IPT=1) or Pressure (IPT=2) iteration
   30   IF (SY.GE.0.99995.AND.SY.LE.1.00005) GOTO 40
        IF (ITP.EQ.1.AND.SY.GT.1.00005) WHI=T
        IF (ITP.EQ.1.AND.SY.LT.0.99995) WLO=T
        IF (ITP.EQ.2.AND.SY.GT.1.00005) WLO=P
        IF (ITP.EQ.2.AND.SY.LT.0.99995) WHI=P
        GOTO 10
   40   RETURN
        END
```

Use of the truncated virial equation and neglect of the Poynting correction limit the routine to low pressures, roughly up to 0.5 MPa. For higher pressures, an improved equation of state and the Poynting correction should both be adopted.

Inputs to the subroutine are:
-   the specified temperature, T, in Kelvin or the specified pressure, P, in Pascal.
-   flag ITP set to 1 for specified pressure or 2 for specified temperature.
-   upper and lower limits for the unknown bubble-point temperature or pressure specified as WLO and WHI in units of Kelvin or Pascal.
-   the number of components of the mixture, IS
-   liquid-phase mole fractions as array X(IS),
-   critical temperature in Kelvin, critical pressure in Pascal and acentric factor for each component, in arrays TC(IS), PC(IS) and OMEG(IS).
-   the constants of the Antoine vapour-pressure equation for each component, as array VP(IS,3)
-   the molar volume ratios V_{ij} and parameters λ_{ij}/R, as arrays V(IS,IS) and SLR(IS,IS) for the subroutine TKWILSON which is called by BUBLVW for the evaluation of activity coefficients.

The output is as follows:
-   the vapour mole fractions, as array Y(IS).
-   the bubble-point temperature T or pressure P (according to flag ITP),
-   the pure-component fugacity coefficients at saturation and (mixture) vapour fugacity coefficients for every component, as arrays FPUR(IS) and F(IS).
-   the pure-component vapour pressures, as array PSAT(I), and
-   the vapourisation equilibrium ratios, as array EK(I).

Examples 8.3, 8.4 and 8.5 demonstrate the application of this subroutine in the estimation of bubble-point temperatures and pressures.

EXAMPLE 8.3

Calculation of the Bubble-Point Temperature and the Vapour Composition for the Liquid Mixture {x_1 Ethanol(1) + x_2 Methylcyclopentane(2) + x_3 Benzene(3)} at P = 0.1 MPa with compositions as follows: a) x_1=0.878, x_2=0.068, x_3=0.054 and b)) x_1=0.569, x_2=0.359, x_3=0.072.

Subroutine BUBLVW is used with ITP=1. The critical constants and acentric factors are obtained from Table A1, Appendix A, while the coefficients of the Antoine equation, are taken from Table A2. The liquid molar volumes were estimated by calculation from the BWR(HS) equation of state with the results $V_1 = 61.72$ cm^3/mol, $V_2 = 120.36$ cm^3/mol, $V_3 = 94.40$ cm^3/mol. Wilson parameters, λ_{ij}/R are obtained from Table A4 but, in this case, alternative values are also available in the literature [4] and these are used too.

The calling program is shown in Display 8.5 for the first liquid composition. The results obtained with both sets of Wilson parameters are given in Table 8.2 together with experimental data [4] for this mixture.

Table 8.2 Comparison of Results.

		x_i	Experimental values [4] T / K	y_i	BUBLVW λ_{ij}/R from [4] T / K	y_i	BUBLVW λ_{ij}/R from Table A4 T / K	y_i
a)	(1)	0.878	340.85	0.594	341.21	0.599	341.89	0.615
	(2)	0.068		0.296		0.287		0.272
	(3)	0.054		0.110		0.114		0.113
b)	(1)	0.569	334.05	0.386	333.84	0.380	334.85	0.382
	(2)	0.359		0.538		0.547		0.545
	(3)	0.072		0.076		0.073		0.073

Although, as expected, the parameters λ_{ij}/R derived in [4] produce slightly better results, both sets work well and the maximum error in the bubble-point temperature is only about 1 K. The maximum error in the vapour mole fractions is about 0.02.

```
                                    DISPLAY 8.5  BUBLVW Sample Input/Output
    PROGRAM EX0803
    PARAMETER (IS=3)
    DIMENSION X(IS),Y(IS),PSAT(IS),EK(IS),FPUR(IS),F(IS),G(IS)
    DIMENSION TC(IS,IS),PC(IS,IS),OMEG(IS,IS),VP(IS,3)
    DIMENSION SLR(IS,IS),V(IS,IS)
    DATA (TC(J,J),J=1,IS)/513.88,532.7,562.06/
```

```
      DATA (PC(J,J),J=1,IS)/6.132E+6,3.780E+6,4.895E+6/
      DATA (OMEG(J,J),J=1,IS)/0.644,0.231,0.212/
      DATA (SLR(1,I),I=1,IS)/0.,1109.61,699.11 /
      DATA (SLR(2,I),I=1,IS)/123.40,0.,6.693/
      DATA (SLR(3,I),I=1,IS)/ 63.06,125.16,0./
      DATA (VP(1,I),I=1,3)/7.24222,1595.811,-46.702/
      DATA (VP(2,I),I=1,3)/5.99178,1188.329,-46.843/
      DATA (VP(3,I),I=1,3)/6.01905,1204.637,-53.081/
      DATA (V(1,I),I=1,IS)/1.,0.5128,0.6538/
      DATA (V(2,I),I=1,IS)/1.9501,1.,1.2750/
      DATA (V(3,I),I=1,IS)/1.5295,0.7843,1./
      DATA ITP,P,WLO,WHI,X/1,0.1E+6,200.,600.,0.878,0.068,0.054/
      CALL BUBLVW
     &(IS,P,T,ITP,WLO,WHI,X,Y,TC,PC,OMEG,SLR,V,PSAT,EK,FPUR,F,G,VP)
      WRITE (*,1100) T,P*1.E-6
1100  FORMAT(3X,/,4X,'>>> T=',F7.2,'K & P=',F6.3,'MPA',/,4X,70('-'),/,
     &11X,'X(I) Y(I)  PSAT    K     FPUR     F      G')
      DO 20 J=1,IS
20    WRITE (*,1200) J,X(J),Y(J),PSAT(J)/1.E+6,EK(J),FPUR(J),F(J),G(J)
1200  FORMAT(4X,'(',I1,')',F8.4,F9.4,5F10.5)
      END
```

```
>>> T= 341.21K & P=0.100MPA,                              SUM OF Y(I)= 1.0000
------------------------------------------------------------------------------
        X(I)      Y(I)       PSAT        K        FPUR       F         G
  (1)  0.8780   0.5988     0.06664   0.68205   0.96907   0.96938   1.02318
  (2)  0.0680   0.2874     0.09014   4.22665   0.95847   0.95908   4.68581
  (3)  0.0540   0.1138     0.06890   2.10721   0.96111   0.96202   3.05551
```

❑

| **EXAMPLE 8.4** | *Calculation of the Bubble-point Pressure and Vapour Composition for the Liquid Mixture {x₁ n-Propanol + x₂ Water} at T = 362.35 K with liquid compositions as follows: (a) x₁=0.179, and (b) x₁=0.712.* |

Calculation of the Bubble-point Pressure and Vapour Composition for the Liquid Mixture $\{x_1$ n-Propanol $+ x_2$ Water$\}$ at $T = 362.35$ K with liquid compositions as follows: (a) $x_1=0.179$, and (b) $x_1=0.712$.

Subroutine BUBLVW is used with ITP = 2. The critical constants and acentric factors are obtained from Table A1, Appendix A, while the coefficients of the Antoine equation are taken from Table A2. The liquid molar volumes estimated from the BWR(HS) equation of state are $V_1 = 81.18$ cm³/mol and $V_2 = 18.73$ cm³/mol. Wilson parameters λ_{ij}/R are obtained from Table A4, Appendix A.

The calling program and sample output for composition (a) are shown in Display 8.6 while, in Table 8.3, all results are shown for both compositions, together with experimental results obtained by Murti and Winkle [5]. The agreement in both the bubble-point pressure and the vapour composition is reasonably good but not outstanding. Note, that a small change to the calling program permits the bubble-point temperature to be determined instead of the pressure.

Table 8.3 Comparison of Results.

	x_i	Expt. values [5] P / MPa	y_i	BUBLVW P / MPa	y_i
(1)	0.179	0.101	0.388	0.095	0.356
(2)	0.821		0.612		0.644
(1)	0.712	0.101	0.560	0.098	0.579
(2)	0.288		0.440		0.421

DISPLAY 8.6 Sample Input/Output

```
      PROGRAM EX0804
      PARAMETER (IS=2)
      DIMENSION X(IS),Y(IS),PSAT(IS),EK(IS),FPUR(IS),F(IS),G(IS)
      DIMENSION TC(IS,IS),PC(IS,IS),OMEG(IS,IS),VP(IS,3)
      DIMENSION SLR(IS,IS),V(IS,IS)
      DATA (TC(J,J),J=1,IS)/536.74,647.3/
      DATA (PC(J,J),J=1,IS)/5.168E+6,22.100E+6/
      DATA (OMEG(J,J),J=1,IS)/0.623,0.344/
      DATA (SLR(1,I),I=1,IS)/0.,529.10/
      DATA (SLR(2,I),I=1,IS)/598.08,0./
      DATA (VP(1,I),I=1,3)/6.87065,1438.587,-74.598/
      DATA (VP(2,I),I=1,3)/7.06252,1650.270,-46.804/
      DATA (V(1,I),I=1,IS)/1.,4.4184/
      DATA (V(2,I),I=1,IS)/0.2307,1./
      DATA ITP,T,WLO,WHI,X/2,361.10,0.06E+6,0.15E+6,0.179,0.821/
C     DATA ITP,P,WLO,WHI,X/1,0.095E+6,350,370,0.179,0.821/
      CALL BUBLVW
     &(IS,P,T,ITP,WLO,WHI,X,Y,TC,PC,OMEG,SLR,V,PSAT,EK,FPUR,F,G,VP)
      WRITE (*,1100) T,P*1.E-6
1100  FORMAT(3X,/,4X,'>>> T=',F7.2,'K  &  P=',F6.3,'MPA'/,4X,70('-'),/,
     &11X,'X(I)    Y(I)    PSAT      K      FPUR      F      G')
      DO 20 J=1,IS
20    WRITE (*,1200) J,X(J),Y(J),PSAT(J)/1.E+6,EK(J),FPUR(J),F(J),G(J)
1200  FORMAT(4X,'(',I1,')',F8.4,F9.4,5F10.5)
      END


95189.21
95183.71

  >>> T= 361.10K  & P=0.095MPA                      SUM OF Y(I)= 1.0000
  ------------------------------------------------------------------------
          X(I)      Y(I)      PSAT       K        FPUR       F        G
  (1)   0.1790   0.3562    0.07070   1.98986    0.96823  0.96781  2.68000
  (2)   0.8210   0.6438    0.06484   0.78418    0.98656  0.98643  1.15135
```

❑

| **EXAMPLE 8.5** | *Construction of the x-y Composition Diagram for the Mixture {x₁ Ethanol + x₂ Benzene} at T = 333 K, 353 K and 373 K.* |

Construction of the x-y Composition Diagram for the Mixture $\{x_1$ Ethanol $+ x_2$ Benzene$\}$ at $T = 333$ K, 353 K and 373 K.

This kind of composition diagram gives a convenient summary of the phase behaviour from which azeotropes may easily be identified. Subroutine BUBLVW is employed with flag ITP = 2 and run for a series of liquid compositions at each temperature.

Critical temperatures and pressures and acentric factors for each component are obtained from Table A1 in Appendix A, while the coefficients for the Antoine equation and the Wilson parameters λ_{ij}/R are obtained from Tables A2 and A4. Liquid molar volumes are estimated from the BWR(HS) equation of state at the approximate pressure PARR. Binary interaction parameters here are set to zero.

The calling program and its output for the isotherm at 353 K are listed in Display 8.7. The results are plotted in Figure 8.15 for the three temperatures.

DISPLAY 8.7 BUBLVW Sample Input/Output

```
      PROGRAM EX0805
      PARAMETER (IS=2)
      DIMENSION X(IS),Y(IS),PSAT(IS),EK(IS),FPUR(IS),F(IS),G(IS)
      DIMENSION TC(IS,IS),PC(IS,IS),DC(IS,IS),OMEG(IS,IS),VP(IS,3)
      DIMENSION SLR(IS,IS),V(IS,IS),DD(IS),BK(IS,IS)
      DATA (TC(J,J),J=1,IS)/513.88,562.06/
      DATA (PC(J,J),J=1,IS)/6.132E+6,4.895E+6/
      DATA (DC(J),J=1,IS)/5984.4,3861.0/
      DATA (OMEG(J,J),J=1,IS)/0.644,0.212/
      DATA (SLR(1,I),I=1,IS)/ 0.,653.13/
      DATA (SLR(2,I),I=1,IS)/66.16,0./
      DATA (VP(1,I),I=1,3)/7.24222,1595.811,-46.702/
      DATA (VP(2,I),I=1,3)/6.01905,1204.637,-53.08/
      DATA BK(1,1),BK(1,2),BK(2,1),BK(2,2)/0.,0.,0.,0./
      DATA ITP,T,PAPR/2,353.,0.15E+6/
C........Components molar volumes through BWR
      CALL BWR
     &(T,PAPR,1,1,1.,TC(1,1),DC(1),OMEG(1,1),Y1,Y2,Y3,Y4,BK,DD(1),H,S,FUG)
      CALL BWR
     &(T,PAPR,1,1,1.,TC(2,2),DC(2),OMEG(2,2),Y1,Y2,Y3,Y4,BK,DD(2),H,S,FUG)
      WRITE (*,1000)
 1000 FORMAT(7X,'P [MPA]',6X,'X(1)',6X,'Y(1)',10X,'X(2)',6X,'Y(2)',8X
     &,'K(1)/K(2)')
      DO 20 I=1,IS
      DO 20 J=1,IS
   20 V(I,J)=(1./DD(I))/(1./DD(J))
      DO 40 K=1,9
      X(1)=0.1*FLOAT(K)
```

```
        X(2)=1.-X(1)
        WLO=0.05E+6
        WHI=0.30E+6
        CALL BUBLVW
        &(IS,P,T,ITP,WLO,WHI,X,Y,TC,PC,OMEG,SLR,V,PSAT,EK,FPUR,F,G,VP)
  40    WRITE (*,1100) P*1.E-6,X(1),X(2),Y(1),Y(2),EK(1)/EK(2)
1100    FORMAT(5X,F8.3,4X,2(F8.4,2X),4X,2(F8.4,2X),4X,F8.4)
        END
```

P [MPA]	X(1)	X(2)	Y(1)	Y(2)	K(1)/K(2)
0.133	0.1000	0.9000	0.3034	0.6966	3.9195
0.144	0.2000	0.8000	0.3892	0.6108	2.5484
0.149	0.3000	0.7000	0.4333	0.5667	1.7841
0.151	0.4000	0.6000	0.4654	0.5346	1.3058
0.151	0.5000	0.5000	0.4955	0.5045	0.9823
0.150	0.6000	0.4000	0.5298	0.4702	0.7510
0.148	0.7000	0.3000	0.5746	0.4254	0.5789
0.142	0.8000	0.2000	0.6413	0.3587	0.4469
0.130	0.9000	0.1000	0.7555	0.2445	0.3434

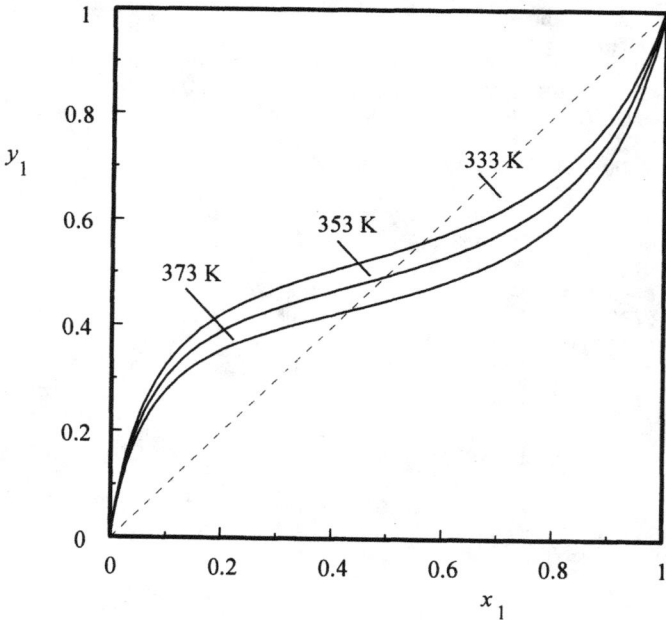

Figure 8.15. Composition diagram for $\{x_1 \text{ ethanol} + x_2 \text{ benzene}\}$.

8.5.2. *Isothermal Flash Calculations*

Subroutine FLASHTP listed in Display 8.8 implements the isothermal flash algorithm of Figure 8.7 and makes use of exactly the same thermodynamic models as described above in connection with the bubble-point routine BUBLVW. The routine is therefore suitable for isothermal flash calculations at low pressures (roughly less than 0.5 MPa) and at temperatures such that $T < T_i^c$ for all components. There is no restriction in principle on the number of components.

Inputs to the subroutine are:
- the flash temperature T and pressure P in units of Kelvin and Pascal.
- the number of components in the feed, IS
- overall composition (mole fractions) of the feed as array Z(IS),
- critical temperature in Kelvin, critical pressure in Pascal and acentric factor for each component, in arrays TC(IS), PC(IS) and OMEG(IS).
- the constants of the Antoine vapour-pressure equation for each component, as array VP(IS,3)
- the molar volume ratios V_{ij} and parameters λ_{ij}/R, as arrays V(IS,IS) and SLR(IS,IS) for the subroutine TKWILSON which is called by BUBLVW for the evaluation of activity coefficients.

The output is as follows:
- the compositions (mole fractions) of the liquid and vapour output streams, as arrays X(IS) and Y(IS).
- the vapour fraction BETA.
- the pure-component fugacity coefficients at saturation and the (mixture) vapour fugacity coefficients for every component, as arrays FPUR(IS) and F(IS).
- the liquid-phase activity coefficients, as array G(IS).
- the pure-component vapour pressures, as array PSAT(I),
- the vapourisation equilibrium ratios, as array EK(I).

Example 8.6 demonstrates the application of this subroutine to the case of a three-component mixture.

DISPLAY 8.8 FLASHTP List

SUBROUTINE FLASHTP
&(IS,P,T,X,Y,Z,TC,PC,OMEG,VP,V,SLR,PSAT,EK,FPUR,F,G,BETA)

C   The subroutine performs a Flash calculation at given P [Pa],T [K]
C   and the composition of the feed Z(I) of a multicomponent mixture
C   of IS components
C   Method adopted is based on

```
C          a) Activity coeficients, G(I),(supplied by another subroutine e.g.TKWILSON), and
C          b) Virial equation truncated to second coefficient (Tsonopoulos
C             expressions with Lorentz-Berthelot mixing rules).
C          - Additional input to the subroutine are critical constants
C          TC(I,I) in [K], and PC(I,I) in [Pa], acentric factor OMEG(I.I),
C          and Antoine equation coeffs. VP(IS,3). For use by the TKWILSON
C          routine the molar volume ratios V(I,J) and the Wilson parameters
C          lamda/R, SLR(I,J) must be supplied.
C          - Output includes vapour pressure PSAT(I), equilibrium constant
C          EK(I),pure fugacity coef.FPUR(I), component fugacity coef. F(I)
C          activity coefficients G(I), and the kmoles of vapour/feed, BETA
C
           DIMENSION X(IS),Y(IS),Z(IS),PSAT(IS),EK(IS),FPUR(IS),F(IS),G(IS)
           DIMENSION TC(IS,IS),PC(IS,IS),OMEG(IS,IS),VP(IS,3),V(IS,IS)
           DIMENSION SLR(ISMIS),B(10,10),VC(10,10),ZC(10,10)
           R=8.3145
C......... Mixture Critical TC(I,J), PC(I,J) (Lorentz-Berthelot Mixing Rules)
           DO 100 I=1,IS
    100    VC(I,I)=(0.291-0.080*OMEG(I,I))*R*TC(I,I)/PC(I,I)
           DO 120 I=1,IS
           DO 120 J=1,IS
           OMEG(I,J)=(OMEG(I,I)+OMEG(J,J))/2
           ZC(I,J)=0.291-0.080*OMEG(I,J)
           VC(I,J)=(1./8.)*((VC(I,I)**(1./3.))+(VC(J,J)**(1./3.)))**3
           TC(I,J)=SQRT(TC(I,I)*TC(J,J))
    120    PC(I,J)=ZC(I,J)*R*TC(I,J)/VC(I,J)
C......... Virial Coefficients B(I,J) (Tsonopoulos Equations)
           DO 140 I=1,IS
           DO 140 J=1,IS
           TRI=TC(I,J)/T
    140    B(I,J)=R*TC(I,J)/PC(I,J)*(0.1445-0.33*TRI-0.1385*(TRI**2)-0.0121*
          &(TRI**3)-0.000607*(TRI**8)+OMEG(I,J)*(0.0637+0.331*(TRI**2)-0.423*
          &(TRI**3)-0.008*(TRI**8)))
C......... Sat.Pressure PSAT(J), Pure Fugac.FPUR(J), first Equil.Const. EK(J)
           DO 160 J=1,IS
           PSAT(J)=1000.*(10.**(VP(J,1)-VP(J,2)/(T+VP(J,3))))
           FPUR(J)=EXP(B(J,J)*P/R/T)
    160    EK(J)=PSAT(J)*FPUR(J)/P
C......... Calculate BETA & first X(I), Y(I)
     20    BETA=1.
     40    ST=0.
           SB=0
           DO 180 I=1,IS
           ST=ST+Z(I)/(1.+BETA*(EK(I)-1.))
    180    SB=SB+Z(I)*(EK(I)-1.)/((1.+BETA*(EK(I)-1.))**2)
           BETAN=BETA+(ST-1.)/SB
           IF (ABS(BETAN-BETA).LE.0.000001) GOTO 60
           BETA=BETAN
```

```
      GOTO 40
  60  SX=0.
      SY=0.
      DO 200 I=1,IS
      X(I)=Z(I)/(1.+BETA*(EK(I)-1.))
      Y(I)=EK(I)*X(I)
      SX=SX+X(I)
 200  SY=SY+Y(I)
      DO 220 I=1,IS
      X(I)=X(I)/SX
 220  Y(I)=Y(I)/SY
C........Activity coefficient G(I) - via TKWILSON subroutine
C        Pure Fug.Coef. FPUR(I), Comp.Fug.Coef. F(I), Sat.Pressure PSAT(I)

      BMIX=0.
      DO 240 I=1,IS
      DO 240 J=1,IS
 240  BMIX=BMIX+Y(I)*Y(J)*B(I,J)
      CALL TKWILSON(IS,X,T,SLR,V,G,1)
      SYDIF=0.
      DO 280 I=1,IS
      FTERM=0.
      DO 260 K=1,IS
 260  FTERM=FTERM+Y(K)*B(K,I)
      F(I)=EXP(P/R/T*(2.*FTERM-BMIX))
      FPUR(I)=EXP(B(I,I)*P/R/T)
      PSAT(I)=1000.*(10.**(VP(I,1)-VP(I,2)/(T+VP(I,3))))
      EK(I)=G(I)*FPUR(I)*PSAT(I)/(P*F(I))
C........New Equilbrium Constant EK(I) - Check if Sum[abs(x*K-y)] = 0

      EK(I)=G(I)*FPUR(I)*PSAT(I)/(P*F(I))
 280  SYDIF=SYDIF+ABS(X(I)*EK(I)-Y(I))
      IF (SYDIF.GT.0.00001) GO TO 20
      RETURN
      END
```

| **EXAMPLE 8.6** | *Calculation of the Vapour Fraction and the Composition of the Liquid and Vapour Phases Resulting from an Isothermal Flash of the Mixture {0.7663 Ethanol(1) + 0.1558 Methyl-cyclopentane(2) + 0.0779 Benzene(3)} at T = 341.21 K and P = 0.1 MPa.* |

The calculation employs subroutine FLASHTP. The system and the thermodynamic model are the same as those considered in Example 8.3; the inputs parameters to the thermodynamic model are therefore identical with those given earlier (Wilson parameters are those from [4]). The calling program and its output are shown in Display 8.9. The temperature, pressure and feed composition were chosen for this example such that the liquid and vapour streams have the same composition as in Example 8.3(a) and the vapour fraction obtained is 0.4.

DISPLAY 8.9   FLASHTP Sample Input/Output

```
      PROGRAM EX0806
      PARAMETER (IS=3)
      DIMENSION X(IS),Y(IS),Z(IS),PSAT(IS),EK(IS),FPUR(IS),F(IS),G(IS)
      DIMENSION TC(IS,IS),PC(IS,IS),OMEG(IS,IS),VP(IS,3)
      DIMENSION SLR(IS,IS),V(IS,IS)
      DATA (TC(J,J),J=1,IS)/513.88,532.7,562.06/
      DATA (PC(J,J),J=1,IS)/6.132E+6,3.780E+6,4.895E+6/
      DATA (OMEG(J,J),J=1,IS)/0.644,0.231,0.212/
      DATA (SLR(1,I),I=1,IS)/0.,1109.61,699.11 /
      DATA (SLR(2,I),I=1,IS)/123.40,0.,  6.693/
      DATA (SLR(3,I),I=1,IS)/ 63.06,125.16,0. /
      DATA (VP(1,I),I=1,3)/7.24222,1595.811,-46.702/
      DATA (VP(2,I),I=1,3)/5.99178,1188.329,-46.843/
      DATA (VP(3,I),I=1,3)/6.01905,1204.637,-53.081/
      DATA (V(1,I),I=1,IS)/1.,0.5128,0.6538/
      DATA (V(2,I),I=1,IS)/1.9501,1.,1.2750/
      DATA (V(3,I),I=1,IS)/1.5295,0.7843,1./
      DATA P,T,Z/0.1E+6,341.21,0.76632,0.15576,0.07792/
      CALL FLASHTP
     &(IS,P,T,X,Y,Z,TC,PC,OMEG,VP,V,SLR,PSAT,EK,FPUR,F,G,BETA)
      WRITE (*,1100) T,P*1.E-6
1100  FORMAT(1X,/,4X,'>>> T=',F7.2,'K & P=',F6.3,'MPA',/,4X,72('-')
     &/,13X,'Z(I)   X(I)   Y(I)',5X,'PSAT',6X,'K',5X,'FPUR',5X,'F',7X,
     &'G')
      DO 20 J=1,IS
20    WRITE (*,1200) J,Z(J),X(J),Y(J),PSAT(J)/1.E+6,EK(J),FPUR(J),
     &F(J),G(J)
1200  FORMAT(4X,'(',I1,') ',3F8.4,2X,5F8.4)
      WRITE (*,1300) (1.-BETA),BETA
1300  FORMAT(3X,'PHASE',/,3X,'KMOLES  1.000 ',2(F6.3,2X))
      END
```

```
>>>  T= 341.21K  &  P= 0.100MPA
```

	Z(I)	X(I)	Y(I)	PSAT	K	FPUR	F	G
(1)	0.7663	0.8779	0.5987	0.0666	0.6820	0.9691	0.9694	1.0232
(2)	0.1558	0.0680	0.2875	0.0901	4.2253	0.9585	0.9591	4.6850
(3)	0.0779	0.0540	0.1138	0.0689	2.1067	0.9611	0.9620	3.0551

```
PHASE
KMOLES 1.000    0.600    0.400
```

❑

8.5.3. Isenthalpic Flash Calculations

Subroutine FLASHHP performs the isenthalpic falsh calculation outlined in Figure 8.10 for the case in which the Peng-Robinson equation of state is adopted. A single-phase (vapour or liquid) feed is assumed.

Inputs to the subroutine are,

- the feed temperature, TF in units of Kelvin, pressure P in units of Pascal, and state (IS=1 liquid, IS=2 vapour, as requested by the PR subroutine),
- the number of components, IS, and their total composition in the mixture Z(I),
- critical temperature, TC(I) in units of Kelvin, critical pressure, PC(I) in units of Pascal and acentric factor, OMEG(I),
- the ideal heat-capacity coefficients Y1(I),Y2(I),Y3(I),Y4(I) for PR subroutine.
- initial and final value for the flash temperature iteration, WLO and WHI

```
                                          DISPLAY 8.10   FLASHHP List
      SUBROUTINE FLASHHP
     &(IS,P,WLO,WHI,TF,IF,Z,TC,PC,OMEG,Y1,Y2,Y3,Y4,BK,X,Y,T,FUGL,FUGV,BETA)
C
C
C     The subroutine performs an Adiabatic Flash calculation at given P
C     in [Pa] for a mixture feed Z(I) of IS components at temperature
C     TF [K] at specified phase (IF=1 liquid,IF=2 vapour).
C        Method adopted is based on the calculation of the liquid and
C     vapour partial fugacity coefficients from the Peng & Robinson EoS
C      - Additional input to the subroutine are critical constants
C     TC(I) in [K], and PC(I) in [Pa], acentric factor OMEG(I).
C     and heat capacity coefficients Y1(I),Y2(I),Y3(I),Y4(I)
C        Also first and last value, WLO and WHI, for the temperature must
C     be provided, as well as the PR binary interaction parameters BK(IS,IS).
C      - Output includes the liquid and vapour mole fractions, X(I), Y(I)
C     the flash temperature T [K], the vaporised fraction BETA and the
C     liquid and vapour partial fugacity coefficients FUGL(I), FUGV(I).
C
      DIMENSION TC(IS),PC(IS),OMEG(IS),Y1(IS),Y2(IS),Y3(IS),Y4(IS)
      DIMENSION X(IS),Y(IS),Z(IS),FUGL(IS),FUGV(IS),BK(IS,IS),EK(10)
      R=8.3145
C     Initial flash temperature & feed enthalpy HF
      CALL PR(TF,P,IF,IS,Z,TC,PC,OMEG,Y1,Y2,Y3,Y4,BK,D,HF,SF,FUGL)
   10 T=(WLO+WHI)/2.
C     Set X(I), Y(I) equal to Z(I) to start
      DO 100 I=1,IS
      X(I)=Z(I)
  100 Y(I)=Z(I)
C     First EK, and enthalpies of 2 phases
   20 CALL PR(T,P,2,IS,Y,TC,PC,OMEG,Y1,Y2,Y3,Y4,BK,D,HV,SV,FUGV)
      CALL PR(T,P,1,IS,X,TC,PC,OMEG,Y1,Y2,Y3,Y4,BK,D,HL,SL,FUGL)
```

```
      CHKT=ABS(FUGV(1)-FUGL(1))*100./FUGV(1)
      IF (CHKT.GT.0.001) GOTO 30
      WRITE (*,*)   NO CONVERGENCE - TRY CLOSER TEMPERATURE LIMITS
      STOP
30    SY=0.
      DO 120 I=1,IS
120   EK(I)=FUGL(I)/FUGV(I)
C     .....BETA & first X(I), Y(I)
40    BETA=1.
50    ST=0.
      SB=0.
      DO 140 I=1,IS
      ST=ST+Z(I)/(1.+BETA*(EK(I)-1.))
140   SB=SB+Z(I)*(EK(I)-1.)/((1.+BETA*(EK(I)-1.))**2)
      BETAN=BETA+(ST-1.)/SB
      IF (ABS(BETAN-BETA).LE.0.000001) GOTO 60
      BETA=BETAN
      GOTO 50
60    SYDIF=0
      DO 160 I=1,IS
160   SYDIF=SYDIF+ABS(X(I)*EK(I)-Y(I))
      SX=0
      SY=0
      DO 180 I=1,IS
      X(I)=Z(I)/(1.+BETA*(EK(I)-1.))
      Y(I)=EK(I)*X(I)
      SX=SX+X(I)
180   SY=SY+Y(I)
      DO 200 I=1,IS
      X(I)=X(I)/SX
200   Y(I)=Y(I)/SY
      IF (SYDIF.GT.0.00001) GO TO 20
C     .....Temperature iteration
      HVL=BETA*HV+(1.-BETA)*HL
      HDIF=HF-HVL
      WRITE (*,*) T
      IF (ABS(HDIF).LE.0.1) GOTO 70
      IF (ABS(HF).GT.ABS(HPROD)) WHI=T
      IF (ABS(HF).LT.ABS(HPROD)) WLO=T
      GOTO 10
70    RETURN
      END
```

The output is as follows:
-   liquid and vapour mole fractions, X(I), Y(I), and vapour fraction, BETA,
-   the liquid and vapour partial fugacity coefficients, FUGL(I) and FUGV(I).
Example 8.7 demonstrates the application of this subroutine in the case of a three-component hydrocarbon mixture.

EXAMPLE 8.7 *Calculation of the Vapour Fraction, Liquid and Vapour Compositions and the Temperature Resulting from an Isenthalpic Flash of the Mixture {x_1 Ethane + x_2 n-Butane + x_3 n-Pentane} with $x_1 = 0.3$, $x_2 = 0.3$ and $x_3 = 0.4$ The Feed Temperature and Pressure are 310 K and 0.7 MPa.*

The calculation employs subroutine FLASHHP. Critical temperature, critical pressure, acentric factor and ideal heat capacity coefficients for ethane, *n*-butane and *n*-pentane are obtained from Table A1 in Appendix A. The Peng - Robinson binary interaction parameters are all set equal to 0.01 [10].

The calling program and its output is shown in Display 8.11. Note that only last few iterations over the temperature are shown.

In Table 8.4 the results obtained are compared with an equivalent flast calculation performed at 294.55 K and 0.7 MPa, by SUPERTRAPP [8], a commercially available FORTRAN package. The results are very similar both for the compositions and the amounts in each phase.

Table 8.4 Comparison of Results.

	Example 8.7		SUPERTRAPP [8]	
	x_i	y_i	x_i	y_i
(1)	0.1921	0.8233	0.1862	0.8280
(2)	0.3363	0.1240	0.3389	0.1198
(3)	0.4716	0.0527	0.4749	0.0522
β	0.171		0.177	

DISPLAY 8.11 FLASHHP Sample Input/Output

```
PROGRAM EX0807
PARAMETER (IS=3)
DIMENSION TC(IS),PC(IS),OMEG(IS),Y1(IS),Y2(IS),Y3(IS),Y4(IS)
DIMENSION X(IS),Y(IS),Z(IS),FUGV(IS),FUGL(IS),BK(IS,IS)
DATA TC(1),PC(1),OMEG(1)/305.33,4.871E+6,0.099/
DATA TC(2),PC(2),OMEG(2)/425.25,3.792E+6,0.199/
DATA TC(3),PC(3),OMEG(3)/469.80,3.375E+6,0.251/
DATA Y1(1),Y2(1),Y3(1),Y4(1)/5.409,0.1781,-6.9385E-5,8.713E-9/
DATA Y1(2),Y2(2),Y3(2),Y4(2)/9.487,0.3313,-1.108E-4,-2.822E-9/
DATA Y1(3),Y2(3),Y3(3),Y4(3)/-3.626,0.4873,-2.580E-4,5.305E-8/
DATA BK(1,1),BK(1,2),BK(1,3)/0.,0.01,0.01/
DATA BK(2,1),BK(2,2),BK(2,3)/0.01,0.,0.01/
DATA BK(3,1),BK(3,2),BK(3,3)/0.01,0.01,0./
DATA TF,IF,Z/310.,1,0.30,0.30,0.40/
DATA P,WLO,WHI/0.7E+6,285.,305./
CALL FLASHHP
```

```
        &(IS,P,WLO,WHI,TF,IF,Z,TC,PC,OMEG,Y1,Y2,Y3,Y4,BK,X,Y,T,FUGL,FUGV,BETA)
        WRITE (*,1100) T,P*1.E-6
 1100   FORMAT(3X,/,4X,'>>>  T=',F7.2,'K  &  P=',F6.3,'MPA',/,4X,70('-'),
        &/,12X,'Z(I)',5X,'X(I)',5X,'Y(I)',8X,'FUG.LIQ.',4X,'FUG.VAP.',8X,'K')
        DO 20 J=1,IS
   20   WRITE (*,1200) J,Z(J),X(J),Y(J),FUGL(J),FUGV(J),FUGL(J)/FUGV(J)
 1200   FORMAT(4X,'(',I1,') ',3F9.4,2X,2(2X,F10.5),3X,F10.5)
        WRITE (*,1300) (1-BETA),BETA
 1300   FORMAT(3X,'PHASE',/,3X,'KMOLES  1.000 ',2(F7.3,2X))
        END
```

294.5459
294.5435
294.5447
294.5453

```
   >>>  T= 294.55K  &  P=0.700MPA
   -----------------------------------------------------------------------
            Z(I)      X(I)      Y(I)       FUG.LIQ.   FUG.VAP.        K
     (1)   0.3000    0.1921    0.8233      4.03968    0.94244     4.28642
     (2)   0.3000    0.3363    0.1240      0.30860    0.83680     0.36879
     (3)   0.4000    0.4716    0.0527      0.08804    0.78806     0.11172
   PHASE
   KMOLES  1.000    0.829     0.171
```

❑

8.5.4. Isothermal Liquid-Liquid Flash Calculations

Subroutine FLASH2L, listed in Display 8.12, implements the liquid-liquid isothermal flash algorithm of Figure 8.13.

Input to the subroutine are,

- the temperature T in units of Kelvin, pressure P in units of Pascal, the number of components, IS, and the overall mole fractions in the mixture $Z(I)$,
- initial values for the mole fractions in phase 1, as $XST(I)$, and the phase fraction β as BST,
- UNIQUAC parameters, $U(I,J)$, $R(I)$, $Q(I)$ and $QPR(I)$

The output is as follows:

- the mole fractions of the two phases, $X1(I)$ and $X2(I)$, and the phase fraction, BETA.

In order for the routine to converge, the initial values given for the mole fractions of phase 1 and for the phase fraction β must be reasonable. Example 8.8, demonstrates the application of subroutine FLASH2L in constructing the isothermal-isobaric triangular phase diagram of a ternary system.

```
      SUBROUTINE FLASH2L
     &(IS,P,T,X1,X2,XST,BST,Z,U,R,Q,QPR,BETA)
C
C     The subroutine calculates equilibrium compositions of the two
C     liquid phases, X1(I) and X2(I), of a ternary mixture (IS=3) of
C     composition Z(I). The fractional amount in phase 1, BETA, is also calculated.
C        The distribution constant is calculated from activity coefficients of the two
C     phases calculated by UNIQUAC.
C     -  Input to the subroutine is the pressure P in [Pa], the tempera-
C     ture T in [K], initial values of mole fractions of phase 1, XST(I),
C     and the fraction of amount in phase 1, BST, mole fractions of the
C     total mixture Z(I), and UNIQUAC parameters, U, R, Q and QPR.
C     -  Output of the subroutine are the mole fractions of the two liquid
C     phases, X1(I) and X2(I) and the fraction of amount in phase 1, BETA.
C
      DIMENSION X1(IS),X2(IS),Z(IS),XST(IS)
      DIMENSION U(IS,IS),R(IS),Q(IS),QPR(IS),G1(3),G2(3)
C........Set initial X1(I) (=XST), BETA (=BST)
      DO 100 I=1,IS
  100 X1(I)=XST(I)
      BETA=BST
C........Calculate X2(I), & activity coefficients
      DO 120 I=1,IS
  120 X2(I)=(Z(I)-BETA*X1(I))/(1.-BETA)
   20 BETAPR=BETA
      CALL UNIQUAC(IS,T,X1,R,Q,QPR,U,G1)
      CALL UNIQUAC(IS,T,X2,R,Q,QPR,U,G2)
C........Calculate BETA and X1(I), X2(I)
   40 ST=0
      SB=0.
      DO 140 I=1,IS
      DK=G1(I)/G2(I)
      ST=ST+Z(I)/(BETA+DK*(1.-BETA))
  140 SB=SB+Z(I)*(1.-DK)/((BETA+DK*(1.-BETA))**2)
      BETAN=BETA+(ST-1.)/SB
      IF (ABS(BETAN-BETA).LE.0.000001) GOTO 60
      BETA=BETAN
      GOTO 40
   60 SX=0.
      SY=0.
      DO 160 I=1,IS
      DK=G1(I)/G2(I)
      X1(I)=Z(I)/(BETA+DK*(1.-BETA))
  160 X2(I)=DK*X1(I)
C........Iteration for BETA
      IF (ABS(BETA-BETAPR).GE.0.00001) GOTO 20
      RETURN
      END
```

EXAMPLE 8.8	*Construction of the Liquid-liquid Phase Diagram for the Mixture {Toluene(1) + 2-Propanone(2) + Water(3)} at T = 283.15 K and P = 0.1 MPa.*

The calculation employs subroutine FLASH2L with values of the UNIQUAC parameters taken from [6]. Display 8.13 shows the calling program and its output. Each line in the table represents a set of equilibrium compositions of the two liquid phases, and hence a tie-line. The seven tie-lines are shown in Figure 8.16, the triangular phase diagram of the system.

DISPLAY 8.13 FLASH2L Sample Input/Output

```
      PROGRAM EX0808
      PARAMETER (IS=3)
      DIMENSION X1(IS),X2(IS),Z(IS),XST(IS)
      DIMENSION U(IS,IS),R(IS),Q(IS),QPR(IS),G1(IS),G2(IS)
      DATA (U(1,I),I=1,IS)/0.,115.14,814.64/
      DATA (U(2,I),I=1,IS)/ 29.585,0.,317.21/
      DATA (U(3,I),I=1,IS)/334.88,-22.287,0./
      DATA R/3.9228,2.5735,0.9200/
      DATA Q/2.968,2.336,1.400/
      DATA QPR/2.968,2.336,1.400/
C.....Note in this particular case Q=QPR
      DATA T,P/283.15,0.101E+6/
      WRITE (*,1000)
 1000 FORMAT(5X,'X1(1)',5X,'X1(2)',5X,'X1(3)',8X,'X2(1)',5X,'X2(2)',
     &5X,'X2(3)',7X,'BETA')
      DO 20 I=1,7
      Z(1)=0.1*FLOAT(I)
      Z(2)=(1.-Z(1))/2
      Z(3)=Z(2)
C.....Set initial values for X1 phase, XST, and BETA, BST
      XST(1)=0.1*FLOAT(I)
      XST(2)=0.9-XST(1)
      XST(3)=0.1
      BST=0.6
      CALL FLASH2L(IS,P,T,X1,X2,XST,BST,Z,U,R,Q,QPR,BETA)
   20 WRITE (*,1100) X1(1),X1(2),X1(3),X2(1),X2(2),X2(3),BETA
 1100 FORMAT(2X,3(F8.4,2X),3X,3(F8.4,2X),F10.4)
      END
```

X1(1)	X1(2)	X1(3)	X2(1)	X2(2)	X2(3)	BETA
0.1772	0.6691	0.1537	0.0013	0.1699	0.8288	0.5611
0.3418	0.5816	0.0766	0.0010	0.1452	0.8538	0.5839
0.4747	0.4801	0.0452	0.0008	0.1272	0.8720	0.6313
0.5829	0.3879	0.0292	0.0007	0.1081	0.8912	0.6858

0.6738	0.3063	0.0199	0.0005	0.0882	0.9113	0.7419
0.7526	0.2335	0.0140	0.0004	0.0685	0.9312	0.7972
0.8224	0.1676	0.0100	0.0003	0.0496	0.9502	0.8511

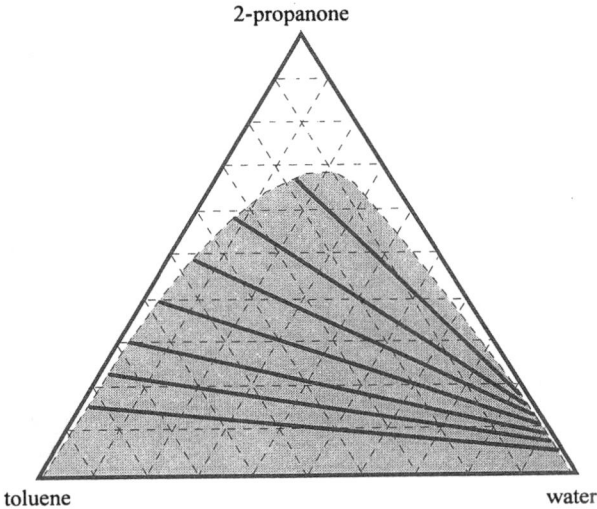

Figure 8.16. Triangular phase diagram of toluene+2-propanone+water at 283.15 K.

◻

References

1. S.M. Walas, *Phase Equilibria in Chemical Engineering* (Butterworth Publishers, Boston, 1985). Ch. 5.
2. J.M. Prausnitz, R.N. Lichtenthaler and E. Gomes de Azevedo, *Molecular Thermodynamics of Fluid-Phase Equilibria*, 2nd Edition (Prentice-Hall, Englewood Cliffs, 1986). Ch. 10.
3. J.S. Rowlinson and F.L. Swinton, *Liquids and Liquid Mixtures*, 3rd Edition (Butterworth, London, 1982).
4. J.E. Sinor and J.H. Weber, *J. Chem. Eng. Data* **5** (1960) 243 (1960).
5. P.S. Murti and M. van Winkle, *Chem. Eng. Data Ser.* **3** (1978) 72.
6. DECHEMA Liquid-Liquid Data Collection, **5** (1980) part 2.
7. D.Y. Peng and D.B. Robinson, *Ind. Eng. Chem. Fundam.* **15** (1976) 59.

8. *SUPERTRAPP. FORTRAN Package for the Calculation of the Transport Properties of Nonpolar Fluids and their Mixtures* (produced and supplied by N.I.S.T., Gaithersburg, U.S.A.).
9. A. Fredenslund and J. Mollerup *J. Chem. Soc. Faraday Trans. I* **70** (1974) 1653.
10. D.L. Katz and A. Firoozabadi, *J. Petr. Tech.* **11** (1978) 1649.

9

Transport Properties: Theory

In the preceding chapters, we have considered in detail prediction of the thermo-dynamic properties of fluids as this is of prime importance in the design of chemical process systems. In many areas of design, quantitative knowledge of the transport properties of the process fluids is also required and it is to the theory and practice of predicting such properties that we now turn.

Unlike thermodynamic quantities, which refer fundamentally to properties of a system at equilibrium, *transport properties* determine the rate at which processes such as heat and mass transfer occur in a system that is not at equilibrium. Practical process situations often involve a system at or near to a steady state but, neverthe-less, far from true equilibrium. In particular, there may be large spatial gradients in temperature, concentration or fluid velocity which give rise to non-linear transfer processes. In contrast, the science of fluid transport properties is concerned almost exclusively with somewhat idealised systems displaced only slightly from equili-brium. The well known linear rate laws of Fourier, Fick and Newton arise in this context and each is associated with a transport coefficient that relates flux to gradient. The connection between these situations is that empirical models which are often, out of necessity, adopted in the description of practical transfer processes involve so-called *transfer coefficients* which depend parametrically on the formal transport properties. Here we shall be concerned only with the latter.

Although for engineering purposes temperature and pressure are usually specified quantities, in the case of the transport properties both theory and expe-rience show that temperature and density (or molar volume) are the fundamental state variables while, in essence, pressure is of no direct importance. Indeed, a transport property X, where X may be the viscosity η, the thermal conductivity λ, the diffusion coefficients D_{ii} and D_{ij}, or the thermal diffusivity α, is written most conveniently as the sum of three contributions:

$$X(\rho_n, T) = X_0(T) + \Delta X(\rho_n, T) + \Delta X_c(\rho_n, T). \qquad (9.1)$$

Here, $X_0(T)$ represents the *dilute-gas* value of the property, $\Delta X(\rho_n,T)$ the *excess* value of the property and $\Delta X_c(\rho_n,T)$ the so-called *critical enhancement*. The sum $\{X_0(T) + \Delta X(\rho_n,T)\}$ is also known as the *background* value of the property. The contributions of the three terms in Eq.(9.1) are illustrated in Figure 9.1 for the case of the thermal conductivity λ of a pure fluid on a near-critical isotherm. We see that the background term is a gently increasing function of density but that, near to $\rho_n = \rho_n^c$ and $T = T^c$, there is a substantial critical enhancement.

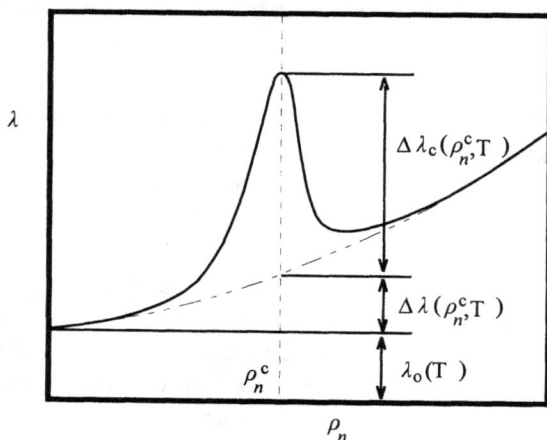

Figure 9.1. Contributions to the thermal conductivity for an isotherm.

The transport coefficients describe the rate at which a fluid relaxes towards its equilibrium state when it is perturbed by the application of a temperature, pressure, density, velocity or composition gradient. The non-equilibrium statistical mechanics describing these processes is known as *kinetic theory* and aims to relate the macroscopic (observable) properties of a system to the microscopic properties and interactions of the individual molecules.

As a result of substantial advances in recent years, it is now possible to describe essentially exactly the low-density transport properties as well as the critical enhancements [1]. The excess contribution is less well understood theoretically although various modifications of the Enskog theory can be used to establish effective correlations of experimental results. In particular, for dense gases and liquids remote from the critical region, methods based on the Enskog theory for hard spheres give an excellent representation of experimental data.

Modern kinetic theory is far from simple. However, in this chapter, we shall consider in detail only the most elementary theory pertaining to the viscosity and thermal conductivity of a gas composed of hard spheres. We then proceed to explain how the results of that simple theory can be modified so as to incorporate the predictions of the essentially-exact kinetic theory for real monatomic gases. We also discuss the theoretical status of real polyatomic systems. The nature of the excess transport properties of dense gases and liquids is then discussed and, finally, a brief description of critical enhancements will be given.

Based on this discussion, we shall describe in the next chapter predictions of the viscosity and thermal conductivity of polyatomic fluids (including mixtures) in both the gaseous and liquid states.

9.1. Dilute Gases and Mixtures

In kinetic theory, the term *dilute gas* refers to a state in which the density is:
(a) low enough that the molecules spend most of the time in free flight and that only binary encounters need be considered; and
(b) high enough to ensure that the effects of molecule-wall collisions can be neglected compared to those of molecule-molecule encounters.
This clearly implies that the mean free path of the molecules must be much smaller than any dimension of the container. It is also worth noting that the terms 'dilute' and 'low-density' refer to the real physical situation defined above whereas the term 'zero-density limit' refers to the results of a mathematical extrapolation of a particular transport property to the limit of zero-density. In most cases, the value of the transport property obtained from such an extrapolation may be properly identified with the theoretical dilute-gas value whereas the true zero-density value, if one exists, corresponds to a situation to which 'normal' kinetic theory cannot be applied.

The transport properties of a dilute gas depend directly on the intermolecular forces that exist between molecules. Thus, on the one hand, knowledge of the intermolecular forces enables the dilute-gas transport properties to be evaluated while, on the other hand, accurate measurements of such quantities may permit the determination of intermolecular forces. Indeed, Maitland *et al.* [2] showed that it was possible to determine the intermolecular pair potential of monatomic species directly from measurements of the viscosity of the dilute gas by a process of iterative data inversion. Recently, attempts have been made to employ gas transport properties in the determination of intermolecular potentials for polyatomic molecules. Unfortunately, application of the exact kinetic theory for such systems requires a large computational effort and such calculations are only just becoming feasible on a routine basis.

Monatomic species interact though a potential-energy function that is spherically symmetric. Furthermore, at thermal energies, atomic collisions do not result in any change in the internal state of the atoms which may therefore be considered as structureless particles. Polyatomic molecules differ not only because they interact through non-spherically symmetric intermolecular pair potentials, but also because they possess internal degrees of freedom in the form of rotational and vibrational modes of motion. The energy required to bring about a change in the vibrational or rotational state of a molecule is within the range of thermal energies. Consequently, molecular collisions may result in an exchange of translational and internal energy and the possibility of such inelastic collisions seriously complicates the kinetic theory of polyatomic gases. To deal with this situation properly requires a fully quantum-mechanical treatment [3-6] which, even today, remains computationally intractable except for the simplest of molecules. However, Wang Chang, Uhlenbeck [7] and de Boer [8] formulated a semi-classical kinetic theory that was later developed by Monchick, Mason and their collaborators [9,10] into a powerful and practical, albeit approximate, treatment. We shall mention some other approximate treatments of the effects of inelastic collisions later in this chapter.

As already stated, a full treatment of the kinetic theory, including polyatomic systems, is beyond the scope of this book. We therefore turn now to a simple treatment of the viscosity and thermal conductivity of a dilute gas composed of elastic hard spheres. The derivation will be based on the mean-free-path approach which, while simple and approximate, contains the essential physics of the problem for dilute monatomic gases. The resulting transport coefficients in fact only differ from those of the exact kinetic theory by numerical factors of order unity.

9.1.1. Mean Free Path

Since the elementary kinetic theory is based on the collision frequency and the mean free path, these quantities will be discussed first. We consider a two-component gas composed of hard spheres of diameter σ_1 and σ_2 and masses m_1 and m_2, that move between collisions in straight lines with velocities $\mathbf{c}_1$ and $\mathbf{c}_2$ respectively. These velocities are distributed in accordance with the Maxwell-Boltzmann velocity distribution function for the gas at temperature T [11] according to which the mean speeds of the molecules (irrespective of direction) are:

$$<c_1> = \left(\frac{8k_{\mathrm{B}}T}{\pi m_1}\right)^{1/2} \quad \text{and} \quad <c_2> = \left(\frac{8k_{\mathrm{B}}T}{\pi m_2}\right)^{1/2}. \tag{9.2}$$

The trajectories of two molecules define a plane and make an angle with each other which we denote by θ. In solving the problem of when two such molecules will collide, it is simplest to adopt a co-ordinate system fixed to the centre of mass of one of the molecules and to determine the trajectory of the second molecule in that frame of reference. If we consider two molecules, the first of type i and the second of type j, then the speed c_{ij} of the first relative to an origin at the second is given by

$$c_{ij}^2 = c_i^2 + c_j^2 - 2c_i c_j \cos\theta . \tag{9.3}$$

In the following derivation, we shall require the mean relative speed $<c_{ij}>$ which is obtained by averaging c_{ij} over all possible speeds and directions, each weighted in accordance with the Maxwell-Boltzmann velocity distribution function. The results of that operation [11] is:

$$< c_{ij} >^2 = < c_i >^2 + < c_j >^2 = \left(\frac{8k_B T}{\pi m_i}\right) + \left(\frac{8k_B T}{\pi m_j}\right) = \left(\frac{8k_B T}{\pi \mu_{ij}}\right), \tag{9.4}$$

where

$$\mu_{ij} = \frac{m_i m_j}{m_i + m_j} \tag{9.5}$$

is the reduced mass of the pair.

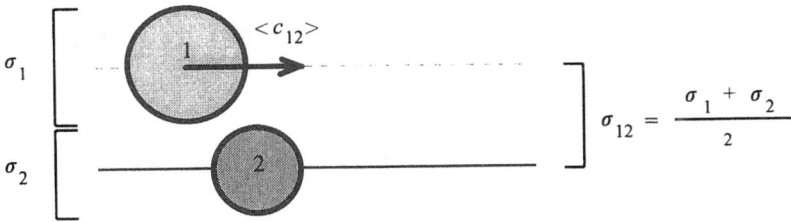

Figure 9.2. Collision between a molecule of species 1 and a molecule of species 2.

Now, a molecule of species 1 moving in a gas composed of molecules of species 1 and 2 will collide with molecules of both species as depicted in Figure 9.2. Let us therefore consider the two types of collisions separately.

(a) Collisions with other molecules of species 1.

A molecule of type 1 moving with mean speed $<c_{11}>$ relative to another molecule of species 1 sweeps out a cylindrical volume $\pi \sigma_1^2 < c_{11} >$ in unit time.

If the centre of the second molecule is present in that volume then it will experience a collision. Since there are n_1 molecules of species 1 per unit volume, the mean number of collisions experienced by a single molecule of that type per unit time will be $\nu_{11} = n_1 \pi \sigma_1^2 < c_{11} >$.

(b)   Collisions with other molecules of species 2.

Similarly, a molecule of type 1 moving with a mean speed $<c_{12}>$ relative to another molecule of species 2 sweeps out a volume $\pi \sigma_{12}^2 < c_{12} >$ in unit time and experiences $n_2 \pi \sigma_{12}^2 < c_{12} >$ collisions with such molecules per unit time, where $\sigma_{12} = (\sigma_1 + \sigma_2)/2$. The mean collision frequency of a single molecule of species 1 with other molecules of species 2 is therefore given by $\nu_{12} = n_2 \pi \sigma_{12}^2 < c_{12} >$.

Thus the total number of collisions that a single molecule of type 1 undergoes per unit time will be $n_1 \pi \sigma_1^2 < c_{11} > + n_2 \pi \sigma_{12}^2 < c_{12} >$. Furthermore, the mean distance travelled by this molecule in unit time is $< c_1 >$.

The mean free path, Λ, is defined as the distance travelled by one molecule in unit time divided by the number of collisions it underwent in that time. Hence the mean free path of molecule 1 in a two-component gas, Λ_1^{mix}, is

$$\Lambda_1^{mix} = \frac{< c_1 >}{n_1 \pi \sigma_1^2 < c_{11} > + n_2 \pi \sigma_{12}^2 < c_{12} >} . \tag{9.6}$$

In the case of a pure gas composed of molecules of species 1, $n_2 = 0$ and the mean free path becomes

$$\Lambda_1 = \frac{< c_1 >}{n_1 \pi \sigma_1^2 < c_{11} >} = \frac{1}{\sqrt{2} \, n_1 \pi \sigma_1^2} . \tag{9.7}$$

To simplify Eq.(9.6) somewhat, we employ Eqs.(9.4) and (9.7) to obtain

$$\Lambda_1^{mix} = \frac{\Lambda_1}{\left[1 + (n_2/n_1) H_{12} \right]} \quad \text{where} \quad H_{12} = \frac{\sigma_{12}^2}{\sigma_1^2} \left[\frac{m_1 + m_2}{2 m_2} \right]^{1/2} . \tag{9.8}$$

By an analogous argument, the mean free path of molecules of species 2 in a gas composed of molecules of species 1 and 2 is

$$\Lambda_2^{\mathrm{mix}} = \frac{\Lambda_2}{[1+(n_1/n_2)H_{21}]} \qquad \text{where} \qquad H_{21} = \frac{\sigma_{12}^2}{\sigma_2^2}\left[\frac{m_1+m_2}{2m_1}\right]^{1/2}. \qquad (9.9)$$

From these expressions, we conclude that the mean free path is a purely geometric quantity, independent of the temperature, and inversely proportional to the density. The usefulness of the mean free path will become apparent in the following sections where the viscosity and the thermal conductivity of dilute-gas mixtures will be discussed. It should be noted that, although we ascribed the same speed to all molecules of the same species, exactly the same results would have been obtained by retaining the full Maxwell-Boltzmann velocity distribution function and then averaging over all possible initial trajectories when we calculated the collision frequency.

EXAMPLE 9.1	*Interstellar space consists largely of atomic hydrogen at $T \approx 10\,K$ and density of the order of 1 atom/cm³. Calculate how often a hydrogen atom has collided with others since the death of Alexander the Great in 332 BC ($\sigma = 0.25$ nm).*

The number of collisions that a hydrogen atom undergoes per unit time is given by $v_{11} = n_1 \pi \sigma_1^2 < c_{11} >$. Employing Eq.(9.4) for $<c_{11}>$ with $\mu = m_1/2$, we obtain

$$v_{11} = \sqrt{2}\,\pi\sigma^2\,n_1\left(\frac{8k_{\mathrm{B}}T}{\pi m_1}\right)^{1/2}$$

and with $n_1 = 1\times10^{-6}$ m⁻³ and $m_1 = (0.002$ kg·mol⁻¹$/6.023\times10^{23}$ mol⁻¹$)$, we find that the collision rate is 9.03×10^{-11} s⁻¹. Since the death of Alexander the Great occurred about $(1996+332)\times365\times24\times3600 = 0.72\times10^{11}$ seconds ago, the number of collisions suffered since then by a hydrogen atom in interstellar space is approximately 7.
□

9.1.2. Transport Properties of Dilute Hard-Sphere Gases

Let us consider a pure dilute gas composed of molecules of type 1 and confined between two parallel plates of area A. We can move the top plate relative to the bottom one at a constant velocity u in the x-direction, as shown in Figure 9.3, by applying a constant force F.

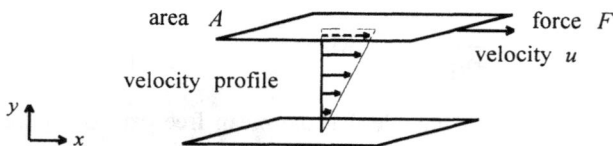

Figure 9.3.

Newton's Law of viscosity states that the force necessary to maintain the upper plate in motion is

$$F = -\eta_1 A \frac{du}{dy} , \qquad (9.10)$$

where η is the coefficient of viscosity. We can extend this picture to the molecular level and define three parallel planes each a distance Λ_1 apart, as shown in Figure 9.4, such that a typical molecule which suffered its most recent collision in one plane will suffer its next collision in an adjacent plane. We further assume that the steady velocity u is very much smaller than the average molecular speed $<c_1>$ so that the underlying velocity distribution function is not perturbed.

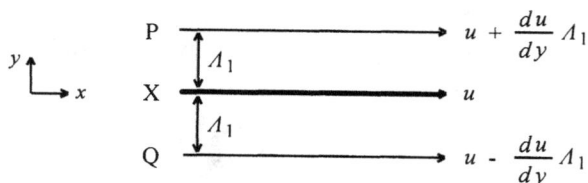

Figure 9.4.

Consider a molecule that enters plane X, having had its most recent collision in plane P. This molecule will carry with it a net momentum characteristic of plane P and equal to $m_1(u+\Lambda_1 du/dy)$ in the x-direction. If there are n_1 molecules per unit volume in the gas, a third of them will move in the y-direction and half of those (i.e. one sixth of the total) in the positive y-direction. Hence the flow of mole-cules from plane P to plane X per unit time will be approximately $(n_1/6)A <c_1>$ and these will therefore transport momentum at the rate $[(n_1/6)A <c_1>] \times [m_1(u+\Lambda_1 du/dy)]$. A similar argument can be developed for the transport of momentum by the molecules entering the plane X from the plane Q.

The net rate at which momentum is transferred in the positive y direction across the plane X is, by definition, equal to the F acting on that plane and is given by

$$F = -\frac{1}{6}n_1 A <c_1> m_1 \left[u + \Lambda_1 \frac{du}{dy} \right] + \frac{1}{6} n_1 A <c_1> m_1 \left[u - \Lambda_1 \frac{du}{dy} \right]$$

$$= -\frac{1}{3} n_1 <c_1> m_1 \Lambda_1 A \frac{du}{dy} \ . \tag{9.11}$$

Comparing this expression with Eq.(9.10), we see that the coefficient of viscosity η_1 for the pure dilute gas composed of molecules of species 1 is

$$\eta_1 = \frac{1}{3} n_1 <c_1> m_1 \Lambda_1 \ . \tag{9.12}$$

Combining this with Eq.(9.2) for the average speed $<c_1>$ and Eq.(9.7) for the mean free path Λ_1, we obtain

$$\eta_1 = \frac{2}{3} \left[\frac{m_1 k_B T}{\pi} \right]^{1/2} \frac{1}{\pi \sigma_1^2} \ . \tag{9.13}$$

This predicts that the dilute-gas viscosity is independent of density, proportional to the square root of the temperature, and inversely proportional to the collision cross-section $\pi \sigma_1^2$. In this approximation, the collision cross-section is a purely geometric factor independent of temperature. This result was first derived by Maxwell in 1859 long before a rigorous solution of Boltzmann's transport equation was achieved by Chapman and Enskog in 1916 [11].

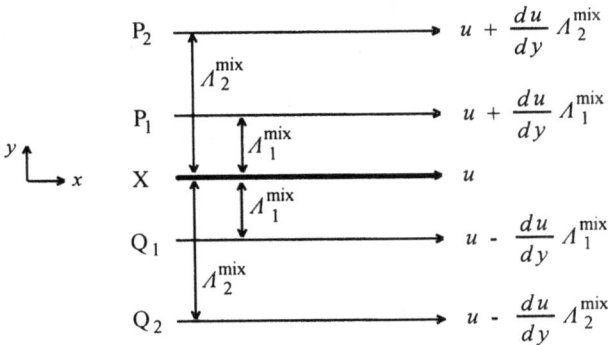

Figure 9.5.

Similar arguments can easily be applied for the case of a binary mixture of species 1 and 2. Having obtained in Section 9.1.1. expressions for the mean free paths of both components, we now set up five parallel planes as shown in Figure 9.5.

The net momentum transferred per unit time across plane X will now result in a force F given by

$$F = -\frac{1}{6} n_1 A < c_1 > m_1 \left[u + \Lambda_1^{mix} \frac{du}{dy} \right] - \frac{1}{6} n_2 A < c_2 > m_2 \left[u + \Lambda_2^{mix} \frac{du}{dy} \right]$$
$$+ \frac{1}{6} n_1 A < c_1 > m_1 \left[u - \Lambda_1^{mix} \frac{du}{dy} \right] + \frac{1}{6} n_2 A < c_2 > m_2 \left[u - \Lambda_2^{mix} \frac{du}{dy} \right] \tag{9.14}$$

This expression reduces to

$$F = -\left[\frac{1}{3} n_1 < c_1 > m_1 \Lambda_1^{mix} + \frac{1}{3} n_2 < c_2 > m_2 \Lambda_2^{mix} \right] A \frac{du}{dy}, \tag{9.15}$$

and, comparing with Newton's Law of viscosity, Eq.(9.10), we obtain the coefficient of viscosity η_{mix} as

$$\eta_{mix} = \frac{1}{3} n_1 < c_1 > m_1 \Lambda_1^{mix} + \frac{1}{3} n_2 < c_2 > m_2 \Lambda_2^{mix}. \tag{9.16}$$

This can be further simplified by expressing the results in terms of the pure-component dilute-gas viscosities and the parameters H_{12} and H_{21} defined above:

$$\eta_{mix} = \frac{\eta_1}{1 + (n_2/n_1) H_{12}} + \frac{\eta_2}{1 + (n_1/n_2) H_{21}}. \tag{9.17}$$

In terms of the mole fractions instead of the number densities, this becomes

$$\eta_{mix} = \frac{y_1 \eta_1}{y_1 + y_2 H_{12}} + \frac{y_2 \eta_2}{y_2 + y_1 H_{21}}. \tag{9.18}$$

Finally, upon generalisation to a mixture of N components one finds

$$\eta_{mix} = \sum_{i=1}^{N} \frac{y_i \eta_i}{\displaystyle\sum_{j=1}^{N} y_j H_{ij}}, \tag{9.19}$$

where

$$H_{ij} = \frac{\sigma_{ij}^2}{\sigma_i^2} \left[\frac{m_i + m_j}{2 m_j} \right]^{1/2}. \tag{9.20}$$

In the case of the thermal conductivity, a similar argument can be developed for the transfer of thermal energy flow across the plane X. For a pure dilute gas composed of hard spherical molecules of species 1 with constant-volume heat capacity per molecule c_{V1} we construct the following diagram.

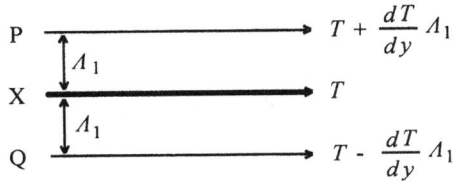

Figure 9.5.

The total thermal energy, Q, transferred per unit time across the plane X will be

$$Q = -\frac{n_1 A <c_1> c_{V1}}{6}\left[T + \Lambda_1\frac{dT}{dy}\right] + \frac{n_1 A <c_1> c_{V1}}{6}\left[T - \Lambda_1\frac{dT}{dy}\right]$$
$$= -\frac{1}{3}n_1 <c_1> c_{V1}\Lambda_1 A\frac{dT}{dy} , \tag{9.21}$$

and comparing this with Fourier's equation,

$$Q = -\lambda_1 A(dT/dy) , \tag{9.22}$$

we see that the coefficient of thermal conductivity for species 1 is given by

$$\lambda_1 = \frac{2}{3}c_{V1}\left[\frac{k_B T}{\pi m_1}\right]^{1/2}\frac{1}{\pi \sigma_1^2} . \tag{9.23}$$

It is worthwhile noting the similarity of this expression to that obtained for the viscosity in terms of the dependence on both temperature and the collision cross-section.

In the case of a dilute-gas mixture, the same arguments leads to the following expression for the thermal conductivity:

$$\lambda_{mix} = \sum_{i=1}^{N}\frac{y_i \lambda_i}{\sum_{j=1}^{N} y_j H_{ij}} . \tag{9.24}$$

The results of this section have been derived under simplifying assumptions which retain most of the essential physics but involve some quantitative approximations. Thus the expressions obtained for the viscosity and thermal conductivity of a pure gas differ from those which follows from an exact treatment of the problem for hard spheres but only by constant numerical factors of order unity. The correct results are in fact given by

$$\eta_1 = \frac{5}{16} \left(\pi m_1 k_B T \right)^{1/2} \frac{f_\eta}{\pi \sigma_1^2} \tag{9.25}$$

and

$$\lambda_1 = \frac{25}{32} c_{V1} \left[\frac{\pi k_B T}{m_1} \right]^{1/2} \frac{f_\lambda}{\pi \sigma_1^2}, \tag{9.26}$$

where $f_\eta = 1.015$ and $f_\lambda = 1.024$. Since the hard-sphere gas has only translational degrees of freedom, the heat capacity per molecule would be $3k_B/2$.

9.1.3. Collision Cross Section

We define the *collision cross section* of a molecule as the area perpendicular to the direction of motion within which the centre of a second molecule must lie if a collision between the two is to occur. For a gas composed of hard spherical molecules, this cross section is $\pi \sigma^2$, the distance of closest approach in a collision is σ and both are independent of the temperature.

For real molecules, the repulsive part of the intermolecular potential is not infinitely steep and a more realistic approach is therefore to replace the hard-sphere potential assumed above with a 'soft-sphere' potential of the form

$$\phi(r) = \frac{8}{\pi} \varepsilon \left(\frac{\sigma}{r} \right)^s. \tag{9.27}$$

Here σ and ε are characteristic length- and energy-scaling parameters which define the extent and strength of the potential and s is an exponent which determines the steepness of the repulsion. For hard spheres $s = \infty$, while typical values used in the modelling of real molecules lie in the range 9 to 13.

In the strictest sense there is no such thing as a mean free path for real molecules as the interaction potential approaches zero only asymptotically and it is therefore impossible to ascribe to each molecule a unique diameter. However, it is a simple matter to determine the distance of closest approach r_0 in a binary collision and this forms a convenient measure upon which to base the molecular diameter and collision cross section. We consider two molecules approaching each other along their line of centres with an initial relative speed $<c_{12}>$. Their initial kinetic energy is converted gradually to potential energy as they approach until they reach the classical turning point r_0 where

$$\frac{1}{2} m_1 <c_{12}>^2 = \frac{8}{\pi} \varepsilon \left(\frac{\sigma}{r_0} \right)^s . \tag{9.28}$$

Employing Eq.(9.4) for the mean relative speed, we obtain the following expression for the effective cross section πr_0^2:

$$\pi r_0^2 = \pi \sigma^2 \left(\frac{\varepsilon}{k_B T} \right)^{2/s} . \tag{9.29}$$

This takes the form of the corresponding hard-sphere cross section modified by a temperature dependent factor. The expressions for the viscosity and thermal conductivity obtained in the previous section may be modified for soft spheres by simply replacing the hard sphere cross section by πr_0^2. This results in coefficients η and λ which vary with temperature like $T^{(s+4)/2s}$ and, with s in the range 9-13, this is close to the temperature dependence observed experimentally for real monatomic gases.

The rôle of the cross section in the elementary kinetic theory is taken on in the rigorous theory by a *collision integral* $\Omega(T)$. This is a temperature-dependent effective cross section which is related to the intermolecular potential through a multidimensional integral and reduces to $\pi \sigma^2$ in the case of hard spheres. These qualities are also exhibited by the effective cross section defined in Eq.(9.29); indeed that quantity is almost identical with the exact collision integral for soft spheres. It is convenient to define a dimensionless reduced collision integral $\Omega^* = \Omega /(\pi \sigma^2)$ and to note that, for a two-parameter intermolecular potential such as the Lennard-Jones potential or the soft-sphere potential, this is a function of the reduced temperature $T^* = k_B T/\varepsilon$ only. For example, the reduced collision integral in the case of soft spheres is, from Eq.(9.29), approximately $(\varepsilon / k_B T)^{2/s}$. For mixtures, the exact kinetic-theory results involve interaction collision integrals $\Omega_{12}(T)$ which take the place of the hard-sphere cross section $\pi \sigma_{12}^2$ and may be obtained from the unlike intermolecular potential in the same way as Ω is obtained from the pair potential of a pure substance. The estimation of collision integrals will be discussed in the next chapter.

9.1.4. Transport Properties of Real Dilute Gases

The exact expressions for the viscosity and thermal conductivity of a dilute monatomic gas were obtained by Chapman and Enskog by rigorous solution of the
Boltzmann transport equation [11]. The results may be written[1]

$$\eta_0 = \frac{5}{16}\left[\frac{m k_B T}{\pi}\right]^{1/2} \frac{f_\eta}{\sigma^2 \Omega^*} \tag{9.30}$$

$$\lambda_0 = \frac{75}{64}\left[\frac{k_B^3 T}{m\pi}\right]^{1/2} \frac{f_\lambda}{\sigma^2 \Omega^*}, \tag{9.31}$$

where the subscript 'o' now denotes a property of the dilute gas.

The Chapman-Enskog solution takes the form of a series of successive
approximations the first of which corresponds to $f_\eta = 1$ and $f_\lambda = 1$. Fortunately, the
series converges rapidly and the true values of f_η and f_λ are very nearly unity.

Comparing these equations and Eq.(9.25-6), with $c_V = 3k_B/2$, it can be seen that
the expressions are identical except for the replacement of the hard-sphere cross
section by the collision integral. In order to calculate the transport properties, it is
necessary to evaluate the collision integral and the correction factors. This is a
rather involved calculation that we will not describe here. Fortunately, the reduced
collision integral has been evaluated as a function of reduced temperature for a
number of two-parameter intermolecular potentials and tables of results are available.

The results of the rigorous kinetic theory for the case of a multicomponent
mixture conform to the composition dependence derived from the elementary kinetic
theory but with coefficients H_{ij} that differ from those given by Eq.(9.20) [1,11]. The
results involve the interaction collision integrals Ω_{12}. Some useful approximations
for H_{ij} will be discussed in the next chapter.

An interesting result of both the elementary and the rigorous theory is that the
thermal conductivity and viscosity share the same dependence on (a) temperature
and (b) the collision integral.

---

[1] In recent years, the kinetic theory relations have increasingly been written in terms of
effective cross sections $\mathfrak{S}$ instead of collision integrals. The relation between Ω as used
here and the cross section $\mathfrak{S}(2000)$ is [1]:

$$\mathfrak{S}(2000) = (4/5)\,\Omega.$$

Note also that our Ω is identical with the collision integral $\Omega^{(2,2)}$ used by other authors.

The ratio

$$f_{\text{Eu}} = \frac{\lambda_0 \, M}{\eta_0 \, C_{V,\text{m}} \, \xi} \,, \tag{9.32}$$

is known as the Eucken factor and, according to the rigorous theory, it has the value $5/2\xi$. Here, $C_{V,\text{m}}$ is the molar heat capacity at constant volume, M is the molar mass, and $\xi = f_\eta / f_\lambda$ differs from unity by no more than 1 or 2 per cent. Knowledge of the Eucken factor is very useful in the estimation of the one property from the other at the same temperature.

The discussion thus far is restricted to dilute monatomic gases. We have already mentioned that the existence of non-spherical intermolecular potentials and the possibility of inelastic collisions seriously complicate the kinetic theory of polyatomic gases. Nevertheless, formal expressions for the viscosity and thermal conductivity of pure gases and mixtures may be derived. These expressions, which are functions of several different types of cross sections, can be found elsewhere [1]. The thermal conductivity, which is found to be especially sensitive to inelastic collisions, is expressed as the sum of two contributions, the one for the transport of translational energy and the other for the transport of internal energy. The problem then reduces to the calculation of the full set of collision cross sections and a number of methods are available by which this may be accomplished starting from a prescribed non-spherical intermolecular potential. At present, essentially-exact calculations based on a quantum-mechanical treatment are possible only for systems having, at the temperature of interest, a small number of populated rotational states [1]. On the other hand, when the number of populated rotational states becomes large, a classical treatment of collisions is appropriate and calculations based on that approach, while computationally demanding, are nevertheless possible. It appears that a judicious combination of quantum and classical calculations can achieve the desired objectives.

It must be said that rigorous methods for polyatomic gases have, at present, no rôle to play in routine property predictions and that approximate methods are of much greater utility. It has been found that the viscosity of polyatomic gases may be correlated quite satisfactorily by assuming an effective intermolecular potential of spherical symmetry. For example, parameters for the Lennard-Jones potential are available for many polyatomic gases.

The difficulties with the thermal conductivity have been addressed through several useful approximations, most of which seek to relate λ to η. Eucken proposed that Eq.(9.32) be modified for polyatomic gases by separating the translational and internal energy contributions so that

$$f_{Eu} = \frac{\lambda_0 M}{\eta_0 C_{V,m}} = f_{tr}\Delta + f_{int}(1-\Delta) ,\qquad (9.33)$$

where f_{tr} and f_{int} are factors of order unity and

$$\Delta = \left(\frac{3}{2(C_{V,m}/R)}\right)\qquad (9.34)$$

is the fraction of the constant-volume heat capacity associated with the translational modes. Assuming ξ to be unity, Eq.(9.32) is recovered by setting $f_{tr}=f_{int}=2.5$. In his original formulation, Eucken in fact chose $f_{tr}=2.5$ and $f_{int}=1.0$ which corresponds to an assumption of free exchange of translational and internal energy in molecular collisions. If one makes the alternative assumption that the exchange of internal and translational energy in collisions is very inefficient, then the transport of internal energy becomes a process of diffusion and one can show that the appropriate choice is to set $f_{int}=1.32$ [12]. The resulting expression for f_{Eu}, often called the *modified Eucken formula*, is

$$f_{Eu} = \frac{\lambda_0 M}{\eta_0 C_{V,m}} = 1.32 + 1.77(R/C_{V,m}) .\qquad (9.35)$$

Experimental results typically lie somewhere in-between the predictions of the Eucken and modified Eucken formulae, although the latter is usually to be preferred. Formulae, such as Eq.(9.35), are useful for estimating the dilute-gas thermal conductivity once the viscosity is known.

More accurate expressions for the factors f_{tr} and f_{int} were obtained by Mason and Monchick [9] from the theory of Wang Chang and Uhlenbeck [7]. However these expressions involve parameters which quantify the efficiency with which internal and translational energies are exchanged in collisions; these parameter are seldom known except for very simple molecules and the method will therefore be considered here no further.

9.2. The Density Dependence

Although at low densities (and hence low pressures), the viscosity at constant temperature is more or less independent of density, this is not the case at higher densities and pressures. The somewhat surprising low-density behaviour, predicted first by Maxwell [13], can be clearly seen if we examine the experimental results for the viscosity of argon. At 298.15 K, η increases by just 0.8 per cent for a tenfold increase in pressure from 0.1 to 1 MPa. However, at high pressures and at about the same temperature, the viscosity increases from 124 to 421 µPa·s, that is by a factor of about 3.5, for a four fold increase of pressure from 200 to 800 MPa [14].

There are two main factors to which this difference in the pressure dependence between the dilute-gas and dense-gas states may be attributed. The first relates to the mechanism of momentum transfer itself. In a dilute gas, the mean free path is long compared with the size of the molecules and free flight of molecules is therefore the primary mechanism of momentum transfer. In contrast, the mean free path may be smaller than the molecular diameter in a dense gas and then momentum is transported only over a relatively small distance through free flight of molecules. However, a collision transports momentum instantly over a distance of the order of the molecular diameter and such transfers become the dominant mechanism at high densities. This phenomenon of collisional transport is exhibited rather nicely by the 'executive toy' known as Newton's cradle. The second important factor is the collision rate which is proportional to density in the dilute gas but increases more rapidly in the dense gas as the free volume is filled up by the molecules themselves.

In Figure 9.6 the viscosity of argon [15] is plotted as a function of pressure and also as a function of density. It is apparent that the dependence upon the latter is the

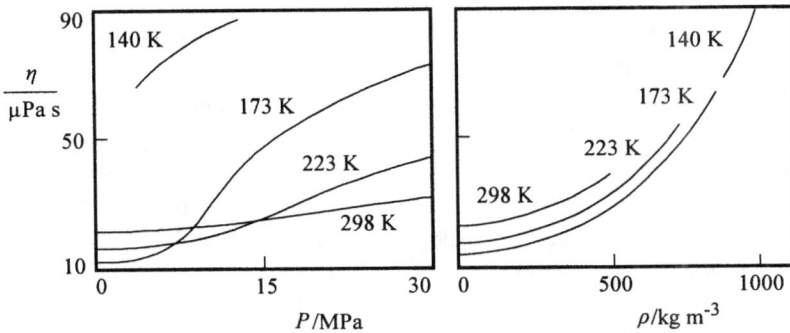

Figure 9.6. Pressure and density dependence of the viscosity of argon [15].
(note: $T^c = 150.8$ K, $P^c = 4.87$ MPa and $\rho^c = 533.4$ kg·m^{-3})

simpler and this is in accord with the theoretical importance of the density. Similar results are obtained for the thermal conductivity far from the critical region.

There is at present no complete theory for the density dependence of the transport properties. It is preferable to separate the problem into three:

(a)   the initial density dependence,
(b)   the intermediate density range, and
(c)   the very dense gas and liquid regions.

9.2.1. The Initial Density Dependence

If only moderate densities are considered, then the background transport property, defined in Eq.(9.1), can be expanded in powers of ρ_n to obtain a virial-type series. For example, the series of the viscosity, truncated after the second viscosity 'virial coefficient', is

$$\eta(\rho_n, T) = \eta_0(T) + \Delta\eta(\rho_n, T) = \eta_0(T)[1 + B_\eta(T)\rho_n] . \qquad (9.36)$$

This linear expansion in sufficient only up to a rather modest density. For the viscosity, the presence of internal degrees of freedom has only a small influence and the second viscosity virial coefficient may therefore be treated in much the same way for both monatomic and polyatomic gases.

In the case of the thermal conductivity, a separate treatment of the translational and internal degrees of freedom is required and one writes

$$\lambda(\rho_n, T) = \lambda_{0,\mathrm{tr}}(T)[1 + B_{\lambda,\mathrm{tr}}(T)\rho_n] + \lambda_{0,\mathrm{int}}(T)[1 + B_{\lambda,\mathrm{int}}(T)\rho_n] , \qquad (9.37)$$

where $B_{\lambda,\mathrm{tr}}$ and $B_{\lambda,\mathrm{int}}$ are second thermal-conductivity virial coefficients for the translational and internal modes.

The most modern theory by which the second transport-property virial coefficients may be calculated is that due to Rainwater and Friend [16,17]. This is a model that treats the moderately dense gas as a mixture of monomers and dimers which interact according to the Lennard-Jones (12,6) potential. Contributions, assumed to be independent, to transport-property virial coefficients arise from the interactions between two monomers, three monomers and between a monomer and a dimer.

Among the limitations of the Rainwater-Friend theory is that it has not yet been extended to describe $B_{\lambda,\mathrm{int}}$. The best that can be done for this term is an approximation based on modifications [18,19] of the Enskog hard-sphere theory [20].

Some efforts have been made to establish a principle of corresponding states for transport-property virial coefficients (see for example Bich & Vogel [21]) and this is an area of current research.

9.2.2. *The Intermediate Density Range*

The truncated virial-type expansion presented in the previous section becomes inadequate as the density increases. However, it appears that the background transport property, defined through Eq. (9.1) as

$$\Delta X(\rho_n, T) = X(\rho_n, T) - X_0(T) - \Delta X_c(\rho_n, T) , \qquad (9.38)$$

acquires a dependence upon density that is nearly independent of temperature. This behaviour, apparent in Figure 9.6, has been the basis of several successful correlation methods, although it must be said that at very high densities there is a noticeable dependence upon both temperature and density.

One of the first theoretical attempts to describe the density dependence was that of Enskog [20] who considered a system of hard spheres and proposed empirical modifications to the Boltzmann theory to account for the finite size of the molecules. The Boltzmann transport equation is based on a consideration of binary collisions only and might therefore be thought totally inappropriate for the dense gas and liquid regions. Enskog justified his use of the theory by means of the fact that exactly synchronous many-body encounters between hard spheres have negligible probability even at high densities. Enskog's expressions for the coefficients of self diffusion, D_E, viscosity, η_E, and thermal conductivity, λ_E, are :

$$\frac{nD_E}{nD_0} = \frac{1}{g(\sigma)} \qquad (9.39)$$

$$\frac{\eta_E}{\eta_0} = \frac{1}{g(\sigma)} + \frac{0.8b}{V_m} + 0.761g(\sigma)\left(\frac{b}{V_m}\right)^2 \qquad (9.40)$$

$$\frac{\lambda_E}{\lambda_0} = \frac{1}{g(\sigma)} + \frac{1.2b}{V_m} + 0.755g(\sigma)\left(\frac{b}{V_m}\right)^2 , \qquad (9.41)$$

where V_m is the molar volume, and $g(\sigma)$ is the value of the radial distribution function, $g(r)$, at contact $(r = \sigma)$, while $b = 2\pi N_A \sigma^3/3$ is the van der Waals co-volume.[2]   The Enskog theory accounts for the enhancement of the collision frequency in a dense gas through the term $1/g(\sigma)$. This is the only term contributing to the departure of the self diffusion coefficient from its dilute gas value. However, the viscosity and thermal conductivity are also affected by the collisional transport mechanism discussed above and this gives rise to the terms containing b/V_m.

For polyatomic molecules, Eq.(9.41) may be extended to account for the transport of internal energy on the assumption that this occurs by diffusion alone. This assumption leads to the result:

$$\lambda_E = \lambda_{0,tr}\left[\frac{1}{g(\sigma)} + \frac{1.2b}{V_m} + 0.755g(\sigma)\left(\frac{b}{V_m}\right)^2\right] + \frac{D_0\,C_{V,int}}{V_m\,M\,g(\sigma)}. \qquad (9.42)$$

The value of the radial distribution function at contact is known from computer simulations for hard spheres but the question then arises as to how the Enskog theory may be applied to real molecules.   One method that has been proposed for the selection of appropriate values of $g(\sigma)$ and b involves consideration of the so-called thermal pressure $P_t = T(\partial P/\partial T)_{\rho_n}$ [22] which, for hard spheres, is given exactly by

$$P_t = T\left(\frac{\partial P}{\partial T}\right)_{\rho_n} = \frac{RT}{V_m}(1 + b\rho_n g). \qquad (9.43)$$

The method for determining $g(\sigma)$ involves evaluating P_t from the equation of state of the real fluid and thence $b \cdot g(\sigma)$ from Eq.(9.43). The co-volume may be obtained conveniently from the van der Waals theory of fluids which leads to the relation

$$b = B + T\left(\frac{dB}{dT}\right) \qquad (9.44)$$

and thus permits evaluation of b from the second virial coefficient.

The use of Eqs (9.43) and (9.44) constitute what is called the *Modified Enskog Theory*. Hanley *et al.* [23] point out that the results obtained are very sensitive to

---

[2] The radial distribution function $g(r)$ is defined such that the local number density (i.e. the number of other molecules per unit volume) at a distance r from the centre of a chosen molecule is $g(r) \cdot (N/V)$. For hard spheres, $g(r) = 0$ for $r < \sigma$ but $g(\sigma) \geq 1$ at all temperatures and densities. Since isotropic fluids lack long-range order, $g(r) \to 1$ as $r \to \infty$.

any inconsistencies in the thermodynamic surface and suggest that both P_t and B should be determined from the same equation of state. Detailed comparisons between experimental data for the transport coefficients and the predictions of the *Modified Enskog Theory* were made by these authors. At densities below twice the critical, agreement to within about ±15 per cent was found. Recently, Vesovic and Wakeham [24-26] have extended these methods to mixtures.

9.2.3. The Very Dense Gas & Liquid Density Range

Although the Enskog theory is certainly not exact for real molecules it might be thought correct for hard spheres. However, computer simulations of hard-sphere systems reveal that there are correlated molecular collisions, of a kind not considered in the Enskog theory, and that these become important at very high densities. Alder *et al.* [27] calculated the corrections to the Enskog theory for the coefficients of self diffusion, viscosity and thermal conductivity based on molecular-dynamics simulations of systems containing 108 or 500 molecules. He showed that the corrections were functions only of the ratio (V_m/V_o), where V_o is the molar volume for close packed spheres $(V_o = N_A \sigma^3/2^{1/2})$. Erpenbeck and Wood [28] confirmed the diffusion results obtained by Alder and extrapolated them to a system of infinite size. However, in the case of the viscosity and thermal conductivity, the correction factors turn out to be associated with a rather large uncertainty and this limits somewhat their usefulness. Nevertheless, these results indicate that the dimensionless transport property X^*, defined by

$$X^* = \left(\frac{X}{X_E}\right)_{MD} \left(\frac{X_E}{X_o}\right) \left(\frac{V_m}{V_o}\right)^{2/3}, \qquad (9.45)$$

is a universal function of (V_m/V_o), independent of T. In Eq.(9.45), the factor $(X/X_E)_{MD}$ is the correction to the Enskog theory computed from molecular dynamics simulations.

In order to apply these results to real molecules without prior knowledge of V_o, Dymond [29] proposed a modified scheme in which the dimensionless transport property is defined simple as

$$X^* = \left(\frac{X}{X_o}\right) \left(\frac{V_m}{V_o}\right)^{2/3}. \qquad (9.46)$$

Here, X_0 is the dilute gas transport property for the hard-sphere model with $\sigma = (2^{1/2} V_0/N_A)^{1/3}$ and V_0 is permitted to be a function of the temperature. Employing Eqs.(9.32) and (9.33), with $\Omega^* = 1$, expressions for the dimensionless viscosity and dimensionless thermal conductivity are obtained as follows:

$$\eta^* = \frac{16}{5}(2N_A)^{1/3}\pi^{1/2}\left(\frac{1}{MRT}\right)^{1/2}\eta V_m^{2/3} = \mathcal{F}_\eta(V_m/V) \tag{9.47}$$

$$\lambda^* = \frac{64}{75}\frac{\pi^{1/2}}{R}(2N_A)^{1/3}\left(\frac{M}{RT}\right)^{1/2}\lambda V_m^{2/3} = \mathcal{F}_\lambda(V_m/V) \tag{9.48}$$

where M is the molar mass and $\mathcal{F}_\eta$ and $\mathcal{F}_\lambda$ denote universal functions of (V_m/V_0). The optimum values of V_0 for a set of real fluids may be obtained by plotting $\ln X^*$ against $\ln V_m$ for the experimental results on various isotherms. The results for different compounds and temperatures form a set of parallel curves and the translation along the molar volume axis that maps each curve on to a chosen reference curve produces the corresponding values of V_0. It is found that V_0 is only a weak function of temperature for a given compound. The reference curve was chosen so that it produced the closest agreement with the simulated results for hard spheres.

As this method is an approximate scheme, it is advisable to optimise V_0 by simultaneous consideration of the three properties: self-diffusion, viscosity and thermal conductivity and this can in fact be accomplished.

In Figure 9.7, the universal curve for viscosity is shown including both experimental values for the monatomic gases and the results computed for hard spheres. The results cover the compressed-gas and liquid ranges but not the critical region where an additional enhancement exists. Having obtained the universal curves, a measurement of η or λ at, say, atmospheric pressure for a particular liquid will permit determination of V_0 at the temperature in question and thence prediction of the property at higher pressure. It is of course necessary to know accurately the molar volume as a function of pressure to accomplish this. The advantage of scheme is that only values of the transport properties at atmospheric pressure are required for the determination of V_0.

Following the successful application of this scheme to simple liquids, the method was applied to non-spherical dense fluids such as hydrocarbons [30], alcohols [31], refrigerants [32] and their mixtures. In order that these systems too

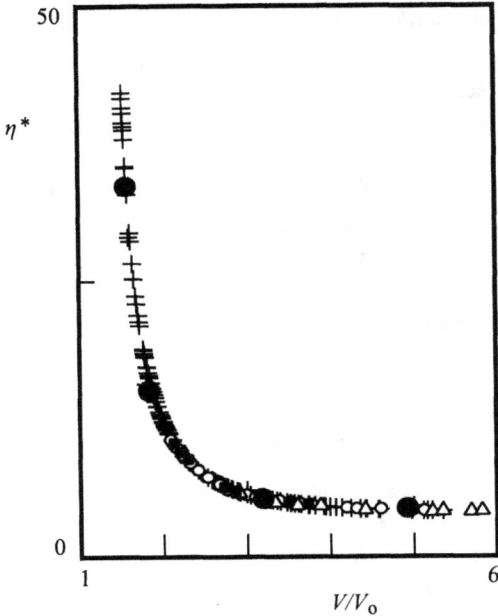

Figure 9.7. Reduced viscosity η^* as a function of (V/V_0).
(■ Ne; + Ar; Δ Kr; ○ Xe; ● computed values for hard spheres)

coincide with the universal curves, the definitions of the dimensionless transport properties were modified through the inclusion of temperature-independent roughness factors R_η and R_λ:

$$\eta^* = \frac{16}{5}(2N_A)^{1/3}\pi^{1/2}\left(\frac{1}{MRT}\right)^{1/2}\frac{\eta V_m^{2/3}}{R_\eta} \tag{9.49}$$

$$\lambda^* = \frac{64}{75}\frac{\pi^{1/2}}{R}(2N_A)^{1/3}\left(\frac{M}{RT}\right)^{1/2}\frac{\lambda V_m^{2/3}}{R_\lambda} \tag{9.50}$$

Experimental results over a range of temperatures at atmospheric pressure are sufficient to obtain both the characteristic volume and the roughness factors: V_0 is obtained from the horizontal translation and R_η or R_λ is obtained from the vertical translation needed to superimpose the experimental curves on the (universal) reference curves for η^* and λ^*.

This powerful scheme permits the prediction of the transport properties of compressed liquids at very high pressures (it has been tested up to $P = 500$ MPa) based on experimental results for a single isobar at atmospheric pressure plus knowledge of the liquid density. The main drawback of the method is that the predictions are very sensitive to the values of (V_m/V_o) and an accurate equation of state for the liquid is therefore required. The application of this scheme to the prediction of the viscosity and thermal conductivity of selected liquids and mixtures will be discussed further in Section 10.3.

9.3. Critical Enhancement

Near to the vapour-liquid critical point of a pure fluid, many of the thermo-dynamic and transport properties of the fluid approach either zero or infinity. The modern theory of the critical region is based on consideration of density fluctuations and their correlation in space and time [1]. It predicts a strong enhancement of the thermal conductivity in the critical region and a weak enhancement of the viscosity; both approach infinity at the critical point itself.

In Figure 9.8, the thermal conductivity of ethane is shown on various isotherms near the critical temperature [33]. The critical enhancement can be seen to affect a wide range of densities around ρ_n^c and to extend to temperatures some way above and below T^c. The corresponding enhancement to the viscosity can be observed only with some difficulty as, on the critical isochore, it exceeds 1 per cent of η only within 5 K of the critical temperature. For carbon dioxide, Vesovic *et al.* [34] showed that even at twice the critical density and one-and-a-half times the critical temperature, the thermal conductivity enhancement still amounts to about 1 per cent of the total value.

Very close to the critical point, the transport properties diverge in accordance with simple power laws. However, it is not possible to represent the critical enhance-ment over the wider critical region in this way. A theoretical formulation that permits accurate correlation of the experimental data for both the thermal; condu-ctivity and the viscosity has been derived by Olchowy and Sengers [35]. Although measurements near to the critical point are very difficult, and experimental results are therefore scarce, the theory provides a good description of the available data although it does require an accurate equation of state for the substance in question.

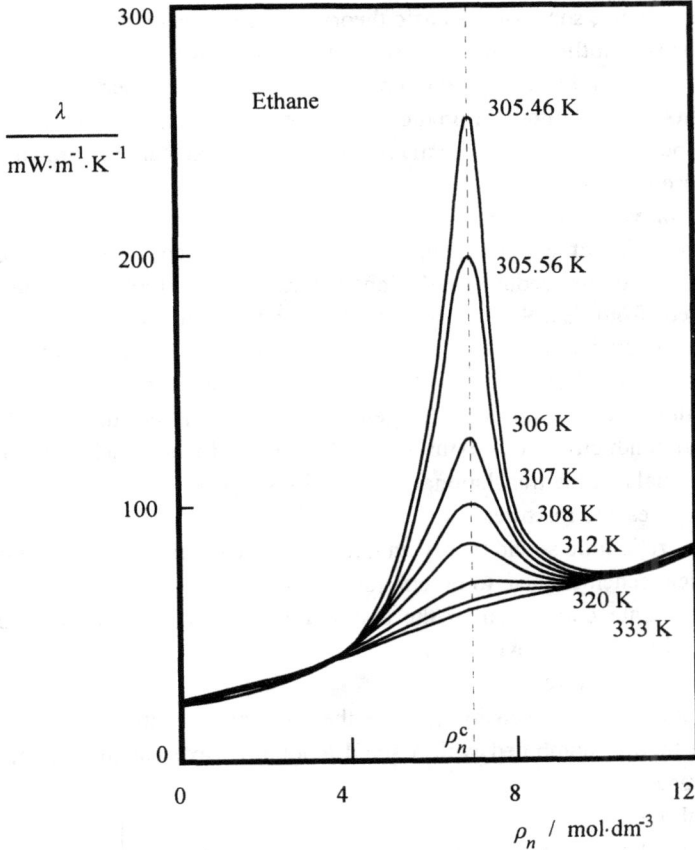

Figure 9.8. Ethane thermal conductivity as function of density along isotherms [33]. (Note that $T^c = 305.33$ K, $P^c = 4.878$ MPa, $\rho_n^c = 6.87$ mol·dm^{-3})

As the details of the modern theory are clearly beyond the scope of an introductory text such as this, the reader is referred to a monograph on transport properties [1] and to the original literature [35].

9.4. Summary

In this chapter, some of the basic theory of the viscosity and thermal conductivity of fluids was outlined. The purpose of the presentation was to indicate the most favourable schemes for the prediction of the transport properties of real fluids including mixtures. Methods of prediction and/or correlation based on the theory are usually to be preferred and the main conclusions that can be drawn from the discussion are as follows:

(1) Dilute gases and mixtures.

The viscosity of a pure component is best obtained from Eq.(9.32) and a correlation of the reduced collision integral Ω^*. Mixture viscosities are best obtained from Eq.(9.19) with H_{ij} from either the rigorous theory or an empirical-estimation scheme (examples of which will be considered in the next chapter). The thermal conductivity of a pure component is best be obtained from the viscosity through the generalised Eucken formula, Eq.(9.33). The thermal conductivity of mixtures must be obtained from a scheme which treats the internal and the translational contributions separately.

(2) Dense gases and mixtures.

Here there is little sound theory and, to cover a wide range of substances, semi-empirical schemes have to be employed. The best approach seems to be based on a separate correlation of the excess contribution together with dilute-gas values obtained as above.

(3) Liquids and mixtures.

A scheme for the viscosity and the thermal conductivity based on the hard-sphere theory, combined with a small amount of experimental data, can give good results.

(4) Critical region

Although there is a well-developed theory for the critical region, its complexity places it beyond the scope of this chapter. In the case of the viscosity, the critical enhancement is of little technological importance as it is restricted to a small region very close to the critical point. For the thermal conductivity, we have seen that the critical enhancement is much larger over an extended region around the critical point and its neglect could give rise to significant errors.

Based on these conclusions, we shall present in the next chapter semiempirical methods for the prediction of the viscosity and the thermal conductivity in all fluid regions except the critical region where special techniques apply.

References

1. J. Millat, J.H. Dymond and C.A. Nieto de Castro (Eds.), "Transport Properties of Fluids. Their Correlation, Prediction and Estimation", Cambridge University Press (1996).
2. G.C. Maitland, M. Rigby, E.B. Smith and W.A. Wakeham, "Intermolecular Forces. Their Origin and Determination", Clarendon Press, Oxford (1981).
3. L. Waldman, *Z. Natuforsch.* **12a**:660 (1957).
4. R.F. Snider, *J. Chem. Phys.* **32**:1051 (1960).
5. F.R.W. McCourt, J.J.M. Beenakker, W.E. Kohler and I. Kuscer, "Noenequilibrium Phenomena in Polyatomic Gases. Volume I", Clarendon Press, Oxford (1990).
6. F.R.W. McCourt, J.J.M. Beenakker, W.E. Kohler and I. Kuscer, "Noenequilibrium Phenomena in Polyatomic Gases. Volume II", Clarendon Press, Oxford (1991).
7. C.S. Wang Chang and G.E. Uhlenbeck, "Transport Phenomena in Polyatomic Gases", *Univ. Michigan Eng. Res. Rep.* CM-681, Ann Arbor, Michigan (1951).
8. J. de Boer and G.E. Uhlenbeck (Eds.), "Studies in Statistical Mechanics. Volume II.", North Holland, Amsterdam (1964).
9. E.A. Mason and L. Monchick, *J. Chem. Phys.* **36**:1622 (1965).
10. L. Monchick, A.N.G. Pereira and E.A. Mason, *J. Chem. Phys.* **42**:3241 (1967).
11. J.H. Ferziger ang H.G. Kaper, "Mathematical Theory of Transport Processes in Gases", North-Holland, Amsterdam (1972).
12. R.C. Reid, J.M. Prausnitz and B.E. Poling, "The Properties of Gases & Liquids", 4th Ed., McGraw-Hill, New York (1988).
13. J.C. Maxwell, *Phil. Trans. Roy. Soc. London* **157**:49 (1867).
14. N.J. Trappeniers, P.S. van der Gulik and H. van den Hooff, *Chem. Phys. Lett.* **70**:438 (1980).
15. W.M. Haynes, *Physica* **67**:440 (1973).
16. D.G. Friend and J.C. Rainwater, *Chem. Phys. Lett.* **107**:590 (1984).
17. J.C. Rainwater and D.G. Friend, *Phys. Rev.* **A36**:4062 (1987).
18. H.J. Hanley and E.G.D. Cohen, *Physica* **A83**:215 (1976).
19. C.A. Nieto de Castro, D.G. Friend, R.A. Perkins and J.C. Rainwater, *Chem. Phys.* **145**:19 (1990).
20. D. Enskog, *Kgl. Svenska Ventensk. Handl.* **63**:4 (1922).
21. E. Bich and E. Vogel, *Int. J. Thermophys.* **12**:27 (1991).
22. S. Chapman and T.G. Cowling, "The Mathematical Theory of Non-Uniform Gases", Cambridge University Press, London (1952).
23. H.J.M. Hanley, R.D. McCarty and E.G.D. Cohen, *Physica* **60**:322 (1972).
24. V. Vesovic and W.A. Wakeham, *Int. J. Thermophys.* **10**:125 (1989).
25. V. Vesovic and W.A. Wakeham, *Chem. Eng. Sci.* **44**:2181 (1989).
26. V. Vesovic and W.A. Wakeham, *High Temp. High Press.* **23**:179 (1991).
27. B.J. Alder, D.M. Gass and T.E. Wainwright, *J. Chem. Phys.* **53**:3813 (1970).
28. J.J. Erpenbeck and W.W. Wood, *Phys. Rev. A.* **43**:4254 (1991).
29. J.H. Dymond, *Proc. 6th Symp. Thermophys. Prop.*, ASME, New York, 143 (1973).
30. M.J. Assael, J.H. Dymond, M. Papadaki and P.M. Patterson, *Int. J. Thermophys.* **13**:269 (1992).

31. M.J. Assael, J.H. Dymond and S.K. Polimatidou, *Int. J. Thermophys.* **15**:189 (1994).
32. M.J. Assael, J.H. Dymond and S.K. Polimatidou, *Int. J. Thermophys.* **16**:761 (1995).
33. R. Mostert, PhD thesis, University of Amsterdam (1991).
34. V. Vesovic, W.A. Wakeham, G.A. Olchowy, J.V. Sengers, J.T. Watson and J. Millat, *J. Phys. Chem. Ref. Data* **19**:763 (1990).
35. G.A. Olchowy and J.V. Sengers, *Phys. Rev. Lett.* **61**:15 (1988).

Transport Properties: Calculation

In the previous chapter, a brief description of the kinetic theory was presented. We now consider appropriate applications of the theory to the prediction of transport properties of pure fluids and mixtures in the gaseous and liquid states. The methods chosen are those that have a reasonable connection with the theory and in which a transport property X is written:

$$X(\rho_n, T) \;=\; X_0(T) + \Delta X(\rho_n, T) + \Delta X_c(\rho_n, T) \;. \tag{10.1}$$

The critical enhancement ΔX_c will not be considered further here. The dilute-gas term X_0 will be discussed first. Then, the estimation of the excess transport properties ΔX for compressed gases by the method of Thodos will be examined while, for liquids, the scheme proposed by Dymond and Assael will be presented. Finally, for the estimation of the viscosity and thermal conductivity of non-polar mixtures, the corresponding-states scheme of Ely and Hanley will be outlined.

In Section 10.6, numerical examples based on these methods will be given.

10.1. Dilute Gases and Mixtures

For a dilute gas, the viscosity is given from Eq.(9.30) as

$$\eta_0 \;=\; \frac{5}{16} \left[\frac{m k_B T}{\pi} \right]^{1/2} \frac{f_\eta}{\sigma^2 \, \Omega^*}, \tag{10.2}$$

f_η is generally taken as unity. This expression, although strictly applicable only to a monatomic gas, is used in practice as a correlating tool for many pure gases including those with polyatomic molecules. The reduced collision integral $\Omega^* = \Omega/(\pi\sigma^2)$ has been tabulated as a function of $T^* = k_B T/\varepsilon$ for a number of model intermolecular potentials. A few experimental values of the viscosity are then required to establish appropriate values of the scaling parameter σ and ε in a chosen model and it is then possible to predict η at other temperatures with reasonable confidence.

The reduced collision integral Ω^* for the Lennard-Jones (12,6) potential has been calculated by Neufeld *et al.* [2] and correlated as a function of T^* over the range $0.3 \leq T^* \leq 100$. This correlation is reproduced in Eq.(10.6) of Table 10.1. Many sets of the parameters σ and ε/k_B for the (12,6) potential have been reported in the literature and a selection of those determined from viscosity data are given in Table A9 of Appendix A [3]. When reliable parameters are not available, rough estimates may be made based on the critical constants [4]:

$$(\varepsilon/k_B)_{LJ} = 0.775 \, T_c \qquad (10.3)$$

and
$$\frac{2}{3} \pi N_A \, \sigma_{LJ}^3 = 0.211 \left(\frac{RT_c}{P_c} \right). \qquad (10.4)$$

For mixtures, the composition dependence of the dilute-gas viscosity is given in leading order by Eq.(9.19) with coefficients H_{ij} that depend upon both pure-component and interaction collision integrals. It turns out in practice to be much more convenient to employ approximations [6] to the 'exact' coefficients as these are easy to calculate and, at the same time, can compensate for higher-order terms omitted from the simple mixture theory. The scheme of Wilkes [7] is particularly useful, being both simple and generally quite accurate. Indeed, although this method was proposed on purely empirical grounds, it may be derived [8] from approximations applied to the formal kinetic theory.

| EXAMPLE 10.1 | *Evaluation of Dilute-Gas Viscosity of n-Butane at T = 293 K.*

For this calculation, Eqs.(10.5) and (10.6) are employed. The parameters σ, and ε/k_B are obtained from Table A9 in Appendix A as 0.4687 nm and 531.4 K respectively. Hence, $T^* = 0.551$ and, from Eq.(10.6), the reduced collision integral is found to be 2.175 (note the very small contribution of the sine term)

The molecular mass of *n*-butane is $m_1 = 9.652 \times 10^{-26}$ kg and, substituting in Eq.(10.5), we obtain $\eta = 7.29$ µPa·s. The experimental value reported by Kestin and Yata [5] is (7.27 ± 0.02) µPa·s and the agreement is therefore excellent.

Had we used Eqs.(10.3) and (10.4) to estimate the Lennard-Jones parameter then the calculated viscosity would have been 7.11 µPa·s, which is still within 2.5 per cent of the experimental value. Such good agreement is not always observed.
❑

Table 10.1   Viscosity of Dilute Gases and their Mixtures[1]

Pure Gas *i*

$$\eta_i \; = \; \frac{5}{16}\left[\frac{m_i\,k_{\mathrm B}T}{\pi}\right]^{1/2}\frac{1}{\sigma_i^2\;\Omega^*(T^*)} \tag{10.5}$$

where

$$\Omega^*(T^*)= 1.16145(T^*)^{-0.14874}+0.52487\,\mathrm{e}^{-0.7732\,T^*} + 2.16178\,\mathrm{e}^{-2.43787\,T^*}$$
$$- \,6.435\mathrm{x}10^{-4}(T^*)^{0.14874}\sin\!\left[18.0323(T^*)^{-0.7683}-7.2371\right] \tag{10.6}$$

Multicomponent Mixtures

$$\eta_{\mathrm{mix}} \; = \; \sum_{i=1}^{N}\frac{y_i\eta_i}{\displaystyle\sum_{j=1}^{N} y_j\,\Phi_{ij}} \tag{10.7}$$

where

$$\Phi_{ij} \; = \; \frac{\left[1+\left(\eta_i/\eta_j\right)^{1/2}\left(M_j/M_i\right)^{1/4}\right]^2}{2\sqrt{2}\left[1+\left(M_i/M_j\right)\right]^{1/2}} \tag{10.8}$$

---

[1] Note that subscript "o" indicating dilute-gas property, has been omitted for brevity.

Wilke's treatment of the mixture viscosity is given in Eqs.(10.7) and (10.8) of Table 10.1. These equations have been tested with great success against experimental data for a large number of non-polar mixtures [6] and are usually not in error by more than 2 per cent. Later studies [9] showed that this scheme can be applied with equal success to polar gases. The only cases in which significant disagreement was found were when $M_i \gg M_j$ and $\eta_i \gg \eta_j$.

A reasonable estimate of the thermal conductivity of a pure dilute gas may be obtained from the viscosity and perfect-gas heat capacity through the modified Eucken correlation, Eq.(9.35), which we repeat here as Eq.(10.9) of Table 10.2. This expression is almost exact for a monatomic gas but it can involve considerable error for polyatomic systems. Usually this error is not larger than 10 per cent although, for highly polar gases, it can reach 20 per cent [6].

Many schemes have been proposed for the prediction of the dilute-gas thermal conductivity of multicomponent mixtures containing polyatomic components. These range from the highly-complex rigorous kinetic theory [1] to purely empirical approximations [6] that, while easy to apply, may lead to large errors. A reasonable practical approach to predicting the mixture thermal conductivity is provided by the scheme initially proposed by Mason and Saxena [10] and later refined by Brokaw [4]. This method, shown in Table 10.2, is based on approximations to the rigorous kinetic theory and expresses the mixture thermal conductivity as the sum of two contributions:

(a)   a term containing the translational parts, λ_i', of the pure component thermal conductivities which are estimated from η_i; and

(b)   a term containing the internal parts, λ_i'', of the pure component thermal conductivities which are estimated as $(\lambda_i - \lambda_i')$ with λ_i from Eq.(10.9).

Brokaw's expressions for the mixture thermal conductivity, given in Table 10.2, also produce reasonable results for non-polar and slightly polar compounds. They should not be used for the calculation of the thermal conductivity of mixtures containing strongly polar compounds.

| EXAMPLE 10.2 | *Calculation of the Dilute-Gas Viscosity and Thermal Conductivity of Methane at $T = 308.15$ K.* |

a) Viscosity. Following the method of Example 10.1, Lennard-Jones parameters are obtained from Table A9, Appendix A, as $\varepsilon/k_B = 148.6$ K and $\sigma = 0.3758$ nm. The reduced collision integral at 308.15 K is found to be 1.161 and, employing Eq.(10.5), with $m = 2.664 \times 10^{-26}$ kg we find $\eta = 11.44$ μPa·s. The value reported by Kestin *et al.* [11] is (11.39 ± 0.03) μPa·s and the agreement is again good.

b) Thermal Conductivity. The perfect-gas heat capacity coefficients are obtained from Table A1, Appendix A, and give at 308.15 K $C_{P,m}^{pg} = 36.12$ J·mol^{-1}·K^{-1}, and hence $C_{V,m}^{pg} = 27.81$ J·mol^{-1}·K^{-1}. Employing the modified Eucken equation, Eq.(10.9), the thermal conductivity is calculated to be 36.67 mW·m^{-1}·K^{-1}. The experimental value obtained by Assael and Wakeham [12] is (35.35 ± 0.11) mW·m^{-1}·K^{-1} and the error in the calculated value is therefore about 4 per cent.

The calculated value of the Eucken factor, f_{Eu}, obtained above is 1.85 and this should be compared with the experimental value of 1.79. Had we used Eq.(9.33) with $f_{tr} = 2.5$ and $f_{int} = 1.0$, then we would have obtained $f_{Eu} = 1.67$ and thus $\lambda = 33.17$ mW·m^{-1}·K^{-1}.

❑

Table 10.2 Thermal Conductivity of Dilute Gases and their Mixtures[1]

Pure Gas *i*

$$\frac{\lambda_i M_i}{\eta_i C_{Vi}} = 1.32 + \frac{1.77}{(C_{Vi}/R)} \tag{10.9}$$

Multicomponent Mixtures

$$\lambda_{mix} = \sum_{i=1}^{N} \frac{y_i \lambda_i'}{\sum_{j=1}^{N} y_j C_{ij}} + \sum_{i=1}^{N} \frac{y_i \lambda_i''}{\sum_{j=1}^{N} y_j A_{ij}} \tag{10.10}$$

where

$$\lambda_i' = \frac{2.5 \eta_i (1.5R)}{M_i}, \qquad \lambda_i'' = \lambda_i - \lambda_i' \tag{10.11}$$

and

$$A_{ij} = \frac{\left[1 + (\lambda_i'/\lambda_j')^{1/2} (M_i/M_j)^{1/4}\right]^2}{2\sqrt{2}\left[1 + (M_i/M_j)\right]^{1/2}} \tag{10.12}$$

$$C_{ij} = A_{ij}\left[1 + 2.41\frac{\left(M_i - M_j\right)\left(M_i - 0.142 M_j\right)}{\left(M_i + M_j\right)^2}\right] \tag{10.13}$$

---

[1] Note that subscript "o" indicating dilute-gas property, has been omitted for brevity.

10.2. Dense Fluids and Mixtures

Since, as we saw in Section 9.2, it is not usually possible to calculate the excess viscosity and thermal conductivity from purely theoretical considerations, it is necessary to rely on mainly empirical methods for the prediction of the effect of pressure on the transport coefficients. A considerable number of schemes have been developed for that purpose and the reader is referred to the book by Reid, Prausnitz and Poling [6] for a comprehensive discussion. Here we shall present only the scheme developed by Thodos and his collaborators [13-15].

Table 10.3   Excess Transport Properties

Fluid	Range	Excess viscosity, $\Delta\eta$ / Pa·s	
Non-polar	$0.1 \leq \rho_r \leq 3$	$\dfrac{1}{\xi_\eta} 1.455 \times 10^{-9} \left\{ \left(1.023 + 0.23364\rho_r + 0.58533\rho_r^2 \right.\right.$ $\left.\left. -0.40758\rho_r^3 + 0.093324\rho_r^4 \right)^4 - 1 \right\}$	(10.14)
Polar	$\rho_r \leq 0.1$	$\dfrac{1}{\xi_\eta} \left\{ 2.409 \times 10^{-9} \ \rho_r^{1.111} \right\}$	(10.15)
	$0.1 \leq \rho_r \leq 0.9$	$\dfrac{1}{\xi_\eta} \left\{ 8.832 \times 10^{-11} \left(9.045\rho_r + 0.63 \right)^{1.739} \right\}$	(10.16)
	$0.9 \leq \rho_r \leq 2.2$	$\dfrac{1}{\xi_\eta} 1.455 \times 10^{-9} \left(10^{4-a}\right) \qquad a = 10^{0.6439 - 0.1005\rho_r}$	(10.17)

Fluid	Range	Excess thermal conductivity, $\Delta\lambda$ / (W·m^{-1}·K^{-1})	
Non-polar	$\rho_r \leq 0.5$	$\dfrac{1}{\xi_\lambda Z_c^5} 9.81 \times 10^{-10} \left\{ e^{0.535\rho_r} - 1 \right\}$	(10.18)
	$0.5 \leq \rho_r \leq 2.0$	$\dfrac{1}{\xi_\lambda Z_c^5} 9.17 \times 10^{-10} \left\{ e^{0.67\rho_r} - 1.069 \right\}$	(10.19)
	$2.0 \leq \rho_r \leq 2.8$	$\dfrac{1}{\xi_\lambda Z_c^5} 2.08 \times 10^{-10} \left\{ e^{1.155\rho_r} + 2.016 \right\}$	(10.20)

$$\text{where} \quad \xi_\eta = \left(\frac{M}{\text{kg} \cdot \text{mol}^{-1}} \right)^{-1/2} \left(\frac{(T^c/\text{K})}{(P^c / \text{Pa})^4} \right)^{1/6}$$

$$\xi_\lambda = \left(\frac{M}{\text{kg} \cdot \text{mol}^{-1}} \right)^{1/2} \left(\frac{(T^c/\text{K})}{(P^c / \text{Pa})^4} \right)^{1/6}$$

(10.21)

Note :   in Eq.(10.18) in the original reference [15], there was an error in the sign in the exponential.

The scheme of Thodos *et al.* is based on the development of generalised equations for the excess transport properties as functions of the reduced density. It therefore follows Eq.(10.1) and can be combined easily with the treatment of the dilute gas discussed in the previous section. Table 10.3 gives the generalised correlations for the excess viscosity, $\Delta\eta = (\eta - \eta_0)$, and the excess thermal conductivity, $\Delta\lambda = (\lambda - \lambda_0)$ in various ranges of reduced density. Separate correlations of the excess viscosity were developed for polar and non-polar systems while, for the thermal conductivity, only non-polar systems were successfully correlated. The following points should be made:

(a) The equation for the excess viscosity of non-polar systems [13], Eq.(10.14), was based primarily on viscosity data for light hydrocarbons and other simple polyatomic molecules and is generally found to predict results with errors not worse than about ±10 per cent.

(b) The equations for the excess viscosity of polar compounds [14], Eqs.(10.15) - (10.17), were based on viscosity data for a selection of compounds such as halides, ethers, freons, alcohols and ammonia. They take no account of the degree of polarity and general statements of the uncertainty cannot be made; they should therefore be used with caution.

(c) The equations for the excess thermal conductivity of non-polar fluids [15], Eqs.(10.18)-(10.20), were based on a large selection of experimental results for simple molecules, normal hydrocarbons up to *n*-nonane, aromatic hydrocarbons and cyclohexane. The expected uncertainty is about ±10 per cent.

These expressions for the excess transport properties can be employed for mixtures. Mixing rules for the mixture critical constants are then required and the VDW-1 scheme may be adopted for that purpose - see Eqs.(5.39)-(5.43).

10.3. Liquid Phase and Mixtures

Although the methods of the previous section are applicable to non-polar liquids, correlations based on the semi-empirical scheme of Dymond and Assael can offer much superior accuracy. According to this scheme, dimensionless transport properties are defined by

$$\eta^* = \frac{16}{5}(2N_A)^{1/3}\pi^{1/2}\left(\frac{1}{MRT}\right)^{1/2}\frac{\eta V_m^{2/3}}{R_\eta} \qquad (10.22)$$

$$\lambda^* = \frac{64}{75}\frac{\pi^{1/2}}{R}(2N_A)^{1/3}\left(\frac{M}{RT}\right)^{1/2}\frac{\lambda V_m^{2/3}}{R_\lambda} \tag{10.23}$$

and were correlated as universal functions of $V_r = (V_m/V_o)$. By employing a large amount of experimental data for dense monatomic gases, the following correlations were developed:

$$
\begin{aligned}
\log_{10}(\eta^*) = {}& 1.0945 - 9.26324V_r^{-1} + 71.0385V_r^{-2} - 301.9012V_r^{-3} \\
&+ 797.69V_r^{-4} - 1221.977V_r^{-5} + 987.5574V_r^{-6} - 319.4636V_r^{-7}
\end{aligned} \tag{10.24}
$$

$$\log_{10}(\lambda^*) = 1.0655 - 3.538V_r^{-1} + 12.120V_r^{-2} - 12.469V_r^{-3} + 4.562V_r^{-4} \tag{10.25}$$

The results for other compounds were analysed within this scheme by fitting roughness factors, R_η and R_λ, and establishing a correlation for the characteristic molar volume V_o as a function of T. The roughness factors are expected to be constants for a given compound but, in the case of the normal alcohols, a satisfactory correlation of the data required values that were weak functions of temperature.

In applying the method to molecular liquids, the parameters (i.e. R_η, R_λ and the coefficients in a correlation of V_o as a function of T) were optimised against a large number of critically-assessed measurements of both η and λ. It should be noted that V_o is a function only of T. Hence, having obtained the parameters from experimental results at low pressures, the scheme can be used to predict values up to very high pressures. Dymond and Assael applied this scheme to the transport properties of a number of pure compounds including simple polyatomic liquids [17], n-alkanes [18], aromatic hydrocarbons [19], n-alcohols [20] and refrigerants [21]. The range of application is generally from about 280 to 400 K with pressures from saturation up to 100 MPa; for the exact ranges, the reader is referred to the original papers. The uncertainty of the correlations is not greater than 5 per cent.

An essential requirement of the scheme is the use of accurate and consistent liquid densities. Correlations, based on the Tait equation, were therefore developed for each of the substances considered [20-22]. These correlations have a mean absolute error of about 0.1 per cent.

Tables 10.4 and 10.5 give the complete expressions for the calculation of the liquid density, viscosity and thermal conductivity for the n-alkanes and n-alcohols. For the n-alkanes, about 2000 critically assessed measurements were employed in the optimisation of the parameters in the correlation for the liquid density [22], Eqs.(10.26)-(10.28). A further 2000 experimental values of the viscosity and thermal

Table 10.4 Dymond and Assael Scheme for *n*-Alkanes, 1992 [18,22]

$$\frac{\rho - \rho_0}{\rho} = 0.200 \; \log_{10}\left[\frac{B + P}{B + P_0}\right], \qquad P_0 = 0.1013 \text{ MPa} \qquad (10.26)$$

where

$$\rho_0 = \rho^c \left\{1 + \sum_{i=1}^{5} a_i (1 - T_r)^{i/3}\right\} \qquad (10.27)$$

and

$$B/\text{MPa} = 331.2083 - 713.86 T_r + 401.61 T_r^2 - F \qquad (10.28)$$

except for methane where $B/\text{MPa} = 175.8 - 314.57 T_r + 134.3 T_r^2$.

For ethane to *n*-heptane: $F = 0$ and for *n*-octane to *n*-dodecane: $F = 0.8 \, (C_n - 7)$.

Fluid	a_1	a_2	a_3	a_4	a_5
CH_4	-2.892170	19.372430	-25.964817	12.151863	-
C_2H_6	-0.463364	8.930828	-11.154088	5.271987	-
C_3H_8	5.042481	-12.390253	16.563348	-6.601835	-
C_4H_{10}	-0.164000	8.968145	-11.902120	5.914836	-
C_5H_{12}	1.177555	3.891572	-5.508958	3.291806	-
C_6H_{14}	1.597561	1.842657	-1.726311	0.494308	0.646314
C_7H_{16}	1.331593	3.300918	-4.509610	2.765491	-
C_8H_{18}	1.969770	-1.100623	6.364172	-8.693475	4.420047
C_9H_{20}	1.927780	0.930219	-1.334128	1.392823	-
$C_{10}H_{22}$	0.329139	7.364340	-9.985096	5.283608	-
$C_{11}H_{24}$	-0.275532	9.285034	-11.992330	5.969567	-
$C_{12}H_{26}$	-0.030495	8.325350	-10.826840	5.551783	-

For methane to *n*-butane ($\theta = T / \text{K}$)

$$V_0/(\text{cm}^3.\text{mol}^{-1}) = 45.822 - 6.18670 \theta^{0.5} + 0.36879 \theta - 0.0072273 \theta^{1.5} + C_n\left(2.17871 \theta^{0.5}\right.$$
$$\left. - 0.185198 \theta + 0.00400369 \theta^{1.5}\right) + C_n^2\left(6.951548 - 52.64360 \theta^{-0.5}\right) \qquad (10.29)$$
$$+ C_n^3\left(-7.801897 + 42.24493 \theta^{-0.5} + 0.4476523 \theta^{0.5} - 0.00957351 2\theta\right)$$

For *n*-pentane $\qquad\qquad V_0/(\text{cm}^3.\text{mol}^{-1}) = 81.1713 - 0.046169 \theta \qquad (10.30)$

For *n*-hexane to *n*-dodecane

$$V_0/(\text{cm}^3.\text{mol}^{-1}) = 117.874 + 0.15(-1)^{C_n} - 0.25275 \theta + 0.0005488 \theta^2 - 4.2464 \times 10^{-7} \theta^3 \qquad (10.31)$$
$$+ (C_n - 6)(1.27 - 0.0009 \theta)(13.27 + 0.025 C_n)$$

$$R_\eta = 0.995 - 0.0008944 C_n + 0.005427 C_n^2 \qquad (10.32)$$
$$R_\lambda = -18.8416 C_n^{-1.5} + 41.461 C_n^{-1} - 30.15 C_n^{-0.5} + 8.6907 + 0.0013371 C_n^{2.5} \qquad (10.33)$$

Table 10.5 Dymond and Assael Scheme for *n*-Alcohols, 1994 [20]

$$\frac{\rho - \rho_0}{\rho} = 0.200 \ \log_{10}\left[\frac{B + P}{B + P_0}\right], \qquad P_0 = 0.1013 \ \text{MPa} \qquad (10.34)$$

where

$$\rho_0 = \rho^c \left\{1 + \sum_{i=1}^{7} a_i (1 - T_r)^{i/3}\right\} \qquad (10.35)$$

and

$$B/\text{MPa} = 520.23 - 1240 T_r + 827 T_r^2 - F \qquad (10.36)$$

For methanol: $F = 11.8$, while for ethanol to *n*-hexanol: $F = 0.015 \ C_n \left(1 + 11.5 C_n\right)$

Alcohol	a_1	a_2	a_3	a_4	a_5	a_6	a_7
CH_3OH	2.627813	-4.04742	15.8343	-22.5066	10.9160	0.30477	-
C_2H_5OH	-0.992640	38.02867	-181.1172	445.9045	-588.5184	393.9196	-104.3443
C_3H_7OH	0.940534	12.94424	-53.9519	113.6388	-113.8656	43.3832	-
C_4H_9OH	-3.573785	50.24497	-175.9335	308.5883	-265.6281	89.4997	-
$C_5H_{11}OH$	-2.611289	42.33751	-149.6396	262.2095	-223.0897	73.7879	-
$C_6H_{13}OH$	-0.155309	17.72617	-57.6007	99.9026	-86.5883	29.6281	-

$$V_0/(\text{cm}^3.\text{mol}^{-1}) = \sum_{i=0}^{1} b_i \ \theta^{-i/2} + \sum_{i=0}^{4} d_i \ C_n^{(i+2)/2} + \sum_{i=0}^{5} g_i \ (C_n\theta)^{(i+1)/2} \qquad (10.37)$$

and

$$R_\eta = \sum_{i=0}^{2} h_i \ \theta^i \qquad (10.38)$$

(for $\theta = T/K$)

$$R_\lambda = 1.493 - 0.09139 C_n + 0.02804 C_n^2 \qquad (10.39)$$

i	b_i	d_i	g_i	Alcohol	h_0	h_1	h_2
0	-28.842	486.505	34.6986	CH_3OH	41.152	-0.2175	3.057×10^{-4}
1	-2759.29	-478.064	-3.944	C_2H_5OH	49.74	-0.2505	3.325×10^{-4}
2	-	217.956	0.16991	C_3H_7OH	119.69	-0.6770	9.817×10^{-4}
3	-	-47.901	-3.6810×10^{-3}	C_4H_9OH	55.07	-0.2950	4.110×10^{-4}
4	-	4.139	3.9402×10^{-5}	$C_5H_{11}OH$	46.11	-0.2580	3.802×10^{-4}
5	-	-	-1.6598×10^{-7}	$C_6H_{13}OH$	3.00	-	-

conductivity were employed to determine the roughness factors and the characteristic molar volume. V_0 was correlated as a function of the temperature and the carbon number C_n; these correlations are given by Eqs.(10.29)-(10.31). Correlations of the roughness factors, R_η and R_λ, in terms of C_n were also developed and these are given by Eqs.(10.32) and (10.33). A similar strategy was adopted for the *n*-alcohols. Results were also obtained for cyclohexane [18] and the remaining *n*-alcohols up to *n*-decanol [20].

The method may be applied to the prediction of the transport properties of liquid mixtures containing either normal hydrocarbons or aromatic hydrocarbons or alcohols (but not to mixtures containing different classes of compounds). In this case:

(a)   the mixture density was calculated from the densities of the pure components, assuming there was no change of volume on mixing; and

(b)   mole-fraction-weighted means were assumed for the characteristic volume and the roughness factors.

The agreement with experimental data was found to be within about 5 per cent [20, 24] within the homologous series.

In Section 10.6 a FORTRAN subroutine incorporating these expressions will be presented and employed in comparative examples (a fuller version of this program, the TransP package, is available for serious calculations).

10.4. Corresponding States

The theoretical foundation of the corresponding-states approach was presented in detail in Section 5.1. It is well understood and defined in its simplest form for a conformal system in equilibrium. In the case of polyatomic molecules and complex mixtures, molecular shape factors may be introduced to map the properties of the substance in question onto those of a reference fluid. This approach was adopted by Ely and Hanley [25,26] in the development of a corresponding states scheme for the viscosity and thermal conductivity of non-polar systems. An improved and extended version of this scheme is available as the well-known computer package SUPERTRAPP [27] which may be applied to about 120 non-polar components and their mixtures.

The basic idea of this scheme for the calculation of the transport properties of non-polar mixtures is relatively straightforward. It is assumed that:

(a)   the transport properties of a single-phase mixture may be equated to those of a hypothetical pure fluid; and

(b) the properties of this hypothetical pure fluid may be related to those of a reference fluid at a scaled temperature and density.

To account for the nonconformality of the fluids, property-correction factors and molecular-shape factors were employed. Thus some measurements of the transport properties were required for each substance so that these factors could be evaluated. In the following sections, the scheme proposed by Ely and Hanley for the calculation of the viscosity and thermal conductivity of non-polar mixtures will be described while, in Section 10.6, a sample computer program based on the method will be employed in comparative examples.

The main advantage of the scheme is that it may be applied with some success over a very wide range of conditions. Only the critical constants and the acentric factor are required for the mixture transport property calculation; pure component transport properties are not required.

This corresponding-states scheme has been extended by Huber and Ely [29,30] to include pure refrigerants and their mixtures in both liquid and vapour phases. A similar corresponding-states scheme for the configurational thermodynamic properties of pure refrigerants and their mixtures has also been developed [31] and is available as the computer package REFPROP [32].

10.4.1. Viscosity of Non-Polar Mixtures

The basic assumption of the corresponding-states scheme is that the mixture viscosity $\eta_{mix}(\rho_n, T)$ at amount-of-substance density ρ_n and temperature T may be equated to that of a hypothetical pure fluid, $\eta_x(\rho_n, T)$, at the same T and ρ_n. The viscosity of this hypothetical pure fluid is then evaluated from the extended principle of corresponding states with respect to the viscosity of a given reference fluid at amount-of-substance density ρ_0 and temperature T_0. Methane was adopted as the reference fluid so that

$$\eta_{mix}(\rho_n, T) = \eta_x(\rho_n, T) = F_\eta \, \eta_{CH_4}(\rho_0, T_0). \tag{10.40}$$

The scaling factor F_η is a function of the molar mass and the *temperature and density scaling parameters*, f and h, defined as

$$T_0 = T/f \tag{10.41}$$

$$\rho_0 = \rho_n \, h \, . \tag{10.42}$$

The scaling parameters f and h in a two-parameter corresponding-states approach (see Section 5.1) are simply the ratio of the critical temperatures and the ratio of the critical molar volumes respectively. In the extended corresponding-states method adopted by Ely and Hanley, they were expressed as a function of the critical constants and the so-called *molecular-shape factors*. Shape factors in essense ensure that the configurational thermodynamic properties of the pure fluids map accurately onto the thermodynamic surface of the reference fluid. In principle, shape factors may be determined exactly point-by-point from experimental data. Ely and Hanley adopted generalised equations for the shape factors, of the form originally proposed by Leach *et al.* [28], and fitted the parameters to a large set of experimental data for the saturated vapour and liquid. The resulting formulae are given in Table 10.6 in terms of the critical constants and acentric factor. We also define in Eqs.(10.45), dimensionless analogues of T_0 and ρ_0 for consistency with the published correlations [25].

Methane was chosen as the reference fluid for this correlation as, at the time, it was the only fluid for which both viscosity and thermal conductivity were known sufficiently well over a wide range of conditions [25]. The viscosity of methane at amount-of-substance density ρ_0 and temperature T_0 was correlated as,

$$\eta_{CH_4}(\rho_0, T_0) = \eta_0(T_0) + \eta_1(T_0)\rho_0 + \Delta\eta(\rho_0, T_0)X_\eta \qquad (10.43)$$

where $\eta_0(T_0)$ and $\eta_1(T_0)$ represent the dilute-gas term and the first density correction respectively, while $\Delta\eta(\rho_0, T_0)$ is the remainder of the excess viscosity and dominates the results at high density. The precise forms of these terms, obtained in a fit to the experimental data, are shown in Table 10.7 as Eqs.(10.52)-(10.55).

In order to account for nonconformality, a viscosity correction factor X_η was introduced through Eq.(10.56). Since this factor is always close to unity, the mixture critical properties required for its evaluation may be estimated as mole-fraction weighted averages of the pure component values. Finally, it should be noted that the scaling factor F_η given in Eq.(10.56) incorporates an effective molar mass for the mixture which differs from the mean molar mass of the mixture and is given by Eqs.(10.57) and (10.66).

The scheme was found to produce good results for a large selection of hydrocarbon systems in both liquid and vapour phases. The critical region should be avoided but otherwise the expected accuracy is in the region of 10 to 20 per cent. The whole scheme is of course based on temperature and amount-of-substance density as the independent variables and to apply it to problems in which tempe-rature and pressure are the given variables, an equation of state is required. Ely and

Hanley [25] proposed an extended corresponding-states method for the mixture density based on an equation of state for methane. In the computational examples presented later in this chapter, we adopt the BWR (HS) equation of state discussed in Section 6.6.1.

10.4.2. Thermal Conductivity of Non-Polar Mixtures

For the mixture thermal conductivity, Ely and Hanley [26] adopted essentially the same approach but divided the property into translational and internal parts. They further made the rather drastic assumptions that the latter is independent of the density and may be calculated from the modified Eucken equation. The translational part was treated by the principle of corresponding states with methane again chosen as the reference fluid. In detail the method involves the following terms :

(1) Internal contribution.

This is calculated for the pure components from the modified Eucken equation, Eq.(10.59), and a quadratic mixing rule, Eq.(10.60), is adopted. The dilute gas viscosity of each component η_{oi} required for this calculation is obtained, as described in Section 10.4.1, from Eq.(10.52).

(2) Translational contribution.

This is calculated in the same manner as described for the viscosity. Thus, $\lambda_{CH_4}^{tr}$ is decomposed into a dilute-gas term, λ_0, the initial density dependence, $\lambda_1 \rho_n$, and the remainder of the excess property, $\Delta\lambda$, which dominates the thermal conductivity at high densities. The term λ_0 is calculated from the dilute-gas viscosity. The precise forms of these three terms are given in Table 10.8, Eqs.(10.61)-(10.64).

It should be noted that, in contrast to the scheme for viscosity, the correction factor X_λ is applied to the whole translational part and not just to the high-density part of the excess contribution. This factor, given by Eq.(10.65), is otherwise identical to X_η.

In Section 10.6 a sample computer program is presented for the calculation of the viscosity and thermal conductivity of non-polar mixtures from the extended corresponding-states method and the BWR(HS) equation of state.

Table 10.6 Ely and Hanley Scaling Parameters, 1981 [25]

Dimensionless Equivalent Temperature and Density

$$\overline{T}_0 = \frac{(T/K)}{f_{mix}} \quad \text{and} \quad \overline{\rho}_0 = \frac{V_0}{V_{mix}} h_{mix} \quad \text{with } V_0=16.04 \text{ cm}^3\cdot\text{mol}^{-1} \quad (10.44)$$

Reduced Temperature and Volume

$$T_{ri} = \min\left[2,(T/T_i^c)\right] \qquad V_{ri} = \min\left[2,\max[(V_{mix}/V_i^c),0.5]\right] \quad (10.45)$$

Equivalent Fluid Reducing Parameters Relative to Methane[1]

$$h_{mix} = \sum_i \sum_j x_i x_j h_{ij} \quad \text{and} \quad f_{mix} = \frac{1}{h_{mix}}\sum_i \sum_j x_i x_j f_{ij} h_{ij} \quad (10.46)$$

where,
$$h_{ij} = \frac{1}{8}\left(h_{ii}^{1/3} + h_{jj}^{1/3}\right)^3 \quad \text{and} \quad f_{ij} = \sqrt{f_{ii} f_{jj}} \quad (10.47)$$

$$h_{ii} = \frac{V_i^c}{99.2 \text{ cm}^3\cdot\text{mol}^{-1}}\left\{1+(\omega_i-0.011)\left[\varphi_1(V_{ri}+\varphi_2)+\varphi_3(V_{ri}+\varphi_4)\log\overline{T}_{ri}\right]\right\}\frac{0.288}{Z_i^c} \quad (10.48)$$

$$f_{ii} = \frac{T_i^c}{190.55K}\left\{1+(\omega_i-0.011)\left[\theta_1+\theta_2\log T_{ri}+\left(V_{ri}-0.5\right)\left(\theta_3+\frac{\theta_4}{T_{ri}}\right)\right]\right\} \quad (10.49)$$

	1	2	3	4
θ_i	0.09057	-0.86276	0.31664	-0.46568
φ_i	0.3949	-1.02355	-0.93281	-0.75464

Derivative of Eq.(10.49)

$$\left(\frac{\partial f}{\partial T}\right)_V = \frac{1}{190.55K}(\omega-0.011)\left[\frac{\theta_2}{T_r}-(V_r-0.5)\frac{\theta_4}{T_r^2}\right] \quad (10.50)$$

Note: (1) $(\partial f/\partial T)_V = 0$, if T_r is constant (=2) - see Eq.(10.45) above.
(2) In connection with Eq.(10.50) for mixtures, use mole fraction average of ω_i, T_i^c and V_i^c.

[1] two equal subscripts, e.g. "*ii*", refer to the pure fluid.

Table 10.7   Ely and Hanley Mixture Viscosity Correlation, 1981 [25]

$$\eta_{mix}(\rho,T) = F_\eta\, \eta_{CH_4}(\rho_o, T_o)$$

$$= F_\eta\Big[\eta_o + \eta_1\bar{\rho}_o + \Delta\eta X_\eta\Big] \tag{10.51}$$

Dilute-Gas Term

$$\eta_o/(10^{-7}\,\text{Pa.s}) = \sum_{i=1}^{9} c_i\, \bar{T}_o^{(i-4)/3} \tag{10.52}$$

First-Density Term

$$\eta_1/(10^{-7}\,\text{Pa.s}) = b_1 + b_2\Big[b_3 - \log(\bar{T}_o/b_4)\Big]^2 \tag{10.53}$$

High-Density Term

$$\Delta\eta/(10^{-7}\,\text{Pa.s}) = (\Delta\eta' - 1)\exp\!\left(a_1 + \frac{a_2}{\bar{T}_o}\right) \tag{10.54}$$

$$\Delta\eta' = \exp\!\left\{\left(a_3 + \frac{a_4}{\bar{T}_o^{1.5}}\right)\bar{\rho}_o^{0.1} + \left(\frac{\bar{\rho}_o}{0.1617} - 1\right)\bar{\rho}_o^{0.5}\left(a_5 + \frac{a_6}{\bar{T}_o} + \frac{a_7}{\bar{T}_o^2}\right)\right\} \tag{10.55}$$

i	a_i	b_i	c_i
1	-1.023916E+1	1.696986E+0	2.907741E+6
2	1.742282E+2	-1.333724E-1	-3.312874E+6
3	1.746055E+1	1.400000E+0	1.608102E+6
4	-2.847633E+3	1.680000E+2	-4.331905E+5
5	1.336850E-1		7.062481E+4
6	1.420724E+2		-7.116621E+3
7	5.002067E+3		4.325174E+2
8			-1.445911E+1
9			2.037119E-1

Scaling Factor F_η, High-Density Correction X_η, and Mixture Effective Molar Mass ($M_o = 0.016043$ kg·mol^{-1})

$$F_\eta = \left(\frac{M_{mix}}{M_o}\right)^{1/2} f_{mix}^{1/2}\, h_{mix}^{-2/3} \qquad X_\eta = \left\{\left[1 - 1.5\frac{T}{f_{mix}}\left(\frac{\partial f}{\partial T}\right)_v\right]\frac{Z_{mix}^c}{0.288}\right\}^{0.5} \tag{10.56}$$

$$M_{mix} = \left[\sum_i\sum_j x_i x_j\, M_{ij}^{1/2} f_{ij}^{1/2} h_{ij}^{4/3}\right]^2 \left(f_{mix}^{-1} h_{mix}^{-8/3}\right) \quad \text{and} \quad M_{ij} = \frac{2\,M_{ii}\,M_{jj}}{M_{ii} + M_{jj}} \tag{10.57}$$

Table 10.8   Ely and Hanley Mixture Thermal Conductivity Correlation, 1983 [26]

$$\lambda_{mix}(\rho,T) = \lambda_{mix}^{int}(T_o) + F_\lambda \, \lambda_{CH_4}^{tr}(\rho,T_o) \tag{10.58}$$

$$= \lambda_{mix}^{int} + F_\lambda \left[\lambda_o + \lambda_1 \bar{\rho}_o + \Delta\lambda \right] X_\lambda$$

Internal term

$$\lambda_{mix}^{int} = \sum_i \sum_j x_i \, x_j \, \lambda_{ij}^{int} \tag{10.59}$$

$$\frac{\lambda_{ii}^{int} M_{ii}}{\eta_{oi}} = 1.32 \left(C_{Pi}^{pg} - \frac{5R}{2} \right) \quad \text{and} \quad \lambda_{ij}^{int} = \frac{2\lambda_{ii}^{int} \lambda_{jj}^{int}}{\lambda_{ii}^{int} + \lambda_{jj}^{int}} \tag{10.60}$$

Dilute-Gas Term

$$\lambda_o = \frac{15}{4} \frac{R}{M_o} \eta_o \tag{10.61}$$

First-Density Term

$$\lambda_1/(W.m^{-1}.K^{-1}) = b_1 + b_2 \left[b_3 - \log(\bar{T}_o/b_4) \right]^2 \tag{10.62}$$

High-Density Term

$$\Delta\lambda/(W.m^{-1}.K^{-1}) = (\Delta\lambda'-1)\exp\left(a_1 + a_2/\bar{T}_o\right) \tag{10.63}$$

$$\Delta\lambda' = \exp\left\{ \left(a_3 + \frac{a_4}{\bar{T}_o^{1.5}} \right) \bar{\rho}_o^{0.1} + \left(\frac{\bar{\rho}_o}{0.1617} - 1 \right) \bar{\rho}_o^{0.5} \left(a_5 + \frac{a_6}{\bar{T}_o} + \frac{a_7}{\bar{T}_o^2} \right) \right\} \tag{10.64}$$

i	a_i	b_i
1	-7.197708E+0	-2.527629E-4
2	8.567822E+1	3.343285E-4
3	1.247183E+1	1.120000E+0
4	-9.846252E+2	1.680000E+2
5	3.594685E-1	
6	6.979841E+1	
7	-8.728833E+2	

Scaling Factor F_λ, Translational Correction X_λ, and Mixture Effective Molar Mass

$$F_\lambda = \left(\frac{M_o}{M_{mix}} \right)^{1/2} f_{mix}^{1/2} h_{mix}^{-2/3} \qquad X_\lambda = \left\{ \left[1 - \frac{T}{f_{mix}} \left(\frac{\partial f}{\partial T} \right)_v \right] \frac{0.288}{Z_{mix}^c} \right\}^{1.5} \tag{10.65}$$

$$M_{mix} = \left[\sum_i \sum_j x_i \, x_j \, M_{ij}^{-1/2} f_{ij}^{1/2} h_{ij}^{-4/3} \right]^{-2} \left(f_{mix} \, h_{mix}^{-8/3} \right) \quad \text{and} \quad M_{ij} = \frac{2 M_{ii} M_{jj}}{M_{ii} + M_{jj}} \tag{10.66}$$

10.5. Summary

Although there have been some new advances recently in the kinetic theory of fluids, progress remains slow. Thus while we have, for the most part, soundly based prediction methods for the dilute-gas properties, only with difficulty can we estimate the density dependence of transport properties. The schemes presented in this chapter have some theoretical basis but are also designed to be useful for a wide range of systems. The corresponding-states scheme proposed by Ely and Hanley is one of the few schemes able to predict with any accuracy the viscosity and thermal conductivity of a large number of non-polar components and their mixtures in both liquid and vapour phases. Unfortunately it does not apply to polar fluids.

Many empirical correlations of transport properties may be found in the literature but such methods should be employed only in the prescribed range of conditions for which they were developed.

10.6. Computational Implementation

In the next four subsections, the procedures described in this chapter for the dilute-gas state, the density dependence and the liquid phase, as well as the wide-ranging corresponding-states approach, will be examined through appropriate computer routines and their application to a number of comparative examples.

10.6.1. Dilute Gases and Mixtures

In Tables 10.1 and 10.2, expressions for the estimation of the viscosity and the thermal conductivity of dilute-gas mixtures were presented. Subroutine TPDILG, shown in Display 10.1, implements those methods for a mixture of IS components.

```
                                           DISPLAY 10.1  TPDILG List
      SUBROUTINE TPDILG(IS,T,Y,EK,SIG,WM,Y1,Y2,Y3,Y4,VIS,VMIX,TCC,TMIX)
C
C     The subroutine calculates the viscosity and thermal conductivity
C     of a mixture of IS gases at low pressures and temperature T (K)
C     Calculations are based on the following methods.
C     Viscosity          - Pure gases: Reduced collision integral (Neufeld)
C                        - Mixtures  : Wilke expressions
C     Th.Conductivity    - Pure gases: Modified Eucken correlation
C                        - Mixtures  : Brokaw expressions
C     Data required for each component are potential parameters e/k (K)
C     sigma (nm), EK,SIG, molecular weight WM, molefractions Y
```

```
C       and perfect-gas heat capacity coefficients, Y1,Y2,Y3,Y4.
C
        DIMENSION EK(IS),SIG(IS),WM(IS),VIS(IS),TCC(IS),Y(IS)
        DIMENSION Y1(IS),Y2(IS),Y3(IS),Y4(I)
        DIMENSION PHI(10,10),A(10,10),C(10,10),TCTR(10),TCINT(10)
        R=8.3145
C.......Pure Components Viscosity VIS(I) & Thermal Conductivity TCC(I)
        DO 100 I=1,IS
        TS=T/EK(I)
        CP=Y1(I)+Y2(I)*T+Y3(I)*T**2+Y4(I)*T**3
        OMEGA=1.16145*(TS**(-0.14874))+0.52487*(EXP(-0.7732*TS))+2.16178*
       &(EXP(-2.43787*TS))-6.435E-4*(TS**0.14874)*SIN(18.0323*(TS**
       &(-0.7683))-7.2371)
        VIS(I)=26.69E-9*((WM(I)*T)**0.5)/(SIG(I)*SIG(I)*OMEGA)
100     TCC(I)=(VIS(I)*(CP-R)/(WM(I)/1000.))*(1.32+1.77/((CP/R)-1))
C.......Mixture Viscosity VMIX
        DO 120 I=1,IS
        DO 120 J=1,IS
        PH=(1.+((VIS(I)/VIS(J))**0.5)*((WM(J)/WM(I))**0.25))**2
120     PHI(I,J)=PH/((8.*(1.+WM(I)/WM(J)))**0.5)
        VMIX=0.
        DO 160 I=1,IS
        VB=0.
        DO 140 J=1,IS
140     VB=VB+Y(J)*PHI(I,J)
160     VMIX=VMIX+Y(I)*VIS(I)/VB
C.......Mixture Thermal Conductivity TCMIX (=TTRM+TINM)
        DO 200 I=1,IS
        TCTR(I)=2.5*VIS(I)*(1.5*R)/(WM(I)/1000.)
200     TCINT(I)=TCC(I)-TCTR(I)
        DO 220 I=1,IS
        DO 220 J=1,IS
        AA=(1.+((TCTR(I)/TCTR(J))**0.5)*((WM(I)/WM(J))**0.25))**2
        A(I,J)=AA/((8.*(1.+WM(I)/WM(J)))**0.5)
        CC=(WM(I)-WM(J))*(WM(I)-0.142*WM(J))/((WM(I)+WM(J))**2)
220     C(I,J)=A(I,J)*(1.+2.41*CC)
        TTRM=0.
        TINM=0.
        DO 260 I=1,IS
        TRB=0.
        TIB=0.
        DO 240 J=1,IS
        TRB=TRB+Y(J)*C(I,J)
240     TIB=TIB+Y(J)*A(I,J)
        TTRM=TTRM+Y(I)*TCTR(I)/TRB
260     TINM=TINM+Y(I)*TCINT(I)/TIB
        TMIX=TTRM+TINM
        RETURN
        END
```

EXAMPLE 10.3   *Calculation of the Dilute-Gas Thermal Conductivity of the Mixture {0.592 CO + 0.408 Ar} at T = 308.15 K.*

For this calculation, subroutine TPDILG is employed. Potential parameters ε/k_B and σ are obtained from Table A9, in Appendix A, while molar masses and the perfect-gas heat-capacity coefficients are taken from Table A1.

In Display 10.2 the calling program and its output are shown. The viscosity of the mixture is also calculated. The thermal conductivity of this particular mixture has been measured with an uncertainty of only ±0.2 per cent [33]. The calculated and measured values for both pure components and the mixture are compared below. The agreement is within 6 per cent which, while not particularly good, is typical for such predictions. The calculated viscosity is expected to be much more accurate as the property is not strongly influenced by inelastic collisions.

System	TPDILG	Ref.[33]	Per Cent Dev.
Carbon Monoxide (1)	27.22	25.68	6.0
Argon (2)	18.18	18.18	0.0
Mixture	23.35	22.39	4.3

Display 10.2   TPDILG Sample Input/Output

```
      PROGRAM EX1003
      PARAMETER (IS=2)
      DIMENSION EK(IS),SIG(IS),WM(IS),VIS(IS),TCC(IS),Y(IS)
      DIMENSION Y1(IS),Y2(IS),Y3(IS),Y4(IS)
      DATA EK(1),SIG(1),WM(1),Y(1)/ 91.7,0.3690,28.0104,0.592/
      DATA Y1(1),Y2(1),Y3(1),Y4(1)/30.87,-1.285E-2,2.789E-5,-1.272E-8/
      DATA EK(2),SIG(2),WM(2),Y(2)/ 93.3,0.3542,39.9480,0.408/
      DATA Y1(2),Y2(2),Y3(2),Y4(2), T/20.8,0.,0.,0.308.15/
      CALL TPDILG(IS,T,Y,EK,SIG,WM,Y1,Y2,Y3,Y4,VIS,VMIX,TCC,TMIX)
      WRITE(*,1000)
 1000 FORMAT(21X,'THERMAL CONDUCTIVITY    VISCOSITY',/,
     &27X,'(mW/M/K)',13X,'(uPa.s)')
      DO 20 I=1,IS
 20   WRITE(*,1100) I,TCC(I)*1.E+3,VIS(I)*1.E+6
 1100 FORMAT(5X,'COMPONENT ',I2,9X,F6.2,15X,F6.2)
      IF (IS.GT.1) WRITE (*,1200) TMIX*1.E+3,VMIX*1.E+6
 1200 FORMAT(7X,'MIXTURE',13X,F6.2,15X,F6.2)
      END
```

	THERMAL CONDUCTIVITY (mW/M/K)	VISCOSITY (uPA.S)
COMPONENT 1	27.22	18.04
COMPONENT 2	18.18	23.28
MIXTURE	23.35	20.32

□

10.6.2. Density Dependence

The density dependence of the transport properties may be estimated from the generalized correlation of the excess functions proposed by Thodos [13-15] and given in Table 10.3. In Example 10.4, these excess functions are employed in the estimation of the transport properties of ethane.

| **EXAMPLE 10.4** | *Calculation of the Viscosity and Thermal Conductivity of Ethane at $T = 350K$ and (a) 10 MPa, (b) 20 MPa.* |

Program EX1004 shown in Display 10.3 performs these calculation. The dilute gas terms are obtained as in Example 10.3 while the excess contribution is calculated according to the correlation given in Table 10.3. Potential parameters ε/k_B and σ are obtained from Table A9, in Appendix A, while the critical constants, acentric factor, molar mass and perfect-gas heat-capacity coefficients are from Table A1. To calculate the density, the Peng-Robinson equation of state is employed (subroutine PR from Section 6.8.2).

The output for $P = 20$ MPa is shown in Display 10.3 while, in the following table, the calculated values are compared with those recommended by IUPAC [34,35]. Bearing in mind that the literature values have an uncertainty of about 2 per cent, it can be seen that the agreement is good for the viscosity but poor for the thermal conductivity (especially at the highest pressure).

P/MPa	$\lambda/(mW\cdot m^{-1}\cdot K^{-1})$			$\eta/(\mu Pa\cdot s)$		
	EX1004	Ref.[34]	Per cent Dev.	EX1004	Ref.[35]	Per cent Dev.
$0^\dagger$	29.88	28.98	3.1	10.90	10.80	0.9
10	59.48	61.68	-3.6	23.87	24.18	-1.3
20	96.14	82.44	16.6	43.42	44.46	-2.3

† Denotes the dilute-gas state.

DISPLAY 10.3 Sample Input/Output

```
PROGRAM EX1004
C
C     This program calculates the viscosity and thermal conductivity of
C     a pure nonpolar gas at temperature T (K) & high pressure P (Pa)
C     in a reduced density range 0.1 to 2.0.
C     Calculations are based on the following methods:
C        Viscosity          - Dilute gas : Reduced collision integral (Neufeld)
C                           - Excess term : Thodos expressions
C        Th Conductivity    - Dilute gas : Modified Eucken correlation
C                           - Excess term : Thodos expressions
```

```
C          Density required is calculated by routine PR (Peng & Robinson EoS)
C          Data required are: potential parameters e/k (K) sigma (nm),
C          EK,SIG, molecular weight WM, critical temperature TC (K), pressure
C          PC (Pa) and density DC (MOL/M3), acentric factor OMEG, and the
C          perfect-gas heat capacity coefficients, Y1,Y2,Y3,Y4.
C
           DATA EK,SIG,WM,R/215.7,0.4443,30.0694,8.3145/
           DATA Y1,Y2,Y3,Y4/5.409,0.1781,-6.938E-5,8.713E-9/
           DATA TC,PC,DC,OMEG/305.33,4.871E+6,6743,,0.099/
           DATA T,P/350.0,10.E+6/
C.......Dilute Gas Properties, VIS & TCC
           TS=T/EK
           CP=Y1+Y2*T+Y3*T**2+Y4*T**3
           OMEGA=1.16145*(TS**(-0.14874))+0.52487*(EXP(-0.7732*TS))+2.16178*
          &(EXP(-2.43787*TS))-6.435E-4*(TS**0.14874)*SIN(18.0323*(TS**
          &(-0.7683))-7.2371)
           VIS=26.69E-9*((WM*T)**0.5)/(SIG*SIG*OMEGA)
           TCC=(VIS*(CP-R)/(WM/1000.))*(1.32+1.77/((CP/R)-1))
           CALL PR(T,P,2,1,1,,TC,PC,OMEG,Y1,Y2,Y3,Y4,D,H,S,FUG)
           DDC=D/DC
C.......Excess Viscosity, DVIS
           XVI=((1000./WM)**(.5))*(TC**(1./6.))/(PC**(2./3.))
           DVIS=1.023+0.22364*DDC+0.58533*DDC*DDC-0.40758*(DDC**3)+0.093324*
          &(DDC**4)
           DVIS=(1.455E-9/XVI)*((DVIS**4)-1.)
           IF (P.LT.0.11E+6) DVIS=0.0
C.......Excess Thermal Conductivity, DTCC
           XTC=((WM/1000.)**(.5))*(TC**(1./6.))/(PC**(2./3.))
           ZC=PC/(R*TC*DC)
           IF (DDC.LE.0.5) DTCC=(1./(XTC*(ZC**5)))*9.81E-10*(EXP(0.535*DDC)-1.0)
           IF (DDC.GT.0.5.AND.DDC.LE.2.0) DTCC=(1./(XTC*(ZC**5)))*9.17E-10*
          &(EXP(0.67*DDC)-1.069)
           IF (P.LT.0.11E+6) DTCC=0.0
           WRITE(*,1000) DDC
 1000   FORMAT(1X,'@ REDUCED DENSITY : ',F6.2)
           WRITE(*,1100) TCC*1.E+3,VIS*1.E+6,DTCC*1.E+3,DVIS*1.E+6,
          &(TCC+DTCC)*1.E+3,(VIS+DVIS)*1.E+6
 1100   FORMAT(27X,'THERMAL CONDUCTIVITY  VISCOSITY',/34X,'(mW/M/K)',
          &'(uPA.S)',/3X,'DILUTE-GAS PROPERTY  ',2(10X,F6.2),/3X,'EXCESS  ',
          &'PROPERTY  ',2(10X,F6.2),/3X,' TOTAL PROPERTY ',2(10X,F6.2))
           END
```

@ REDUCED DENSITY : 1.04

	THERMAL CONDUCTIVITY (mW/M/K)	VISCOSITY (uPA.S)
DILUTE-GAS PROPERTY	29.88	10.90
EXCESS PROPERTY	29.60	12.97
TOTAL PROPERTY	59.48	23.87

❑

10.6.3. *Liquid-Phase Transport Properties*

For the transport properties of certain well-studied liquids, one can use the accurate scheme presented in Section 10.3 which is based on the hard-spheres theory with experimentally-determined parameters. In Example 10.5, we apply the method to an alcohol.

| **EXAMPLE 10.5** | *Calculation of the Viscosity and Thermal Conductivity of n-Propanol at $T = 323.15K$ and $P = 6$ MPa.* |

Program EX1005 shown in Display 10.4 is used. The only input data required are the temperature, pressure and the carbon number. The density and the viscosity obtained are compared in the table with recent experimental values and the agreement, as expected, is within 0.2 per cent for the density and within 2 per cent for the viscosity.

	EX1005	Experimental Values	Reference	Per cent Dev.
$\rho/(\text{kg·m}^{-3})$	784.3	784.2	[36]	0.01
$\eta/(\mu\text{Pa·s})$	1193.4	1172	[37]	1.8

DISPLAY 10.4 Sample Input/Output

```
PROGRAM EX1005
COMMON /RVI/RV1(6),RV2(6),RV3(6)

This program calculates the density, the viscosity and the thermal
conductivity of one of the n-alcohols, methanol to n-hexanol,
based on the algorithm developed by Dymond & Assael (Section 10.3)
Data required  temperature T (K), pressure P (Pa), alcohol carbon number CN

DATA T,P,CN/323.15,6,6.E+6,3/

Molecular Weight WM, characteristic volume V0, density DENS
WM=CN*0.012011+(2.*CN+1.)*0.0010079+0.0170073
V0=1.E-6*(-28.842-2759.29*(T**(-0.5))+486.505*CN-478.064*(CN**1.5)
&+217.9562*(CN**2.)-47.901*(CN**2.5)+4.139*(CN**3.)-34.6986*
&((CN*T)**0.5)-3.944*(CN*T)+0.169911*((CN*T)**1.5)-3.681E-3*
&((CN*T)**2.)+3.94017E-5*((CN*T)**2.5)-1.6598E-7*((CN*T)**3.))
V=WM/DENS(CN,T,P)
VR=V/V0

Roughness Factor RV & Viscosity VIS
IC=INT(CN)
RV=RV1(IC)+RV2(IC)*T+RV3(IC)*T*T
VST=10.**(1.0945-9.26324/VR+71.0385/(VR**2.)-301.9012/(VR**3.)
&+797.69/(VR**4.)-1221.977/(VR**5.)+987.5574/(VR**6.)-319.4636/(VR**7.))
VIS=RV*VST*((WM*8.314*T)**0.5)/(6.035E+8*(V)**(2./3.))

Roughness Factor RL & Thermal Conductivity TCC
```

```
      RL=1.493-0.09139*CN+0.02804*CN*CN
      TST=10.**(1.0655-3.538/VR+12.121/(VR**2.)-12.469/(VR**3.)+4.562/(VR**4.))
      TCC=RL*TST*((8.314*T/WM)**0.5)/(1.936E+7*(V)**(2./3.))
      WRITE(*,1000) T,P*1.E-6,DENS(CN,T,P),TCC*1.E+3,VIS*1.E+6
1000  FORMAT(3X,'@',F7.2,' K  &',F8.2,' MPA',//5X,'DENSITY       : ',F6.1,' KG/M3',
     &/5X,'TH.CONDUCTIVITY : ',F6.1,' mW/M/K',/5X,'VISCOSITY      : ',F6.1,' uPA.S')
      END

      FUNCTION DENS(CN,T,P)
      COMMON /COE/TC(6),DC(6),A1(6),A2(6),A3(6),A4(6),A5(6),A6(6),A7(6)
      IC=INT(CN)
      TF=(1.-(T/TC(IC))**(1./3.)
      D0=DC(IC)*(1.+A1(IC)*TF+A2(IC)*(TF**2)+A3(IC)*(TF**3)+A4(IC)*
     &(TF**4)+A5(IC)*(TF**5)+A6(IC)*(TF**6)+A7(IC)*(TF**7))
      TR=T/TC(IC)
      BB=0.015*FLOAT(IC)*(1.+11.5*FLOAT(IC))
      IF (IC.EQ.1) BB=11.8
      B=520.23-1240*TR+827*TR*TR-BB
      BK=(LOG((B+P*1.E-6)/(B+0.101))/LOG(10.))*0.2
      DENS=D0/(1.-BK)
      RETURN
      END

      BLOCK DATA
      COMMON /COE/TC(6),DC(6),A1(6),A2(6),A3(6),A4(6),A5(6),A6(6),A7(6)
      COMMON /RVI/RV1(6),RV2(6),RV3(6)
      DATA (TC(I),I=1,6)/512.60,513.88,536.74,563.01,588.11,610.70/
      DATA (DC(I),I=1,6)/272.,276.,274.,271.,270.,268./
      DATA (A1(I),I=1,6)/2.6278130,-0.992640,0.9405336,-3.673785,
     &-2.611289,-1553091/
      DATA (A2(i),I=1,6)/-4.047421,36.028670,12.944240,50.244970,
     &42.337510,17.726170/
      DATA (A3(I),I=1,6)/15.834250,-181.1172,-53.95190,-175.9335,
     &-149.6396,-57.60072/
      DATA (A4(I),I=1,6)/-22.50656,445.90450,113.63880,308.58830,
     &262.20950,99.902590/
      DATA (A5(I),I=1,6)/10.915970,-588.5184,-113.8656,-265.6281,
     &-223.0897,-86.58833/
      DATA (A6(I),I=1,6)/0.3047744,393.91960,43.383240,89.499660,
     &73.787930,29.628400/
      DATA (A7(I),I=1,6)/0.0,-104.3443,0.0,0.0,0.0,0.0/
      DATA (RV1(I),I=1,6)/41.152,49.740,119.690,55.070,46.110,3.0/
      DATA (RV2(I),I=1,6)/-0.2175,-0.2505,-0.6770,-0.2950,-0.2578,0./
      DATA (RV3(I),I=1,6)/3.057E-4,3.325E-4,9.817E-4,4.11E-4,3.802E-4,0./
      END
```

```
@ 323.15 K  &  6.00 MPA
DENSITY         :    784.3    KG/M3
TH.CONDUCTIVITY :    153.0    mW/M/K
VISCOSITY       :   1193.4    uPA.S
```

□

10.6.4. Corresponding States

In Section 10.4, the corresponding-states scheme of Ely and Hanley was described. The scheme is specified in detail in Tables 10.6-10.8.

Subroutine ELYHANL, shown in Display 10.5, performs calculations by this method with fluid densities calaculated from the BWR(HS) equation of state. The calculations adhere to the method as originally published and do not thus include any revisions subsequently devised for the computer package SUPERTRAPP.

```
                                          DISPLAY 10.5  ELYHANL List
        SUBROUTINE ELYHANL
        &(T,P,IS,IP,X,TC,PC,VC,WM,OMEG,Y1,Y2,Y3,Y4,D,VIS,TCC)
C
C
C       The subroutine calculates the following properties of a mixture of
C       IS components  at state IP (liquid=1, vapour=2)  at temperature,
C       T in [K]  and pressure, P in [PA]
C          1) Viscosity                 (VIS)  in [PA.S]
C          2) Thermal Conductivity      (TCC)  in [W/M/K]
C          3) Density (BWR subroutine)  (D)  in [MOL/M3]
C       based on the corresponding-states scheme developed by Ely & Hanley
C
C       - Input to the subroutine, for each component, are: the molefractions X,
C       the critical constants TC in [K], PC in [Pa], VC in [cm3/mol], the molecular
C       weights WM, the acentric factors OMEG, and the four coefficients of the
C       perfect-gas heat-capacity Y1,Y2,Y3,Y4
C
        COMMON /PROPCON/ VIA(7),VIB(4),VIC(9),TCA(7),TCB(4)
        DIMENSION X(IS),TC(IS),PC(IS),VC(IS),OMEG(IS),WM(IS)
        DIMENSION Y1(IS),Y2(IS),Y3(IS),Y4(IS)
        DIMENSION DC(10),H(10,10),F(10,10),WMM(10,10),TCI(10,10)
C    ...Mixture Density D
        DO 100 I=1,IS
100     DC(I)=1.E+6/VC(I)
        CALL BWR(T,P,IP,IS,X,TC,DC,OMEG,Y1,Y2,Y3,Y4,D,H,S,FUG)
        VMIX=1.E+6/D
C    ...Pure Components (I,I) Equivalent Fluid Reducing Parameters
        DO 120 I=1,IS
        TR=MIN(2.,(T/TC(I)))
        VR=MIN(2.,MAX((VMIX/VC(I)),0.5))
        THI=1.+(OMEG(I)-0.011)*(0.09057-0.86276*LOG(TR)+(0.31664-0.46568/
        &TR)*(VR-0.5))
        PHI=(1.+(OMEG(I)-0.011)*(0.39490*(VR-1.02355)
        &-0.93281*(VR-0.75464)*LOG(TR)))*(0.288*8.314*
        &TC(I)/(PC(I)*VC(I)*1.E-6))
        F(I,I)=THI*TC(I)/190.55
        H(I,I)=PHI*VC(I)/99.2
120     WMM(I,I)=WM(I)*1.E-3
```

```
C..........Binary (I,J) Equivalent Fluid Reducing Parameters
        DO 140 I=1,IS
        DO 140 J=1,IS
        H(I,J)=(1./8.)*((H(I,I)**(1./3.))+(H(J,J)**(1./3.)))**3
        F(I,J)=(F(I,I)*F(J,J))**0.5
   140  WMM(I,J)=2.*WMM(I,I)*WMM(J,J)/(WMM(I,I)+WMM(J,J))
C..........Mixture Equivalent Fluid Reducing Parameters
          and Molecular Weight
        HMIX=0.
        FMIX=0.
        WMIXV=0.
        WMIXT=0.
        DO 160 I=1,IS
        DO 160 J=1,IS
        HMIX=HMIX+X(I)*X(J)*H(I,J)
        FMIX=FMIX+X(I)*X(J)*F(I,J)*H(I,J)
        WMIXV=WMIXV+X(I)*X(J)*(WMM(I,J)**(+0.5))*(F(I,J)**0.5)*(H(I,J)**
       &(+4./3.))
   160  WMIXT=WMIXT+X(I)*X(J)*(WMM(I,J)**(-0.5))*(F(I,J)**0.5)*
       &(H(I,J)**(-4./3.))
        FMIX=FMIX/HMIX
        WMIXV=(WMIXV**(+2))*(FMIX**(-1))*(HMIX**(-8./3.))
        WMIXT=(WMIXT**(-2))*FMIX*(HMIX**(-8./3.))
C..........Equivalent Temperature, T0, and Density, D0
        T0=T/FMIX
        D0=(16.043/VMIX)*HMIX
C..........X correction for Viscosity & Thermal Conductivity
        TCMIX=0.
        VCMIX=0.
        PCMIX=0.
        OMMIX=0.
        DO 170 I=1,IS
        TCMIX=TCMIX+X(I)*TC(I)
        VCMIX=VCMIX+X(I)*VC(I)
        PCMIX=PCMIX+X(I)*PC(I)
   170  OMMIX=OMMIX+X(I)*OMEG(I)
        ZCMIX=PCMIX*VCMIX*1.E-6/(8.314*TCMIX)
        VR=MIN(2.,MAX((VMIX/VCMIX),0.5))
        DFDT=(OMMIX-0.011)*(-0.86276/T+0.46568*TCMIX*
       &(VR-0.5)/(T**2))
        IF ((T/TCMIX).GE.2.) DFDT=0.
        DFDT=(TCMIX/190.4)*DFDT
        XVIS=((1.-1.5*T*DFDT/FMIX)*(ZCMIX/0.288))**0.5
        XTCC=((1.-1.0*T*DFDT/FMIX)*(0.288/ZCMIX))**1.5
C..........Mixture Viscosity VIS
        V0=0.
        DO 180 I=1,9
   180  V0=V0+VIC(I)*(T0**(FLOAT(I-4)/3.))*1.E-7
```

```
      V1=(VIB(1)+VIB(2)*(VIB(3)-LOG(T0/VIB(4))))*D0*1.E-6
      DV=(D0**0.1)*(VIA(3)+VIA(4)/(T0**1.5))+((D0/.1617)-1.)*(D0**0.5)
     &*(VIA(5)+VIA(6)/T0+VIA(7)/(T0**2))
      DV=EXP(VIA(1)+VIA(2)/T0)*(EXP(DV)-1.)*1.E-7
      FVI=((WMIXV/0.016043)**0.5)*(FMIX**0.5)/(HMIX**(2./3.))
      VIS=FVI*(V0+V1+DV*XVIS)
C........Mixture Thermal Conductivity TCTRANS : Translational part
      TC0=(15./4.)*(8.314/0.016043)*V0
      TC1=(TCB(1)+TCB(2)*((TCB(3)-LOG(T0/TCB(4)))**2))*D0
      DTC=(D0**0.1)*(TCA(3)+TCA(4)/(T0**1.5))+((D0/.1617)-1.)*(D0**0.5)
     &*(TCA(5)+TCA(6)/T0+TCA(7)/(T0**2))
      DTC=EXP(TCA(1)+TCA(2)/T0)*(EXP(DTC)-1.)*1.E-3
      FTC=((0.016043/WMIXT)**0.5)*(FMIX**0.5)/(HMIX**(2./3.))
      TCTRANS=FTC*(TC0+TC1+DTC)*XTCC
C........Mixture Thermal Conductivity TCINT : Internal part
      DO 220 J=1,IS
      T0=T/F(J,J)
      V0=0.
      DO 200 I=1,9
  200 V0=V0+VIC(I)*(T0**(FLOAT(I-4)/3.))*1.E-7
      FVI=((WM(J)/16.043)**0.5)*(F(J,J)**0.5)*(H(J,J)**(-2./3.))
      CV=Y1(J)+Y2(J)*T+Y3(J)*T*T+Y4(J)*(T**3)
  220 TCI(J,J)=1.32*(V0*FVI/(WM(J)*1.E-3))*(CV-2.5*8.3145)
      TCINT=0.
      DO 240 I=1,IS
      DO 240 J=1,IS
      TCI(I,J)=2.*TCI(I,I)*TCI(J,J)/(TCI(I,I)+TCI(J,J))
  240 TCINT=TCINT+X(I)*X(J)*TCI(I,J)
C........Total Thermal Conductivity TCC
      TCC=TCTRANS+TCINT
      RETURN
      END


      BLOCK DATA
      COMMON /PROPCON/ VIA(7),VIB(4),VIC(9),TCA(7),TCB(4)
      DATA VIA/    -1.023916E+1,  1.742282E+2,  1.746055E+1, -2.847633E+3,
     &1.336850E-1,  1.420724E+2,  5.002067E+3/
      DATA VIB/     1.696986E+0, -1.333724E-1,  1.400000E+0,  1.680000E+2/
      DATA VIC/     2.907741E+6, -3.312874E+6,  1.608102E+6, -4.331905E+5,
     &7.062481E+4, -7.116621E+3,  4.325174E+2, -1.445911E+1,  2.037119E-1/
      DATA TCA/    -7.197708E+0,  8.567822E+1,  1.247183E+1, -9.846252E+2,
     &3.594685E-1,  6.979841E+1, -8.728833E+2/
      DATA TCB/  -2.527629E-4,  3.343285E-4,  1.120000E+0,  1.680000E+2/
      END
```

EXAMPLE 10.6	*Calculation of the Viscosity of the Mixture {0.7 n-Heptane +* *0.3 n-Decane} at T = 303.15K and (a) P = 0.1 MPa and (b)* *P = 25.64 MPa.*

Critical parameters, acentric factors, ideal-gas heat-capacity coefficients and molar masses for *n*-heptane and *n*-decane are all obtained from Table A1 in Appendix A. The calling program is shown in Display 10.6 for the first pressure and the results

P/MPa	η/(μPa·s)	
	ELYHANL	Assael *et al.* [38]
0.10	467.3	462.6
25.64	561.4	594.2

for both conditions are given in the table above. The viscosity of this particular mixture under these conditions has been measured by Assael *et al.* [38] with an uncertainty of ±0.5 per cent and their values are also shown in the table. It can be seen that the values predicted by the corresponding-states method do not differ by more than 5 per cent from the experimental values. In this particular case, the agreement is therefore reasonable.

```
                                    DISPLAY 10.6  ELYHANL Sample Input/Output
        PROGRAM EX1006
        PARAMETER (IS=2)
        DIMENSION X(IS),TC(IS),PC(IS),VC(IS),OMEG(IS),WM(IS)
        DIMENSION Y1(IS),Y2(IS),Y3(IS),Y4(IS)
        DATA TC(1),PC(1),VC(1),OMEG(1)/540.15,2.735E+6,432.,0.349/
        DATA Y1(1),Y2(1),Y3(1),Y4(1)/-5.146,6.762E-1,-3.651E-4,7.658E-8/
        DATA X(1),WM(1)/0.7,100.2034/
        DATA TC(2),PC(2),VC(2),OMEG(2)/617.65,2.105E+6,603.,0.489/
        DATA Y1(2),Y2(2),Y3(2),Y4(2)/-7.913,9.609E-1,-5.288E-4,1.131E-7/
        DATA X(2),WM(2),WMIX/0.3,142.2838,0./
        DATA IP,T,P/1,303.15,0.1E+6/
        CALL ELYHANL(T,P,IS,IP,X,TC,PC,VC,WM,OMEG,Y1,Y2,Y3,Y4,D,VIS,TCC)
        DO 20 I=1,IS
  20    WMIX=WMIX+X(I)*WM(I)
        WRITE(*,1000) T,P*1.E-6,D*WMIX*1.E-3,VIS*1E+6,TCC*1.E+3
 1000   FORMAT(3X,'@ T =',F8.2,' K  &  P =',F8.2,' MPA'/,8X,'DENSITY
       &,F8.1,' KG/M3'/,8X,'VISCOSITY        ',F8.1,' uPA.S'/,8X,
       &'TH.CONDUCTIVITY : ',F8.1,' mW/M/K')
        END

  @ T = 303.15 K  &  P = 0.10 MPA
        DENSITY           :   687.6   KG/M3
        VISCOSITY         :   467.3   uPA.S
        TH. CONDUCTIVITY  :   119.7   mW/M/K
```

EXAMPLE 10.7	*Calculation of the Thermal Conductivity of the Mixture {0.5 n-Heptane + 0.5 n-Undecane} at P = 0.1 MPa and (a) T = 287.95 K and (b) T = 343.65 K.*

Critical constants, acentric factors, ideal-gas heat-capacity coefficients and molar masses for *n*-heptane and *n*-decane are all obtained from Table A1 in Appendix A. The calling program is shown in Display 10.7 for the first pressure and the results

T/K	λ/(mW·m^{-1}·K^{-1})	
	ELYHANL	Wada *et al.* [39]
287.95	126.9	130.4
343.65	112.8	116.0

for both conditions are given in the table above. The thermal conductivity of this particular mixture under these conditions has been measured by Wada *et al.* [39] with an uncertainty of ±1.5 per cent and their values are also shown in the table. It can be seen that the values predicted by the corresponding-states method do not differ by more than 3 per cent from the experimental values. Thus, in this particular case, the agreement is also good.

DISPLAY 10.7 ELYHANL Sample Input/Output

```
PROGRAM EX1007
PARAMETER (IS=2)
DIMENSION X(IS),TC(IS),PC(IS),VC(IS),OMEG(IS),WM(IS)
DIMENSION Y1(IS),Y2(IS),Y3(IS),Y4(IS)
DATA TC(1),PC(1),VC(1),OMEG(1)/540.15,2.735E+6,432.,0.349/
DATA Y1(1),Y2(1),Y3(1),Y4(1)/-5.146,6.762E-1,-3.651E-4,7.658E-8/
DATA X(1),WM(1)/0.5,100.2034/
DATA TC(2),PC(2),VC(2),OMEG(2)/638.85,1.955E+6,660.,0.535/
DATA Y1(2),Y2(2),Y3(2),Y4(2)/-8.395,1.054E+0,-5.799E-4,1.237E-7/
DATA X(2),WM(2),WMIX/0.5,156.3106,0./
DATA IP,T,P/1,287.95,0.1E+6/
CALL ELYHANL(T,P,IS,IP,X,TC,PC,VC,WM,OMEG,Y1,Y2,Y3,Y4,D,VIS,TCC)
DO 20 I=1,IS
20 WMIX=WMIX+X(I)*WM(I)
WRITE(*,1000) T,P*1.E-6,D*WMIX*1.E-3,VIS*1E+6,TCC*1.E+3
1000 FORMAT(3X,'@ T =',F8.2,' K & P =',F8.2,' MPA'/,8X,'DENSITY
&,F8.1,' KG/M3'/,8X,'VISCOSITY ',F8.1,' uPA.S'/,8X,
&'TH.CONDUCTIVITY ',F8.1,' mW/M/K')
END

@ T = 287.95 K & P = 0.10 MPA
    DENSITY             726.0    KG/M3
    VISCOSITY           742.9    uPA.S
    TH. CONDUCTIVITY    126.9    mW/M/K
```

| EXAMPLE 10.8 |

Calculation of the Viscosity of n-Hexane at T = 601.15 K and P = 0.14 MPa.

Critical parameters, acentric factor, perfect-gas heat-capacity coefficients and molar mass for *n*-heptane are all obtained from Table A1 in Appendix A.

The calling program is shown in Display 10.8 The viscosity of *n*-hexane vapour under these conditions has been measured by Vogel and Strehlow [40] as (12.46 ± 0.06) µPa·s. The viscosity predicted by the corresponding states method differs from the experimental value by less than 3 per cent. If the routine TPDILG (Section 10.6.1) was employed instead then the calculated viscosity would have been 13.09 µPa·s and this is still in reasonable agreement with experiment..

```
DISPLAY 10.8 ELYHANL Sample Input/Output
      PROGRAM EX1008
      PARAMETER (IS=1)
      DIMENSION X(IS),TC(IS),PC(IS),VC(IS),OMEG(IS),WM(IS)
      DIMENSION Y1(IS),Y2(IS),Y3(IS),Y4(IS)
      DATA TC(1),PC(1) VC(1),OMEG(1)/507.90,3.035E+6,370.,0.299/
      DATA Y1(1),Y2(1),Y3(1),Y4(1)/-4.413,5.820E-1,-3.119E-4,6.494E-8/
      DATA X(1),WM(1),WMIX/1.0,86.1766,0./
      DATA IP,T,P/2,601.15,.14E+6/
      CALL ELYHANL(T,P,IS,IP,X,TC,PC,VC,WM,OMEG,Y1,Y2,Y3,Y4,D,VIS,TCC)
      DO 20 I=1,IS
   20 WMIX=WMIX+X(I)*WM(I)
      WRITE(*,1000) T,P*1.E-6,D*WMIX*1.E-3,VIS*1E+6,TCC*1.E+3
 1000 FORMAT(3X,'@ T =',F8.2,' K   &   P =',F8.2,' MPA',/,8X,'DENSITY
     &,F8.1,' KG/M3',/,8X,'VISCOSITY       ',F8.1,' uPA.S',/,8X,
     &'TH.CONDUCTIVITY ',F8.1,' mW/M/K')
      END


  @ T = 601.15 K   &   P = 0.14 MPA
      DENSITY                  2.44  KG/M3
      VISCOSITY               12.84  uPA.S
      TH. CONDUCTIVITY        50.29  mW/M/K
```

❑

References

1. G.C. Maitland, M. Rigby, E.B. Smith and W.A. Wakeham, "Intermolecular Forces. Their Origin and Determination", Clarendon Press, Oxford (1981).
2. P.D. Neufeld, A.R. Janzen and R.A. Aziz, *J. Chem. Phys.* **57**:1100 (1972).
3. R.A. Svehla, *NASA Tech. Rep.* R-132, Lewis Research Center, Cleveland, Ohio (1962), in [6].

4. R.S. Brokaw, *NASA Technical Report* TR R-81, Washington (1960).
5. J. Kestin and J. Yata, *J. Chem. Phys.* **49**:4780 (1968).
6. R.C. Reid, J.M. Prausnitz and B.E. Poling, "The Properties of Gases & Liquids", 4th Ed., McGraw-Hill, New York (1988).
7. C.R. Wilke, *J. Chem. Phys.* **18**:517 (1950).
8. R.S. Brokaw, *J. Chem. Phys.* **29**:391 (1958).
9. R.C. Reid and L.I. Belenyessy, *J. Chem. Eng. Data* **5**:150 (1960).
10. E.A. Mason and S.C. Saxena, *Phys. Fluids* **1**:361 (1958).
11. J. Kestin, S.T. Ro and W.A. Wakeham, *Trans. Faraday Soc.* **67**:2308 (1971).
12. M.J. Assael and W.A. Wakeham, *J. Chem. Soc., Faraday Trans. 1* **77**:697 (1981).
13. J.A. Jossi, L.I. Stiel and G. Thodos, *A.I.Ch.E. J.* **8**:59 (1962).
14. L.I. Stiel and G. Thodos, *A.I.Ch.E. J.* **10**:275 (1964).
15. L.I. Stiel and G. Thodos, *A.I.Ch.E. J.* **10**:26 (1964).
16. "PPDS - Physical Properties Data Service", Package for the Calculation of the Thermophysical Properties of Fluids and Mixtures, Produced and Supplied by N.E.L., East Kilbride, U.K.
17. M.J. Assael, J.H. Dymond, M. Papadaki and P.M. Patterson, *Fluid Phase Equil.* **75**:245 (1992).
18. M.J. Assael, J.H. Dymond, M. Papadaki and P.M. Patterson, *Int. J. Thermophys.* **13**:269 (1992).
19. M.J. Assael, J.H. Dymond, M. Papadaki and P.M. Patterson, *Int. J. Thermophys.* **13**:895 (1992).
20. M.J. Assael, J.H. Dymond and S.K. Polimatidou, *Int. J. Thermophys.* **15**:189 (1994).
21. M.J. Assael, J.H. Dymond and S.K. Polimatidou, *Int. J. Thermophys.* **16**:761 (1995).
22. M.J. Assael, J.H. Dymond and D. Exadaktilou, *Int. J. Thermophys.* **15**:155 (1994).
23. I. Cibulka, *Fluid Phase Equil.* **89**:1 (1993).
24. M.J. Assael, J.H. Dymond, M. Papadaki and P.M. Patterson, *Int. J. Thermophys.* **13**:659 (1992).
25. J.F. Ely and H.J.M. Hanley, *Ind. Eng. Chem. Fundam.* **20**:323 (1981).
26. J.F. Ely and H.J.M. Hanley, *Ind. Eng. Chem. Fundam.* **22**:90 (1983).
27. "SUPERTRAPP", FORTRAN Package for the Calculation of the Transport Properties of Nonpolar Fluids and their Mixtures, Produced and Supplied by N.I.S.T., Gaithersburg, U.S.A.
28. J.W. Leach, P.S. Chappelear and T.W. Leland, *A.I.Ch.E. J.* **14**:568 (1968).
29. M.L. Huber and J.F. Ely, *Fluid Phase Equil.* **80**:239 (1992).
30. M.L. Huber, D.G. Friend and J.F. Ely, *Fluid Phase Equil.* **80**:249 (1992).
31. M.L. Huber and J.F. Ely, *Rev. Int. Froid* **17**:18 (1994).
32. "REFPROP", FORTRAN Package for the Calculation of the Thermodynamic Properties of Refrigerants and Refrigerant Mixtures, Produced and Supplied by N.I.S.T., Gaithersburg, U.S.A.
33. M.J. Assael and W.A. Wakeham, *J. Chem. Soc., Faraday Trans. 1* **78**:185 (1982).
34. V. Vesovic, W.A. Wakeham, J. Luettmer-Stratmann, J.V. Sengers, J. Millat, E. Vogel and M.J. Assael, *Int. J. Thermophys.* **15**:33 (1994).
35. S. Hendl, J. Millat, E. Vogel, V. Vesovic, W.A. Wakeham, J. Luettmer-Strathmann, J.V. Sengers and M.J. Assael, *Int. J. Thermophys.* **15**:1 (1994).

36. H. Kubota, Y. Tanaka and T. Makita, *Int. J. Thermophys.* **8**:47 (1987).
37. M.J. Assael and S.K. Polimatidou, *Int. J. Thermophys.* **15**:95 (1994).
38. M.J. Assael, J.H. Dymond and M. Papadaki, *Fluid Phase Equilib.* **75**:287 (1992).
39. Y. Wada, Y. Nagasaka and A. Nagashima, *Int. J. Thermophys.* **6**:251 (1985).
40. E. Vogel and T. Strehlow, *Z. Phys. Chemie Leipzig* **269**:57 (1988).

A

Tables of Property Values

Table A1   Physical Constants and Perfect-Gas Heat-Capacity Coefficients

Compound Name	Formula	T^b	T^c	P^c	V^c	Dipm	ω
		K	K	MPa	$cm^3{\cdot}mol^{-1}$	debye	
1. Alkanes							
methane	CH_4	111.6	190.55	4.599	99.2	0.0	0.011
ethane	C_2H_6	184.6	305.33	4.871	148.3	0.0	0.099
propane	C_3H_8	231.1	369.85	4.247	203.	0.0	0.153
n-butane	C_4H_{10}	272.7	425.25	3.792	255.	0.0	0.199
isobutane	C_4H_{10}	261.4	408.2	3.65	263.	0.1	0.183
n-pentane	C_5H_{12}	309.2	469.80	3.375	304.	0.0	0.251
2-methylbutane	C_5H_{12}	301.0	460.4	3.39	306.	0.1	0.227
n-hexane	C_6H_{14}	341.9	507.90	3.035	370.	0.0	0.299
2-methylpentane	C_6H_{14}	333.4	497.5	3.01	367.		0.278
3-methylpentane	C_6H_{14}	336.4	504.5	3.12	367.		0.272
2,2-dimethylbutane	C_6H_{14}	322.8	488.8	3.08	359.		0.232
2,3-dimethylbutane	C_6H_{14}	331.1	500.0	3.13	358.		0.247
n-heptane	C_7H_{16}	371.6	540.15	2.735	432.	0.0	0.349
2-methyl hexane	C_7H_{16}	363.2	530.4	2.73	421.	0.0	0.329
2,2-dimethylpentane	C_7H_{16}	352.4	520.5	2.77	416.	0.0	0.287
n-octane	C_8H_{18}	398.8	568.95	2.49	492.	0.0	0.398
n-nonane	C_9H_{20}	424.0	594.90	2.29	548.	0.0	0.445
n-decane	$C_{10}H_{22}$	447.3	617.65	2.105	603.	0.0	0.489
n-undecane	$C_{11}H_{24}$	469.1	638.85	1.955	660.	0.0	0.535
n-dodecane	$C_{12}H_{26}$	489.5	658.65	1.83	713.	0.0	0.575
2. Alkenes							
ethylene	C_2H_4	169.3	282.4	5.04	130.4	0.0	0.089
propylene	C_3H_6	225.5	364.9	4.60	181.	0.4	0.144
1-butene	C_4H_8	266.9	419.6	4.02	240.	0.3	0.191
1-pentene	C_5H_{10}	303.1	464.8	3.53	300.	0.4	0.233
2-methyl-1-butene	C_5H_{10}	304.3	465.	3.45		0.5	0.236
1-hexene	C_6H_{12}	336.6	504.0	3.17	350.	0.4	0.285
2-methyl-2-pentene	C_6H_{12}	340.5	518.	3.28	351.		0.229
1-heptene	C_7H_{14}	366.8	537.3	2.83	440.	0.3	0.358
1-octene	C_8H_{16}	394.4	566.7	2.62	464.	0.3	0.386
3. Alkadienes							
propadiene	C_3H_4	238.7	393.	5.47	162.	0.2	0.313
1,2-butadiene	C_4H_6	284.0	443.7	4.49	219.	0.4	0.255
1,2-pentadiene	C_5H_8	318.0	503.	4.07	276.		0.173

Table A1(con/d) Physical Constants and Perfect-Gas Heat-Capacity Coefficients

$C_{P,0}$ $J \cdot mol^{-1} \cdot K^{-1}$	$C_{P,1}$ $J \cdot mol^{-1} \cdot K^{-2}$	$C_{P,2}$ $J \cdot mol^{-1} \cdot K^{-3}$	$C_{P,3}$ $J \cdot mol^{-1} \cdot K^{-4}$	Molecular Weight	Compound Name
					1. Alkanes
1.925E+1	5.213E-2	1.197E-5	-1.132E-8	16.0436	methane
5.409E+0	1.781E-1	-6.938E-5	8.713E-9	30.069	ethane
-4.224E+0	3.063E-1	-1.586E-4	3.215E-8	44.096	propane
9.487E+0	3.313E-1	-1.108E-4	-2.822E-9	58.123	*n*-butane
-1.390E+0	3.847E-1	-1.846E-4	2.895E-8	58.123	isobutane
-3.626E+0	4.873E-1	-2.580E-4	5.305E-8	72.150	*n*-pentane
-9.525E+0	5.066E-1	-2.729E-4	5.723E-8	72.150	2-methylbutane
-4.413E+0	5.820E-1	-3.119E-4	6.494E-8	86.177	*n*-hexane
-1.057E+1	6.184E-1	-3.573E-4	8.085E-8	86.177	2-methylpentane
-2.386E+0	5.690E-1	-2.870E-4	5.033E-8	86.177	3-methylpentane
-1.663E+1	6.293E-1	-3.481E-4	6.850E-8	86.177	2,2-dimethylbutane
-1.461E+1	6.150E-1	-3.376E-4	6.820E-8	86.177	2,3-dimethylbutane
-5.146E+0	6.762E-1	-3.651E-4	7.658E-8	100.203	*n*-heptane
-3.939E+1	8.642E-1	-6.289E-4	1.836E-7	100.203	2-methylhexane
-5.010E+1	8.956E-1	-6.360E-4	1.736E-7	100.203	2,2-dimethylpentane
-6.096E+0	7.712E-1	-4.195E-4	8.855E-8	114.230	*n*-octane
-8.374E+0	8.729E-1	-4.823E-4	1.031E-7	128.257	*n*-nonane
-7.913E+0	9.609E-1	-5.288E-4	1.131E-7	142.284	*n*-decane
-8.395E+0	1.054E+0	-5.799E-4	1.237E-7	156.311	*n*-undecane
-9.328E+0	1.149E+0	-6.347E-4	1.359E-7	170.337	*n*-dodecane
					2. Alkenes
3.806E+0	1.566E-1	-8.348E-5	1.755E-8	28.054	ethylene
3.710E+0	2.345E-1	-1.160E-4	2.205E-8	42.080	propylene
-2.994E+0	3.532E-1	-1.990E-4	4.463E-8	56.107	1-butene
-0.134E+0	4.329E-1	-2.317E-4	4.681E-8	70.134	1-pentene
1.057E+1	3.997E-1	-1.946E-4	3.314E-8	70.134	2-methyl-1-butene
-1.746E+0	5.309E-1	-2.903E-4	6.054E-8	84.161	1-hexene
-1.475E+1	5.669E-1	-3.341E-4	7.963E-8	84.161	2-methyl-2-pentene
-3.303E+0	6.297E-1	-3.512E-4	7.607E-8	98.188	1-heptene
-4.099E+0	7.239E-1	-4.036E-4	8.675E-8	112.214	1-octene
					3. Alkadienes
9.906E+0	1.977E-1	-1.182E-4	2.782E-8	40.065	propadiene
1.120E+1	2.724E-1	-1.468E-4	3.089E-8	54.091	1,2-butadiene
8.826E+0	3.880E-1	-2.280E-4	5.246E-8	68.118	1,2-pentadiene

Table A1(con/d)   Physical Constants and Perfect-Gas Heat-Capacity Coefficients

Compound Name	Formula	T^b	T^c	P^c	V^c	Dipm	ω
		K	K	MPa	cm³·mol⁻¹	debye	
4. Alkyl Cyclopentanes							
cyclopentane	C_5H_{10}	322.4	511.7	4.51	260.	0.0	0.196
methylcyclopentane	C_6H_{12}	345.0	532.7	3.78	319.	0.0	0.231
ethylcyclopentane	C_7H_{14}	376.6	569.5	3.40	375.		0.271
n-propylcyclopentane	C_8H_{16}	404.1	603.	3.0	425.		0.335
5. Alkyl Cyclohexanes							
cyclohexane	C_6H_{12}	353.8	553.5	4.07	308.	0.3	0.212
methylcyclohexane	C_7H_{14}	374.1	572.2	3.47	368.	0.0	0.236
ethylcyclohexane	C_8H_{16}	404.9	609.	3.0	450.	0.0	0.243
n-propylcyclohexane	C_9H_{18}	429.9	639.0	2.80			0.258
6. Alkyl Cycloalkenes							
cyclopentene	C_5H_8	317.4	506.0			0.9	0.196
cyclohexene	C_6H_{10}	356.1	560.5	4.34		0.6	0.210
7. Acetylenes							
acetylene	C_2H_2	188.4	308.3	6.14	112.7	0.0	0.190
methylacetylene	C_3H_4	249.9	402.4	5.63	164.	0.7	0.215
1-pentyne	C_5H_8	313.3	493.5	4.05	278.	0.9	0.164
8. Alkyl Benzenes							
benzene	C_6H_6	353.2	562.06	4.895	259.	0.0	0.212
toluene	C_7H_8	383.8	591.75	4.108	316.	0.4	0.263
ethylbenzene	C_8H_{10}	409.3	617.16	3.609	374.	0.4	0.302
o-xylene	C_8H_{10}	417.6	630.3	3.732	369.	0.5	0.310
m-xylene	C_8H_{10}	412.3	617.0	3.541	376.	0.3	0.325
p-xylene	C_8H_{10}	411.5	616.2	3.511	379.	0.1	0.320
cumene	C_9H_{12}	177.1	425.6	3.21			0.326
mesitylene	C_9H_{12}	228.4	637.3	3.127	275	0.1	0.399
9. Alkyl Halides							
methyl bromide	CH_3Br	276.6	464.0	6.61		1.8	
methyl chloride	CH_3Cl	249.1	416.3	6.70	138.9	1.9	0.153
methyl fluoride	CH_3F	194.7	315.0	5.6	113.2	1.8	0.187
ethyl bromide	C_2H_5Br	311.5	503.9	6.23	215.	2.0	0.229
ethyl chloride	C_2H_5Cl	285.5	460.4	5.27	199.	2.0	0.191
ethyl fluoride	C_2H_5F	235.5	375.3	5.02	169.	2.0	0.215
propyl chloride	C_3H_7Cl	320.4	503.0	4.58	254.	2.0	0.235
1-chlorobutane	C_4H_9Cl	351.6	542.0	3.68	312.	2.0	0.218

Table A1(con/d) Physical Constants and Perfect-Gas Heat-Capacity Coefficients

$\dfrac{C_{P,0}}{\text{J·mol}^{-1}\text{·K}^{-1}}$	$\dfrac{C_{P,1}}{\text{J·mol}^{-1}\text{·K}^{-2}}$	$\dfrac{C_{P,2}}{\text{J·mol}^{-1}\text{·K}^{-3}}$	$\dfrac{C_{P,3}}{\text{J·mol}^{-1}\text{·K}^{-4}}$	Molecular Weight	Compound Name
					4. Alkyl Cyclopentanes
-5.362E+1	5.426E-1	-3.031E-4	6.485E-8	70.134	cyclopentane
-5.011E+1	6.381E-1	-3.642E-4	8.014E-8	84.161	methylcyclopentane
-5.531E+1	7.511E-1	-4.396E-4	1.004E-7	98.188	ethylcyclopentane
-5.597E+1	8.449E-1	-4.924E-4	1.117E-7	112.214	*n*-propylcyclopentane
					5. Alkyl Cyclohexanes
-5.454E+1	6.113E-1	-2.523E-4	1.321E-8	84.161	cyclohexane
-6.192E+1	7.842E-1	-4.438E-4	9.366E-8	98.188	methylcyclohexane
-6.389E+1	8.893E-1	-5.108E-4	1.103E-7	112.214	ethylcyclohexane
-6.252E+1	9.889E-1	-5.795E-4	1.291E-7	126.241	*n*-propylcyclohexane
					6. Alkyl Cycloalkenes
-4.151E+1	4.631E-1	-2.579E-4	5.434E-8	68.118	cyclopentene
-6.865E+1	7.252E-1	-5.414E-4	1.644E-7	82.145	cyclohexene
					7. Acetylenes
2.682E+1	7.578E-2	-5.007E-5	1.412E-8	26.038	acetylene
1.471E+1	1.864E-1	-1.174E-4	3.224E-8	40.065	methylacetylene
1.807E+1	3.511E-1	-1.913E-4	4.098E-8	68.118	1-pentyne
					8. Alkyl Benzenes
-3.392E+1	4.739E-1	-3.017E-4	7.130E-8	78.113	benzene
-2.435E+1	5.125E-1	-2.765E-4	4.911E-8	92.140	toluene
-4.310E+1	7.072E-1	-4.811E-4	1.301E-7	106.167	ethylbenzene
-1.585E+1	5.962E-1	-3.443E-4	7.528E-8	106.167	*o*-xylene
-2.917E+1	6.297E-1	-3.747E-4	8.478E-8	106.167	*m*-xylene
-2.509E+1	6.042E-1	-3.374E-4	6.820E-8	106.167	*p*-xylene
-3.936E+1	7.842E-1	-5.087E-4	1.291E-7	120.194	cumene
-1.959E+1	6.724E-1	-3.692E-4	7.700E-8	120.194	mesitylene
					10. Alkyl Halides
1.443E+1	1.091E-1	-5.401E-5	1.000E-8	94.939	methyl bromide
1.388E+1	1.014E-1	-3.889E-5	2.567E-9	50.488	methyl chloride
1.382E+1	8.616E-2	-2.071E-5	-1.985E-9	34.033	methyl fluoride
6.657E+0	2.348E-1	-1.472E-4	3.804E-8	108.966	ethyl bromide
-0.553E+0	2.606E-1	-1.840E-4	5.548E-8	64.515	ethyl chloride
4.346E+0	2.180E-1	-1.166E-4	2.410E-8	48.060	ethyl fluoride
-3.345E+0	3.626E-1	-2.508E-4	7.448E-8	78.541	propyl chloride
-2.613E+0	4.497E-1	-2.937E-4	8.081E-8	92.568	1-chlorobutane

Table A1(con/d)   Physical Constants and Perfect-Gas Heat-Capacity Coefficients

Compound Name	Formula	T^b	T^c	P^c	V^c	Dipm	ω
		K	K	MPa	cm^3·mol^{-1}	debye	
10. Refrigerants							
R11	CCl$_3$F	296.4	471.15	4.403	245.7	0.4	0.189
R12	CCl$_2$F$_2$	243.0	384.95	4.129	215.9	0.5	0.204
R22	CHClF$_2$	232.3	369.3	4.988	168.5	1.4	0.221
R32	CH$_2$F$_2$	221.4	351.56	5.83	121.	2.0	0.271
R123	C$_2$HCl$_2$F$_3$	301.0	456.94	3.674	278.1	1.4	
R124	C$_2$HClF$_4$	261.1	395.65	3.634	243.7	1.5	0.281
R125	C$_2$HF$_5$	224.6	339.4	3.631	209.8	1.6	0.299
R134a	C$_2$H$_2$F$_4$	247.0	374.21	4.056	198.1	2.1	0.326
R141b	C$_2$H$_3$Cl$_2$F		477.26	4.23	253.7	2.0	0.223
R142b	C$_2$H$_3$ClF$_2$	263.9	410.25	4.246		2.1	0.251
R143a	C$_2$H$_3$F$_3$	225.8	346.25	3.811	193.6	2.3	
R152a	C$_2$H$_4$F$_2$	249.0	386.44	4.52	179.5	2.3	0.256
11. 1-Alkanols							
methanol	CH$_4$O	337.7	512.6	8.092	118.	1.7	0.556
ethanol	C$_2$H$_6$O	351.4	513.88	6.132	167.1	1.7	0.644
1-propanol	C$_3$H$_8$O	370.3	536.74	5.168	219.	1.7	0.623
1-butanol	C$_4$H$_{10}$O	390.9	563.01	4.424	275.	1.8	0.593
1-pentanol	C$_5$H$_{12}$O	411.1	588.11	3.909	326.	1.7	0.579
1-hexanol	C$_6$H$_{14}$O	430.2	610.7	3.47	381.	1.8	0.560
1-heptanol	C$_7$H$_{16}$O	449.8	632.5	3.135	435.	1.7	0.560
1-octanol	C$_8$H$_{18}$O	468.3	652.5	2.86	490.	2.0	0.587
1-nonanol	C$_9$H$_{20}$O	486.7	671.5	2.63	546.	1.7	
1-decanol	C$_{10}$H$_{22}$O	506.1	689.0	2.41	600.	1.8	
1-undecanol	C$_{11}$H$_{24}$O						
1-dodecanol	C$_{12}$H$_{26}$O	533.1	679.	1.92	718.	1.6	
12. Alkyl Acetates							
methyl acetate	C$_3$H$_6$O$_2$	330.4	506.8	4.69	228.	1.7	0.326
ethyl acetate	C$_4$H$_8$O$_2$	350.3	523.2	3.83	286.	1.9	0.362

Table A1(con/d) Physical Constants and Perfect-Gas Heat-Capacity Coefficients

$\dfrac{C_{P,0}}{\text{J·mol}^{-1}\text{·K}^{-1}}$	$\dfrac{C_{P,1}}{\text{J·mol}^{-1}\text{·K}^{-2}}$	$\dfrac{C_{P,2}}{\text{J·mol}^{-1}\text{·K}^{-3}}$	$\dfrac{C_{P,3}}{\text{J·mol}^{-1}\text{·K}^{-4}}$	Molecular Weight	Compound Name
					11. Refrigerants
3.209E+1	2.162E-1	-2.334E-4	8.869E-8	137.368	R11
2.352E+1	2.263E-1	-2.335E-4	8.586E-8	120.914	R12
1.864E+1	1.443E-1	-3.115E-5	-6.659E-8	86.469	R22
3.679E+1	-6.290E-2	3.754E-4	-3.216E-7	52.024	R32
0.602E+1	5.052E-1	-7.632E-4	5.183E-7	152.931	R123
3.277E+1	2.189E-1	5.751E-5	-1.840E-7	136.477	R124
2.360E+1	2.837E-1	-1.230E-4	-5.672E-8	120.022	R125
1.678E+1	2.864E-1	-2.273E-4	1.133E-7	102.031	R134a
				116.950	R141b
1.709E+1	2.638E-1	-1.369E-4	-1.871E-8	100.496	R142b
1.791E+1	2.156E-1	-1.492E-6	-1.359E-7	84.041	R143a
2.845E+1	8.474E-2	2.449E-4	-2.837E-7	66.050	R152a
					11. 1-Alkanols
2.115E+1	7.092E-2	2.587E-5	-2.852E-8	32.042	methanol
9.014E+0	2.141E-1	-8.390E-5	1.373E-9	46.069	ethanol
2.470E+0	3.325E-1	-1.855E-4	4.296E-8	60.096	1-propanol
3.266E+0	4.180E-1	-2.242E-4	4.685E-8	74.122	1-butanol
3.869E+0	5.045E-1	-2.639E-4	5.120E-8	88.149	1-pentanol
4.811E+0	5.891E-1	-3.010E-4	5.426E-8	102.176	1-hexanol
4.907E+1	6.778E-1	-3.447E-4	6.046E-8	116.203	1-heptanol
6.171E+0	7.607E-1	-3.797E-4	6.263E-8	130.230	1-octanol
1.280E+0	8.817E-1	-4.791E-4	9.801E-8	144.256	1-nonanol
1.457E+1	8.947E-1	-3.921E-4	3.451E-8	158.283	1-decanol
				172.310	1-undecanol
9.224E+0	1.103E+0	-5.338E-4	7.779E-8	186.337	1-dodecanol
					12. Alkyl Acetates
1.655E+1	2.245E-1	-4.342E-5	2.914E-8	74.079	methyl acetate
7.235E+0	4.072E-1	-2.092E-4	2.855E-8	88.106	ethyl acetate

Table A1(con/d)   Physical Constants and Perfect-Gas Heat-Capacity Coefficients

Compound Name	Formula	T^b	T^c	P^c	V^c	Dipm	ω
		K	K	MPa	$cm^3 \cdot mol^{-1}$	debye	
13. Monatomics							
argon	Ar	87.3	150.8	4.87	74.9	0.0	0.001
helium	He	3.19	3.31	0.114	72.9	0.0	-0.473
krypton	Kr	119.9	209.4	5.5	91.2	0.0	0.005
neon	Ne	27.1	44.4	2.76	41.6	0.0	-0.029
xenon	Xe	165.0	289.7	5.84	118.4	0.0	0.008
14. Diatomics							
chlorine	Cl_2	239.2	416.9	7.98	123.8	0.0	0.090
hydrogen (normal)	H_2	20.4	33.2	1.3	65.1	0.0	-0.218
nitrogen	N_2	77.4	126.2	3.39	89.8	0.0	0.039
oxygen	O_2	90.2	154.6	5.04	73.4	0.0	0.025
15. Inorganic Oxides							
carbon dioxide	CO_2		304.1	7.38	93.9	0.0	0.239
carbon monoxide	CO	81.7	132.9	3.5	93.2	0.1	0.066
nitrogen dioxide	NO_2	294.3	431.	10.1	167.8	0.4	0.834
16. Other Compounds							
ammonia	NH_3	239.8	405.5	11.35	72.5	1.5	0.250
hydrogen sulfide	H_2S	213.5	373.2	8.94	98.6	0.9	0.081
water	H_2O	373.2	647.3	22.12	57.1	1.8	0.344

Table A1(con/d)   Physical Constants and Perfect-Gas Heat-Capacity Coefficients

$\dfrac{C_{P,0}}{\text{J·mol}^{-1}·\text{K}^{-1}}$	$\dfrac{C_{P,1}}{\text{J·mol}^{-1}·\text{K}^{-2}}$	$\dfrac{C_{P,2}}{\text{J·mol}^{-1}·\text{K}^{-3}}$	$\dfrac{C_{P,3}}{\text{J·mol}^{-1}·\text{K}^{-4}}$	Molecular Weight	Compound Name
					13. Monatomics
2.080E+1				39.948	argon
2.080E+1				3.017	helium
2.080E+1				83.800	krypton
2.080E+1				20.179	neon
2.080E+1				131.300	xenon
					14. Diatomics
2.693E+1	3.384E-2	-3.869E-5	1.547E-8	70.906	chlorine
3.224E+1	1.924E-3	1.055E-5	-3.596E-9	2.016	hydrogen (normal)
3.115E+1	-1.357E-2	2.680E-5	-1.168E-8	28.013	nitrogen
2.811E+1	-3.680E-6	1.746E-5	-1.065E-8	32.000	oxygen
					15. Inorganic Oxides
1.980E+1	7.344E-2	-5.602E-5	1.715E-8	44.010	carbon dioxide
3.087E+1	-1.285E-2	2.789E-5	-1.272E-8	28.010	carbon monoxide
2.423E+1	4.836E-2	-2.081E-5	0.293E-9	46.006	nitrogen dioxide
					16. Other Compounds
2.731E+1	2.383E-2	1.707E-5	-1.185E-8	17.030	ammonia
3.194E+1	1.436E-3	2.432E-5	-1.176E-8	34.076	hydrogen sulfide
3.224E+1	1.924E-3	1.055E-5	-3.596E-9	18.015	water

Table A2   Vapour-Pressure Coefficients of Antoine Equation

Compound Name	Formula	A	$\dfrac{B}{K}$	$\dfrac{C}{K}$	$\dfrac{T_{min}}{K}$	$\dfrac{T_{max}}{K}$
1. Alkanes						
methane	CH_4	5.77884	397.847	-61.890	91	127
ethane	C_2H_6	5.94631	662.404	-16.387	136	200
propane	C_3H_8	6.95075	1467.362	71.564	310	370
n-butane	C_4H_{10}	5.95358	945.089	-33.256	195	273
isobutane	C_4H_{10}	5.91911	901.914	-30.947	188	263
n-pentane	C_5H_{12}	5.99028	1071.187	-40.384	269	300
2-methylbutane	C_5H_{12}	6.04913	1081.748	-33.333	218	295
n-hexane	C_6H_{14}	5.72763	1031.938	-64.846	300	321
2-methylpentane	C_6H_{14}	5.99479	1152.210	-44.579	285	334
3-methylpentane	C_6H_{14}	5.99139	1162.069	-44.870	288	338
2,2-dimethylbutane	C_6H_{14}	5.87001	1080.723	-43.308	273	320
2,3-dimethylbutane	C_6H_{14}	5.94371	1132.099	-43.656	287	331
n-heptane	C_7H_{16}	6.02701	1267.592	-56.354	298	373
2-methylhexane	C_7H_{16}	6.00310	1238.864	-53.283	288	363
2,2-dimethylpentane	C_7H_{16}					
n-octane	C_8H_{18}	6.04394	1351.938	-64.030	325	400
n-nonane	C_9H_{20}	6.06280	1430.630	-71.323	340	424
n-decane	$C_{10}H_{22}$	6.08321	1504.405	-78.319	269	450
n-undecane	$C_{11}H_{24}$	6.10213	1572.831	-85.088	283	470
n-dodecane	$C_{12}H_{26}$	6.10740	1627.417	-92.661	400	490
2. Alkenes						
ethylene	C_2H_4	5.87310	584.293	-18.288	150	190
propylene	C_3H_6	5.95357	788.339	-25.705	166	226
1-butene	C_4H_8	5.64740	806.813	-45.514	213	270
1-pentene	C_5H_{10}	5.97160	1045.272	-39.552	285	305
2-methyl-1-butene	C_5H_{10}					
1-hexene	C_6H_{12}	5.99426	1154.952	-47.068	288	340
2-methyl-2-pentene	C_6H_{12}					
1-heptene	C_7H_{14}	6.04107	1266.473	-52.948	293	369
1-octene	C_8H_{16}	6.06241	1356.472	-60.051	315	400
3. Alkadienes						
propadiene	C_3H_4	6.56901	1030.052	-12.571	140	190
1,2-butadiene	C_4H_6					
1,2-pentadiene	C_5H_8	6.07423	1130.531	-40.846	213	245

Table A2(con/d) Vapour-Pressure Coefficients of Antoine Equation

Compound Name	Formula	A	$\dfrac{B}{K}$	$\dfrac{C}{K}$	$\dfrac{T_{min}}{K}$	$\dfrac{T_{max}}{K}$
4. Alkyl Cyclopentanes						
cyclopentane	C_5H_{10}	6.02877	1133.199	-40.735	288	325
methylcyclopentane	C_6H_{12}	5.99178	1188.320	-46.843	288	345
ethylcyclopentane	C_7H_{14}	6.00807	1296.209	-52.755	300	380
n-propylcyclopentane	C_8H_{16}	6.03046	1385.284	-59.904	325	405
5. Alkyl Cyclohexanes						
cyclohexane	C_6H_{12}	6.00569	1223.273	-48.061	290	355
methylcyclohexane	C_7H_{14}	5.95366	1273.962	-51.395	298	374
ethylcyclohexane	C_8H_{16}	5.98475	1348.000	-55.894	316	395
n-propyl cyclohexane	C_9H_{18}	6.01886	1465.252	-64.759	345	430
6. Alkyl Cycloalkenes						
cyclopentene	C_5H_8					
cyclohexene	C_6H_{10}	5.99698	1221.700	-50.001	309	365
7. Acetylenes						
acetylene	C_2H_2	6.19114	697.810	-21.688	193	207
methylacetylene	C_3H_4					
1-pentyne	C_5H_8					
8. Alkyl Benzenes						
benzene	C_6H_6	6.01905	1204.637	-53.081	294	378
toluene	C_7H_8	6.08436	1347.620	-53.363	308	385
ethylbenzene	C_8H_{10}	6.08206	1425.305	-59.735	298	365
o-xylene	C_8H_{10}	5.94220	1387.336	-66.741	280	330
m-xylene	C_8H_{10}	6.46290	1641.628	-42.251	285	333
p-xylene	C_8H_{10}	6.11376	1452.215	-57.992	330	412
cumene	C_9H_{12}	6.06528	1464.366	-64.915	330	424
mesitylene	C_9H_{12}	6.16365	1592.442	-66.245	360	450
9. Alkyl Halides						
methyl bromide	CH_3Br	6.22430	1049.898	-27.831	203	277
methylchloride	CH_3Cl	6.11875	902.201	-29.961	183	250
methyl fluoride	CH_3F	5.10201	406.764	-62.797	165	217
ethyl bromide	C_2H_5Br	6.11352	1121.957	-38.409	300	348
ethyl chloride	C_2H_5Cl	6.11833	1033.161	-34.194	218	285
ethyl fluoride	C_2H_5F	6.34446	947.816	-17.076	170	255
ethyl iodide	C_2H_5I	6.06765	1222.418	-44.782	300	330
propyl chloride	C_3H_7Cl	6.07206	1120.381	-44.089	245	320
1-chlorobutane	C_4H_9Cl	5.94963	1167.428	-55.725	257	351

Table A2(con/d)   Vapour-Pressure Coefficients of Antoine Equation

Compound Name	Formula	A	$\dfrac{B}{K}$	$\dfrac{C}{K}$	$\dfrac{T_{min}}{K}$	$\dfrac{T_{max}}{K}$
10. Refrigerants						
R11	CCl_3F	6.01505	1043.600	-36.566	243	335
R12	CCl_2F_2					
R22	$CHClF_2$	6.096617	821.095	-31.640	210	240
R32	CH_2F_2					
R123	$C_2HCl_2F_3$					
R124	C_2HClF_4					
R125	C_2HF_5					
R134a	$C_2H_2F_4$	6.17338	874.648	-37.217	210	260
R141b	$C_2H_3Cl_2F$	6.00691	1049.840	-42.820	250	330
R142b	$C_2H_3ClF_2$					
R143a	$C_2H_3F_3$					
R152a	$C_2H_4F_2$					
11. 1-Alkanols						
methanol	CH_4O	7.24693	1605.615	-31.317	274	336
ethanol	C_2H_6O	7.24222	1595.811	-46.702	292	365
1-propanol	C_3H_8O	6.87065	1438.587	-74.598	330	377
1-butanol	$C_4H_{10}O$	6.76666	1460.309	-83.939	296	390
1-pentanol	$C_5H_{12}O$	6.31559	1292.273	-111.313	347	420
1-hexanol	$C_6H_{14}O$	6.20107	1305.984	-119.249	325	430
1-heptanol	$C_7H_{16}O$	6.10824	1323.566	-126.909	336	449
1-octanol	$C_8H_{18}O$	6.80512	1752.302	-99.080	327	386
1-nonanol	$C_9H_{20}O$	5.83497	1297.059	-147.711	364	486
1-decanol	$C_{10}H_{22}O$	5.76028	1315.019	-154.022	378	504
1-undecanol	$C_{11}H_{24}O$					
1-dodecanol	$C_{12}H_{26}O$	5.70652	1388.338	-163.157	400	538
12. Alkyl Acetates						
methyl acetate	$C_3H_6O_2$	6.25449	1189.608	-50.035	260	351
ethyl acetate	$C_4H_8O_2$	6.20229	1232.542	-56.563	271	373

Table A2(con/d)   Vapour-Pressure Coefficients of Antoine Equation

Compound Name	Formula	A	B K	C K	T_{min} K	T_{max} K
13. Monatomics						
argon	Ar	6.99579	448.390	-3.130	72	84
helium	He					
krypton	Kr	6.20961	540.493	-9.141	127	208
neon	Ne					
xenon	Xe	5.13001	395.205	-38.582	161	163
14. Diatomics						
chlorine	Cl_2	6.07922	867.371	-26.253	206	271
hydrogen (normal)	H_2					
nitrogen	N_2	5.65650	260.222	-6.069	63	85
oxygen	O_2	5.81534	319.165	-6.409	55	101
15. Inorganic Oxides						
carbon dioxide	CO_2	7.52161	1384.861	74.840	267	303
carbon monoxide	CO					
nitrogen dioxide	NO_2					
16. Other Compounds						
ammonia	NH_3					
hydrogen sulfide	H_2S					
water	H_2O	7.06252	1650.27	-46.804	333	403

Table A3   Wu-Stiel Parameter Y

Compound Name	Y
water	1.000
ammonia	0.739
acetone	0.652
ethylene oxide	0.565
hydrogen chloride	0.565
Alcohols	
methanol	0.652
ethanol	0.213
n-propanol	-0.052
i-propanol	0.052
Alkyl Halides	
methyl chloride	0.304
methyl fluoride	0.739
Refrigerants	
R11	0.015
R12	0.021
R22	0.164
R32	0.704
R123	
R124	-0.001
R125	-0.001
R134a	0.180
R141b	0.073
R142b	
R143a	
R152a	0.483

Table A4 Wilson Parameters λ_{12}/R and λ_{21}/R

Components		λ_{12}/R	λ_{21}/R	P	T
1	2	K	K	MPa	K
Alcohol mixtures					
methanol	acetone	334.18	-108.17	0.1	-
		335.64	-102.17	-	328
	benzene	815.39	77.43	0.1	-
	chloroform	857.32	-187.85	0.1	-
	2,3-dimethylbutane	1394.84	225.98	0.1	-
	ethanol	301.15	-257.34	0.1	-
	ethyl acetate	496.02	-100.82	0.1	-
		605.66	-159.48	-	313
		435.86	10.23	-	323
		518.39	-87.28	-	333
	methyl acetate	419.71	-39.66	0.1	-
	2-propanol	44.29	-15.19	0.1	-
ethanol	acetone	210.83	19.21	0.1	-
	benzene	653.13	66.16	0.1	-
	cyclohexane	1082.43	152.69	0.1	-
	ethyl acetate	425.06	-89.98	0.1	-
		413.66	-31.42	-	313
		374.80	-26.24	-	333
	n-heptane	1055.00	310.77	0.1	-
	n-hexane	1148.34	142.73	0.1	-
	methylcyclopentane	1117.88	81.28	0.1	-
	toluene	623.34	126.78	0.1	-
1-propanol	benzene	689.57	-37.19	0.1	-
		614.97	33.79	-	348
	ethyl acetate	332.75	-100.00	0.1	-
		281.00	21.33	-	313
		261.51	-12.63	-	333
	n-heptane	681.35	159.13	-	348
	n-hexane	408.95	420.11	0.1	-
2-propanol	acetone	143.41	64.12	0.1	-
		26.87	215.96	-	328
	benzene	507.21	80.78	0.1	-
		536.90	137.05	0.07	-
	carbon tetrachloride	620.44	55.91	0.1	-
	cyclohexane	872.64	34.73	0.1	-
		800.37	112.28	0.07	-

Table A4(con/d)   Wilson Parameters λ_{12}/R and λ_{21}/R

Components		λ_{12}/R	λ_{21}/R	P	T
1	2	K	K	MPa	K
2- propanol	ethyl acetate	145.77	30.69	0.1	-
		334.35	20.01	-	313
		210.40	23.01	-	333
	methylcyclohexane	921.77	105.55	0.07	-
	n-octane	700.02	212.56	0.05	-
	2,2,4-trimethylpentane	619.81	92.15	0.1	-
1-butanol	benzene	411.47	80.58	0.1	-
	toluene	446.76	52.68	0.1	-
Aromatic hydrocarbon mixtures					
benzene	acetone	-84.50	249.06	0.1	-
	carbon tetrachloride	-52.04	103.07	0.1	-
	chloroform	71.27	-102.77	0.1	-
	cyclohexane	94.22	40.27	0.1	-
	cyclopentane	134.14	-12.17	0.1	-
	n-heptane	49.99	147.41	0.1	-
		37.05	183.49	-	348
	n-hexane	87.52	85.51	0.1	-
	methyl acetate	115.36	-12.00	0.1	-
	methylcyclohexane	-2.088	181.62	0.1	-
	methylcyclopentane	81.24	48.98	0.1	-
toluene	acetone[1]	304.16	-100.44	0.1	308
	cyclohexane	457.61	-208.67	0.1	-
	n-hexane[1]	118.77	-39.29	-	303
	methylcyclopentane	481.86	-227.92	0.1	-
ethylbenzene	ethylcyclohexane	199.28	-121.24	0.05	-
	hexylene glycol	26.38	805.67	0.05	-
	n-octane	153.13	-67.87	0.1	-
Other mixtures					
acetone	carbon tetrachloride	327.98	-6.376	0.1	-
	chloroform	-36.33	-167.18	0.1	-
	2,3-dimethylbutane	477.20	118.24	0.1	-
	heptane[1]	550.00	79.24	0.1	338
chloroform	2,3-dimethylbutane	107.63	112.56	0.1	-
	ethyl acetate	-184.93	-46.55	0.1	-
	methyl acetate	-227.00	56.98	0.1	-
	methyl ethylketone	-116.55	-118.32	0.1	-
cyclohexane	methyl acetate	173.67	348.05	0.1	-

Table A4(con/d)   Wilson Parameters λ_{12}/R and λ_{21}/R

Components		λ_{12}/R	λ_{21}/R	P	T
1	2	K	K	MPa	K
n-hexane	*n*-heptane[1]	-259.08	626.09	-	365
	hexene-1	208.93	-140.83	0.1	-
	methylcyclopentane	136.92	-88.42	0.1	-
	n-octane[1]	-233.85	151.76	-	422
	1,2,3-trichloropropane	58.57	556.83	0.1	-
hexene-1	1,2,3-trichloropropane	78.97	286.99	0.1	-
hexylene glycol	ethylcyclohexane	1807.76	38.72	0.05	-
water	acetone	707.27	221.23	0.1	-
	acetonitrile	810.21	349.27	0.1	-
	ethanol	480.80	192.38	0.1	-
	ethanol[1]	482.67	-95.50	-	283
		444.72	237.04	-	343
	methanol	242.63	103.31	0.1	-
	1-propanol	646.44	511.17	0.1	-
		575.68	977.43	-	313
		598.08	529.10	-	333

[1] see References

Table A5   T-K-Wilson Parameters λ_{12}/R and λ_{21}/R

Components		λ_{12}/R	λ_{21}/R	V_{12}	P	T
1	2	K	K		MPa	K
Water mixtures						
water	acetonitrile[1]	1098.8	-516.04	0.104	0.1	333
	n-butanol	316.11	273.05	0.208	0.1	298
	ethyl acetate[1]	1343.3	-751.71	0.063	0.1	333
	triethylamine	1120.9	-682.23	0.125	0.1	338
Other mixtures						
methanol	cyclohexane	61.91	994.00	2.674	0.1	298
ethyl acetate	acetonitrile[1]	-14.91	105.02	1.653	0.1	333

[1] see References

Table A6 NRTL (Renon) Parameters g_{12}/R, g_{21}/R and α_{12}

Components		g_{12}/R	g_{21}/R	α_{12}	P	T
1	2	K	K		MPa	K
Alcohol mixtures						
ethanol	ethyl acetate	151.47	162.03	0.300	0.1	343
methanol	*n*-heptane[1]	565.80	345.19	0.200	0.1	303
Aromatic hydrocarbon mixtures						
toluene	acetone[1]	-50.46	446.09	0.3017	0.1	308
	2-propanone[2]	211.69	14.07	0.200	0.1	283
	n-hexane[1]	86.16	-6.934	0.3001	0.1	303
Other mixtures						
acetone	cyclohexane[1]	693.79	881.97	0.4841	0.1	298
	n-heptane[1]	553.74	545.25	0.2911	0.1	338
	n-hexane[1]	632.43	583.83	0.2913	0.1	293
ethyl acetate	acetonitrile[2]	224.87	618.29	0.300	0.1	333
n-hexane	*n*-heptane[1]	616.19	-363.99	0.316	0.01	-
	n-octane[1]	453.85	-175.17	1.8262	0.02	-
water	acetonitrile[2]	618.29	224.88	0.300	0.1	333
	ethanol[1]	112.43	245.98	0.2978	0.1	283
		-61.03	673.23	0.2974	0.1	343
	ethanol	491.14	44.28	0.300	0.1	343
	ethyl acetate[2]	1451.90	48.77	0.200	0.1	333
	ethyl acetate	671.80	1263.08	0.400	0.1	343
	2-propanone[2]	250.69	366.44	0.200	0.1	283
	toluene[2]	1643.20	1057.60	0.200	0.1	283

[1,2] see References

Table A7   UNIQUAC Parameters u_{12}/R and u_{21}/R

Components 1	Components 2	u_{12}/R K	u_{21}/R K	T K
Alcohol mixtures				
methanol	acetone	-96.90	359.10	328-337
	benzene	-56.05	988.60	308
		-38.37	902.56	318
		-50.58	912.41	331-343
	chloroform	-143.50	926.31	326-336
	2,3-dimethylbutane	-7.18	1463.90	317-333
	ethanol	660.19	-292.39	338-349
	ethyl acetate	-107.54	579.61	335-347
	methyl acetate	-83.96	508.39	308
		-95.31	516.12	323
	n-heptane	2.49	1419.33	306
		4.84	1325.89	332
	n-hexane	-2.66	1636.05	322-336
ethanol	acetone	-119.36	414.46	305
		-131.25	404.49	321
	acetonitrile	68.72	430.51	293
	benzene	-123.54	920.39	324-329
		-138.90	947.20	340-351
	chloroform	-235.47	1315.02	328
	cyclohexane	-68.00	1349.20	278
		-90.10	1369.41	308
		-113.70	1269.49	338
	ethyl acetate	-148.29	594.60	313
		-154.11	574.22	328
		-167.61	571.73	343
	methylcyclopentane	-118.27	1383.93	333-349
	methyl acetate	-130.78	573.43	320
	n-decane	-127.48	1254.65	352-433
	n-heptane	-110.15	1356.90	333
	n-hexane	-108.93	1441.57	331-349
	n-octane	-103.04	1425.20	318
		-123.57	1425.20	348
	n-propanol	210.95	-67.70	333
	toluene	-132.12	1083.75	308
		-141.16	1009.48	328

Table A7(con/d) UNIQUAC Parameters u_{12}/R and u_{21}/R

Components		u_{12}/R	u_{21}/R	T
1	2	K	K	K
n-propanol	benzene	-155.10	928.50	318
	cyclohexane	-136.92	1328.49	298
		-173.42	1284.75	338
	ethyl acetate	-190.31	539.64	333
	n-decane	-201.82	1137.20	363
	n-hexane	-144.11	1326.05	318
	toluene	-195.40	818.34	369-377
n-butanol	benzene	-181.24	928.90	318
	cyclohexane	-175.51	1299.00	298
		-198.09	1591.50	318
	n-hexane	-159.24	1370.74	298
	n-octane	-236.21	1098.91	382-388
Aromatic hydrocarbon mixtures				
benzene	acetone	147.12	-90.40	303
		174.00	-108.79	318
	acetonitrile	247.13	-51.54	318
	carbon tetrachloride	43.39	-37.52	313
	chloroform	-121.79	83.58	336-352
	cyclohexane	4.33	65.28	283
		-11.20	70.13	313
		-16.69	69.48	343
	cyclopentane	33.15	15.19	322-352
	ethyl acetate	-181.49	233.81	323
		-192.00	248.71	343
	methylcyclopentane	-6.47	56.47	344-352
	methyl acetate	-143.88	203.46	323
		-207.37	278.33	330-350
	n-heptane	19.07	31.35	318
		-32.03	87.50	333-347
	n-hexane	27.92	31.94	298
		-77.13	132.43	341-352
	toluene	-220.57	330.69	354-378
toluene	2-butanone	123.57	-82.85	352-383
	cyclohexane	-44.04	83.67	354-381
	ethyl acetate	-214.26	309.41	351-382
	methylcyclopentane	-48.05	89.77	345-381
	n-heptane	-72.96	108.24	371-383
	n-hexane	4.30	34.27	298-329

Table A7(con/d)   UNIQUAC Parameters u_{12}/R and u_{21}/R

Components 1	2	u_{12}/R K	u_{21}/R K	T K
ethylbenzene	ethyl acetate	-105.50	137.11	333
Other mixtures				
acetone	*n*-hexane	-33.08	261.51	318
	n-pentane	-22.83	266.31	305-322
n-heptane	acetonitrile	545.79	23.71	318
n-hexane	cyclohexane	-145.56	172.73	343-352
	methylcyclopentane	-138.84	162.13	333
water	acetone	-117.85	572.52	309-339
	acetonitrile	61.92	294.10	350-364
	benzene	115.13	2057.42	298
	ethyl acetate	-107.58	877.04	313
		80.91	569.86	323
		-222.02	1081.40	343
	ethanol	284.81	-27.38	313
		380.68	-64.56	328
		387.38	-71.06	351-372
	methanol	148.27	-50.82	338-368
	n-butanol	1097.59	38.69	365-384
	n-propanol	583.03	78.37	333
	toluene	305.71	1371.36	323

Table A8   UNIQUAC Parameters r, q and q'

	r	q	q'
Alcohols			
methanol	1.43	1.43	0.96
ethanol	2.11	1.97	0.92
n-propanol	2.78	2.51	0.89
n-butanol	3.45	3.05	0.88
Aromatic hydrocarbons			
benzene	3.19	2.40	2.40
toluene	3.92	2.97	2.97
ethylbenzene	4.60	3.51	3.51
o-,m-,p-xylene	4.66	3.54	3.54
Alkanes			
ethane	1.80	1.70	1.70
propane	2.48	2.24	2.24
n-butane	3.15	2.78	2.78
n-pentane	3.82	3.31	3.31
n-hexane	4.50	3.86	3.86
n-heptane	5.17	4.40	4.40
n-octane	5.85	4.94	4.94
n-decane	7.20	6.02	6.02
Other compounds			
acetone	2.57	2.34	2.34
acetonitrile	1.87	1.72	1.72
2-butanone	3.25	2.88	2.88
carbon tetrachloride	3.33	2.82	2.82
chloroform	2.70	2.34	2.34
cyclohexane	3.97	3.01	3.01
cyclopentane	3.30	2.47	2.47
ethyl acetate	3.48	3.12	3.12
methyl acetate	2.80	2.58	2.58
methylcyclohexane	4.64	3.55	3.55
methylcyclopentane	3.97	3.01	3.01
water	0.92	1.40	1.00

Note: parameter $q' = q$, in all cases except water and alcohols

Table A9 Lennard-Jones Potential Parameters Determined from Viscosity Data

Compound Name	σ nm	ε/k_B K	Compound Name	σ nm	ε/k_B K
1. Alkanes			*7. 1-Alkanols*		
methane	0.3758	148.6	methanol	0.3626	481.8
ethane	0.4443	215.7	ethanol	0.4530	362.6
propane	0.5118	237.1	1-propanol	0.4549	576.7
n-butane	0.4687	531.4	1-butanol		
isobutane	0.5278	330.1	1-pentanol		
n-pentane	0.5784	341.1	1-hexanol		
n-hexane	0.5949	399.3	1-heptanol		
n-heptane			*8. Alkyl Acetates*		
n-octane[1]	0.7451	320.	methyl acetate	0.4936	469.8
2. Alkenes			ethyl acetate	0.5205	521.3
ethylene	0.4163	224.7	*9. Monatomics*		
propylene	0.4678	298.9	argon	0.3542	93.3
1-butene[1]	0.5198	319.	helium	0.2551	10.22
2-pentene[1]	0.6476	204.	krypton	0.3655	178.9
3. Alkyl Cyclohexanes			neon	0.2820	32.8
cyclohexane	0.6182	297.1	xenon	0.4047	231.
methylcyclohexane			*10. Diatomics*		
ethylcyclohexane			chlorine	0.4217	316
4. Acetylenes			hydrogen (normal)	0.2827	59.7
acetylene	0.4033	231.8	nitrogen	0.3798	71.4
methyl acetylene	0.4761	251.8	oxygen	0.3467	106.7
5. Alkyl Benzenes			*11. Inorganic Oxides*		
benzene	0.5349	412.3	carbon dioxide	0.3941	195.2
toluene[1]	0.5932	377	carbon monoxide	0.3690	91.7
ethylbenzene			nitrous oxide	0.3828	232.4
o-xylene			*12. Other Compounds*		
m-xylene			ammonia	0.2900	558.3
p-xylene			hydrogen bromide	0.3353	449.
6. Alkyl Halides			hydrogen chloride	0.3339	344.7
methyl bromide	0.4118	449.2	hydrogen fluoride	0.3148	330.
methyl chloride	0.4182	350.	hydrogen sulfide	0.3623	301.1
ethyl chloride	0.4898	300.	water	0.2641	809.1

Table A10 Liquid-Viscosity Coefficients of Selected Compounds
at Atmospheric Pressure (accuracy better than ±2%)

Compound Name	$\dfrac{T}{K}$	$\eta(298.15)$ μPa·s	A	B	C	D
1. Alkanes						
n-hexane[1]	280 - 390	294.9	-5.873	11.047	-7.384	2.211
n-heptane[1]	290 - 350	389.0	-6.265	12.022	-8.411	2.655
n-octane[1]	280 - 390	509.2	-6.496	12.682	-9.404	3.219
n-nonane	280 - 340	657.6	-4.187	4.513	-0.327	-
n-decane[1]	280 - 430	849.8	-7.156	14.258	-11.396	4.294
n-undecane	280 - 350	1071.9	-2.784	0.752	2.032	-
n-dodecane[1]	280 - 430	1358.5	-7.807	16.011	-13.863	5.659
2. Alkyl Benzenes						
benzene	280 - 400	603.7	-3.995	4.399	-1.067	0.663
toluene	220 - 380	554.5	-5.349	9.343	-6.216	2.223
ethylbenzene	280 - 350	628.3	-3.558	3.603	-0.047	-
o-xylene	290 - 350	751.0	-3.077	2.176	0.901	-
m-xylene	290 - 350	583.0	-2.885	2.276	0.608	-
p-xylene	290 - 350	603.3	-3.593	3.606	-0.014	-
mesitylene	290 - 390	659.3	-3.260	2.824	0.436	-
3. 1-Alkanols						
methanol	280 - 340	545.1	-29.905	84.017	-82.539	28.429
ethanol	290 - 350	1080.9	-11.794	23.390	-17.155	5.563
1-propanol	290 - 340	1928.3	10.613	-49.640	60.329	-21.288
1-butanol	290 - 340	2471.8	-31.898	83.208	-78.605	27.325
1-pentanol	290 - 330	3344.4	-147.37	448.53	-464.62	163.47
4. Other Compounds						
water	274 - 372	890.5	-10.728	26.279	-27.208	11.656

Table A11 Liquid Thermal-Conductivity Coefficients of Selected Compounds at Atmospheric Pressure (accuracy better than ±2%)

Compound Name	$\dfrac{T}{K}$	$\dfrac{\lambda(298.15)}{mW \cdot m^{-1} \cdot K^{-1}}$	A	B	C
1. Alkanes					
n-hexane	290 - 330	120.7	3.110	-3.329	1.219
n-heptane[1]	190 - 370	122.8	1.730	-0.730	-
n-octane	280 - 350	128.9	2.217	-1.619	0.403
n-nonane	280 - 360	131.6	1.853	-0.998	0.146
n-decane	300 - 380	132.2	1.088	0.302	-0.391
n-undecane	280 - 380	133.5	1.667	-0.715	0.049
n-dodecane	290 - 380	135.1	1.531	-0.417	-0.114
2. Alkyl Benzenes					
benzene[2]	290 - 350	141.1	1.696	-0.696	-
toluene[1]	190 - 360	131.1	1.452	-0.224	-0.226
ethylbenzene					
o-xylene	290 - 360	129.9	1.452	-0.362	-0.090
m-xylene	290 - 360	130.2	1.458	-0.341	-0.117
p-xylene	290 - 360	127.8	1.603	-0.580	-0.023
mesitylene					
3. 1-Alkanols					
methanol	270 - 330	199.2	3.163	-4.029	1.866
ethanol	280 - 340	165.6	5.357	-7.967	3.610
1-propanol	290 - 370	149.2	1.040	0.093	-0.133
1-butanol	290 - 360	147.6	1.694	-1.117	0.423
1-pentanol	290 - 340	145.6	0.769	0.635	-0.404
4. Other Compounds					
water[1]	274 - 370	606.7	-1.265	3.705	-1.439

REFERENCES

Table A1 Physical Constants and Perfect-Gas Heat-Capacity Coefficients
Data   :   R.C. Reid J.M. Prausnitz and B.E. Poling, *The Properties of Gases and Liquids*, 4th Ed. (McGraw Hill, 1988).
Except :   1) *n*-alkanes, alkyl benzenes and *n*-alcohols critical constants:
- I. Cibulka, *Fluid Phase Equil.* (in press, 1995).
2) refrigerants (except R11, R12) critical constants, boiling temperatures and perfect-gas heat-capacity coefficients:
- M.O. McLinden, *Int. J. Refrig.* **13** (1990) 149.
3) R11 and R12 perfect-gas heat-capacity coefficients:
- C.L. Yaws, H.M. Ni and P.Y. Chiang, *Chem. Eng.* May 9 (1988) 91.
4) Refrigerants dipole moments:
- C.W.Meyer and G.Morrison, *J.Phys.Chem.* **95** (1991) 3860.
5) Pitzer's parameter calculated from McLinden's values (see above).

Table A2 Vapour-Pressure Coefficients of Antoine Equation
Data   :   T. Boublik, V. Fried and E. Hala, *The Vapour Pressures of Pure Compounds*, 2nd Ed. (Elsevier, Amsterdam, 1984).
Except :   1) R22, R134a:
- A.R.H. Goodwin, D.R. Defibaugh and L.A. Weber, *Int. J. Thermophys.* **13** (1992) 837.
2) R141b:
- D.R. Defibaugh, A.R.H. Goodwin, G. Morrison and L.A. Weber, *Fluid Phase Equil.* **85** (1993) 271.

Table A3 Wu-Stiel parameter Y
Data   :   G.Z.A. Wu and L.I. Stiel, *AIChE J.* **31** (1985) 1632.
Except :   Refrigerants values calculated from McLinden (above) saturation pressure values

Table A4 Wilson Parameters λ_{12}/R and λ_{21}/R
Data   :   M.J. Holmes and M. van Winkle, *Ind. Eng. Chem.* **62** (1970) 21.
Except :   [1] DECHEMA *Vapour-Liquid Equilibrium Data Collection.*

Table A5 T-K-Wilson Parameters λ_{12}/R and λ_{21}/R
Data   :   T. Tsuboka and T. Katayama, *J. Chem. Eng. Japan* **8** (1975) 181.
Except :   [1] DECHEMA *Vapour-Liquid & Liquid-Liquid Euilibrium Data Collection.*

Table A6 NRTL (Renon) Parameters g_{12}/R, g_{21}/R and a_{12}
Data   :   H. Renon and J.M. Prausnitz, *I&EC Proc. Des. Dev.* **8**:413(1969).
Except :   [1] DECHEMA *Vapour-Liquid & Liquid-Liquid Equilibrium Data Collection*
[2] S.M. Walas, *Phase Equilibria in Chemical Engineering* (Butterworth Publishers, London, 1985).

Table A7   UNIQUAC Parameters u_{12}/R and u_{21}/R

Data   :   J. Prausnitz, T. Anderson, E. Grens, C. Eckert, R. Hsieh and J. O'Connel, *Computer Calculations for Multicomponent Vapour-Liquid and Liquid-Liquid Equilibria* (Prentice Hall, London, 1980).

Table A8   UNIQUAC Parameters r, q and q'

Data   :   J. Prausnitz, T. Anderson, E. Grens, C. Eckert, R. Hsieh and J. O'Connel, *Computer Calculations for Multicomponent Vapour-Liquid and Liquid-Liquid Equilibria* (Prentice Hall, London, 1980).

Table A9   Lennard-Jones Potential Parameters Determined from Viscosity Data

Data   :   R.A. Svehla, *NASA Tech. Rep. R-132* (Lewis Research Center, Cleveland, 1962), in R.C. Reid, J.M. Prausnitz and B.E. Poling, *The Properties of Gases & Liquids* (McGraw-Hill, New York, 1988).

Except :   [1] J.O. Hirschfelder, C.F. Curtiss and R.B. Bird, *Molecular Theory of Gases and Liquids* (John Wiley & Sons, New York, 1964).

Table A10 Liquid-Viscosity Coefficients of Selected Compounds

Data   :   Calculated by authors, from critical collection of accurate measurements.

Except :   [1] IUPAC recommended values: J.H. Dymond and H.A. Oye, *J. Phys. Chem. Ref. Data* **23** (1994) 41.

Table A11 Liquid-Thermal Conductivity Coefficients of Selected Compounds

Data   :   Calculated by authors, from critical collection of accurate measurements.

Except :   [1] IUPAC recommended values: C.A.Nieto de Castro, S.F.Y. Li, A. Nagashima, R.D. Trengove and W.A. Wakeham, *J. Phys. Chem. Ref. Data* **15** (1986) 1073.
[2] IUPAC recommended values: M.J. Assael, M.L.V. Ramires, C.A.Nieto de Castro and W.A. Wakeham, *J. Phys. Chem. Ref. Data* **19** (1990) 113.

B
Supplementary Information

In this Appendix we will elaborate on the definitions of molecular, perfect-gas, configurational, residual and excess properties of pure fluids and mixtures. We also review the thermodynamic relations required in the calculation of the fugacity of a component in a mixture.

B.1. Configurational Properties

In Chapter 1, we saw that the partition function Q of a system containing N molecules in volume V at temperature T may be expressed as the product of the molecular partition function, Q^{mol}, and the configuration integral, $\mathcal{L}$. The thermodynamic properties of the bulk system depend upon

$$\ln Q = \ln Q^{mol} + \ln \mathcal{L} \qquad (B.1)$$

and may therefore be expressed as the sum of a molecular part and a configurational part. We therefore have for a property X

$$X = X^{mol} + X', \qquad (B.2)$$

where X^{mol} is called the *molecular property* and X' is the *configurational property*. The relation, given in Chapter 2, between the molecular properties and the *perfect-gas* properties can easily be derived. The partition function for a perfect gas, Q^{pg}, is obtained by setting the potential energy $\mathcal{U}$ to zero. The configuration integral, Eq.(1.31), then reduces to V^N. Consequently, the molecular partition function and the perfect-gas partition function are related simply by

$$Q^{pg} = Q^{mol} V^N \qquad (B.3)$$

and

$$\ln Q^{pg} = \ln Q^{mol} + N \ln V. \qquad (B.4)$$

It follows from Eq.(B.4) that the perfect-gas part of a thermodynamic property may be expressed in terms of the molecular contribution plus terms in N and V. For example, the Helmholtz free energy of the perfect gas, obtained from Eq.(1.18) with $Q = Q^{pg}$, may be written

$$A^{pg} = -k_B T \ln Q^{pg} = -k_B T \ln Q^{mol} - N k_B T \ln V$$
$$= A^{mol} - N k_B T \ln V \tag{B.5}$$

Similarly, the perfect-gas entropy is related to the molecular contribution as follows:

$$S^{pg} = k_B T (\partial \ln Q^{pg}/\partial T)_V + k_B \ln Q^{pg}$$
$$= k_B T (\partial \ln Q^{mol}/\partial T)_V + k_B \ln Q^{mol} + N k_B \ln V \tag{B.6}$$
$$= S^{mol} + N k_B \ln V$$

and it is easy to show that

$$U^{pg} = U^{mol} \quad \text{and} \quad C_V^{pg} = C_V^{mol}. \tag{B.7}$$

The form of Eqs.(B.7) is as expected as these properties of the perfect gas are independent of volume.

B.2. Residual Properties

A residual property X^{res} is defined as the difference between the actual value of a thermodynamic property X and the value X^{pg} that would prevail in a hypothetical perfect gas at either the same values of T and V_m or at the same values of T and P. The residual property therefore gives a measure of the effects of intermolecular forces in the system. Rather than writing X as the sum of configurational and molecular parts, one can instead write X as the sum of perfect-gas and residual terms:

$$X(V_m, T) = X^{pg}(V_m, T) + X^{res}(V_m, T). \tag{B.8}$$

This separation is often useful because quite different techniques apply to the prediction of these different kinds of property.

Although the variables T and V_m are natural in statistical mechanics based on the canonical ensemble, practical considerations usually dictate the use of T and P. It is important to realise that, while the actual value of a thermodynamic property is of course unaffected by our choice of (T, V_m) or (T, P) as the independent variables, its decomposition into perfect-gas and residual terms may depend upon the choice of independent variables. This is because the values of (T, V_m, P) which characterise the actual state of the fluid are not related by the perfect-gas equation of state and so the hypothetical perfect gas with the given values of (T, V_m) differs from that with the given values of (T, P). In fact, when residual properties are quoted it is almost always with the choice of temperature and pressure as the independent variables and that convention has been followed throughout this book. Consequently, the residual properties given in Tables 4.4 (for the virial equation of state), 5.1 (for the Lee-Kesler corresponding states model) and 6.1 and 6.5-6.7 (for other equations of state) all refer to residual properties at specified T and P even though T and $\rho_n = 1/V_m$ appear in the expressions.

Given that most thermodynamic models are explicit functions of T and V_m (or of T and ρ_n but that distinction is not important here) it is easiest to derive directly from the model expressions for a residual property at T and V_m, $X^{res}(T, V_m)$, and then incorporate an additional term to obtain the residual property at T and P, $X^{res}(T, P)$. The connection between $X^{res}(T, V_m)$ and $X^{res}(T, P)$ is established as follows. We have that

$$X(T, V_m) = X(T, P) \tag{B.9}$$

where

$$X(T, V_m) = X^{pg}(T, V_m) + X^{res}(T, V_m)$$
$$X(T, P) = X^{pg}(T, P) + X^{res}(T, P) \quad . \tag{B.10}$$

The relation between $X^{pg}(T, V_m)$ and $X^{pg}(T, P)$ is given by the identities

$$\{X^{pg}(T, V_m) - X^{pg}(T, P)\} = \int_{RT/P}^{V_m} (\partial X^{pg}/\partial V_m)_T \, dV_m$$
$$= \int_{P}^{RT/V_m} (\partial X^{pg}/\partial P)_T \, dP \tag{B.11}$$

and we therefore deduce that

$$X^{\text{res}}(T,P) = X^{\text{res}}(T,V_{\text{m}}) + \int_{RT/P}^{V_{\text{m}}}(\partial X^{\text{pg}}/\partial V_{\text{m}})_T \, dV_{\text{m}}$$

$$= X^{\text{res}}(T,V_{\text{m}}) + \int_{P}^{RT/V_{\text{m}}}(\partial X^{\text{pg}}/\partial P)_T \, dP. \tag{B.12}$$

The integrals that appears in these equations are easy to evaluate in terms of the equation of state of the perfect gas (a Maxwell equation is sometimes needed to obtain the partial derivative in terms of the equation of state). As an example, consider the molar Helmholtz free energy for which $(\partial A_{\text{m}}/\partial V_{\text{m}})_T = -P$ and hence

$$(\partial A_{\text{m}}^{\text{pg}}/\partial V_{\text{m}})_T = -(RT/V_{\text{m}}). \tag{B.13}$$

Carrying out the integration, we then find that

$$A_{\text{m}}^{\text{res}}(T,P) = A_{\text{m}}^{\text{res}}(T,V_{\text{m}}) - RT \ln Z, \tag{B.14}$$

where

$$Z = (PV_{\text{m}}/RT) = (P/\rho_n RT) \tag{B.15}$$

is the compression factor. The key relations between the two kinds of residual property are summarised in Table B1.

Table B1 Relations Between Residual Properties

$A_{\text{m}}^{\text{res}}(T,P)$	$= A_{\text{m}}^{\text{res}}(T,V_{\text{m}}) - RT \ln Z$
$S_{\text{m}}^{\text{res}}(T,P)$	$= S_{\text{m}}^{\text{res}}(T,V_{\text{m}}) + R \ln Z$
$G_{\text{m}}^{\text{res}}(T,P)$	$= G_{\text{m}}^{\text{res}}(T,V_{\text{m}}) - RT \ln Z$
$H_{\text{m}}^{\text{res}}(T,P)$	$= H_{\text{m}}^{\text{res}}(T,V_{\text{m}})$

B.3. Excess Properties

A molar residual thermodynamic property of a mixture at specified temperature and pressure may be written

$$X_m^{res} = X_m - X_m^{pg} = X_m - \left(\sum_i x_i X_i^{*pg} + \Delta_{mix} X_m^{pg} \right), \qquad (B.16)$$

where X_i^{*pg} is the molar perfect-gas property for the pure component i and $\Delta_{mix} X_m^{pg}$ is the so-called ideal mixing term equal to the change in X_m^{pg} on mixing the components at constant temperature and pressure. It is sometimes convenient to consider a different quantity, the *excess molar property* X_m^E, defined by

$$X_m^E = X_m - X_m^{id} = X_m - \left(\sum_i x_i X_i^* + \Delta_{mix} X_m^{pg} \right). \qquad (B.17)$$

Here, X_m^{id} is the value of the molar property for an ideal mixture and X_i^* is the actual (rather than perfect-gas) value of the molar property for the pure component i at the specified temperature and pressure. This excess property is then the difference between the actual value of the property for the mixture and that which would prevail in an ideal mixture at the same temperature and pressure. The ideal mixing terms may all be derived from the change in molar Helmholtz free energy on mixing the perfect gases at constant temperature and volume.[1] Starting from Eq.(2.25), it is easy to show that

$$\Delta_{mix} A_m^{pg} = RT \sum_i x_i \ln x_i. \qquad (B.18)$$

All other ideal mixing terms follow from this expression and the results are given in Table B2. Since $\Delta_{mix} U_m^{pg} = 0$, $\Delta_{mix} A_m^{pg}$ is most closely related to the change in molar entropy on mixing:

$$\Delta_{mix} A_m^{pg} = - T \Delta_{mix} S_m^{pg}. \qquad (B.19)$$

The ideal mixing term for molar entropy is derived below in Example B.1.

---

[1] Since there is no change in pressure when perfect gases are mixed at constant temperature and volume, there is no distinction between that situation and mixing at constant temperature and pressure.

Table B2   Ideal Mixing Terms

Property (X_m^{pg})	Ideal mixing term ($\Delta_{mix} X_m^{pg}$)
A_m^{pg}	$RT\sum_i x_i \ln x_i$
S_m^{pg}	$-R\sum_i x_i \ln x_i$
G_m^{pg}	$RT\sum_i x_i \ln x_i$
μ_i^{pg}	$RT \ln x_i$
$H_m^{pg}, U_m^{pg}, C_{V,m}^{pg}, C_{P,m}^{pg}$	0

| **EXAMPLE B.1** | *Derivation of the Change in Molar Entropy on Mixing Perfect Gases at Constant T and P.* |

The mixing process is depicted in Figure B.1. Amounts n_1 and n_2 of two perfect gases are initially contained separately at temperature T and pressure P in a closed vessel. The partition separating the two gases is then removed so that they may mix adiabatically and, being perfect gases, this process involves no change in pressure

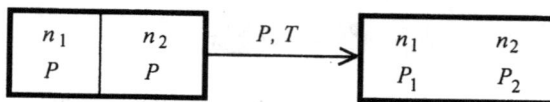

Figure B.1

or temperature. Since for an adiabatic change we have, from Eq.(1.9),

$$dS_m = (P/T)\, dV_m$$

and, in a perfect gas, $(P/T)\, dV_m = -R\, d\ln P$, the change in entropy on mixing the perfect gases is given by

$$(n_1 + n_2)\Delta S_m^{pg} = -R\left[n_1 \ln(P_1/P) + n_2 \ln(P_2/P)\right],$$

where $P_i = x_iP$ is the partial pressure of component i. It therefore follows that

$$\Delta S_m^{pg} = -R\big[x_1 \ln(x_1) + x_2 \ln(x_2)\big]$$

and, upon generalisation to a multicomponent system, one finds

$$\Delta S_m^{pg} = -R\sum_i x_i \ln x_i \,.$$

❏

B.4. Mixture Fugacity

The calculation of the fugacity of a component in a mixture under conditions of specified composition, temperature and pressure is of fundamental importance. In this section, we review the basic thermodynamic relations that make this possible. We begin with the definition of the mixture fugacity f_{mix} and the associated mixture fugacity coefficient $\varphi_{mix} = (f_{mix}/P)$ in terms of the residual molar Gibbs free energy:

$$RT \ln(f_{mix}/P) = RT \ln \varphi_{mix} = G_m^{res}. \tag{B.20}$$

The residual molar Gibbs free energy is itself given by the identity

$$G_m^{res} = \int_0^P \big[(\partial G_m/\partial P)_T - (\partial G_m^{pg}/\partial P)_T\big] dP \tag{B.21}$$

and, since $(\partial G_m/\partial P)_T = V_m$ we have

$$G_m^{res} = \int_0^P \big[V_m - (RT/P)\big] dP. \tag{B.22}$$

In order to convert this relation into one that can be evaluated readily from an equation of state that gives P as a function of T and V_m, we change the variable of integration from P to the total (not molar) volume V, making use of the relation $V(\partial P/\partial V)_{T,n} = -n(\partial P/\partial n)_{T,V}$, and, after some working, we obtain:

$$G_m^{res} = \int_V^\infty \big[(\partial P/\partial n)_{T,V} - (RT/V)\big] dV - RT \ln Z. \tag{B.23}$$

We therefore deduce that

$$\ln \phi_{\text{mix}} = (1/RT)\int_V^\infty \left[(\partial P/\partial n)_{T,V} - (RT/V)\right] dV - \ln Z \tag{B.24}$$

and this is the relation used directly to obtain expressions for the mixture fugacity in terms of an equations of state. Since $G_{\text{m}}^{\text{res}} = H_{\text{m}}^{\text{res}} - TS_{\text{m}}^{\text{res}}$, the mixture fugacity coefficient may also be obtained from the residual enthalpy and residual entropy:

$$\ln \phi_{\text{mix}} = (H_{\text{m}}^{\text{res}}/RT) - (S_{\text{m}}^{\text{res}}/R). \tag{B.25}$$

An analogous derivation may be followed to obtain the fugacity f_i of a component in a mixture and the associated fugacity coefficient $\phi_i = (f_i/x_i P)$. These quantities are related by definition to the residual chemical potential μ_i^{res} at the given temperature, pressure and composition by the relations

$$RT \ln(f_i/x_i P) = RT \ln \phi_i = \mu_i^{\text{res}}. \tag{B.26}$$

It then follows that

$$\ln \phi_i = (1/RT)\int_V^\infty \left[(\partial P/\partial n_i)_{T,V,n_{j\neq i}} - (RT/V)\right] dV - \ln Z \tag{B.27}$$

and this equation may be used to obtain ϕ_i in terms of an equation of state.

The fugacity coefficient may also be obtained in another way by noting that property X, molar property X_{m} and partial molar property X_i are related by

$$\begin{aligned}
X_i &= (\partial X/\partial n_i)_{T,P,n_{j\neq i}} \\
&= X_{\text{m}} - \sum_{k\neq i} x_k (\partial X_{\text{m}}/\partial x_k)_{T,P,x_{j\neq i}} .
\end{aligned} \tag{B.28}$$

Since μ_i^{res} is identical with the partial molar residual Gibbs free energy, it follows from (B.20) and (B.26) that $RT \ln \phi_i$ is the partial molar analogue of $RT \ln \phi_{\text{mix}}$ and hence that

$$\ln \phi_i = \ln \phi_{\text{mix}} - \sum_{k\neq i} x_k (\partial \ln \phi_{\text{mix}}/\partial x_k)_{T,P,x_{j\neq i}}. \tag{B.29}$$

This relation was used in Section 5.4 together with expressions for the partial derivatives of $\ln \phi_{\text{mix}}$ with respect to T and P obtained by operating on Eq.(B.25).

In the case of a liquid mixture, the fugacity of component i may be expressed in terms of an activity coefficient γ_i which is related most easily to μ_i^E rather then μ_i^{res}:

$$RT \ln\gamma_i = \mu_i^E. \tag{B.30}$$

To prove this, recall Eq.(7.2): $\gamma_i = f_i/x_i f_i^\circ$, where f_i° is the standard-state fugacity which is equal to the fugacity of the pure liquid at pressure P°. We then write

$$RT \ln(f_i/x_i P^\circ) = \mu_i^{res} = \mu_i - \mu_i^{pg} \tag{B.31}$$

and

$$RT \ln(f_i^\circ/P^\circ) = \mu_i^{*res} = \mu_i^* - \mu_i^{*pg} \tag{B.32}$$

where * denotes a property of the pure liquid. Subtracting these two equations we obtain

$$
\begin{aligned}
RT \ln\gamma_i &= RT \ln(f_i/x_i f_i^\circ) = \mu_i - \mu_i^* - (\mu_i^{pg} - \mu_i^{*pg}) \\
&= \mu_i - (\mu_i^* + \Delta_{mix}\mu_i^{pg}) \\
&= \mu_i^E,
\end{aligned}
\tag{B.33}
$$

which is the desired relation.

Index

* 9 7 8 1 8 6 0 9 4 0 1 9 4 *